MW01356179

Aquatic Ecology Series

Volume 13

Aquatic ecology is an extraordinarily broad and interesting field. It investigates the interplay between aquatic organisms and their physical, chemical, and biological environment. Aquatic ecology encompasses all freshwater and marine ecosystems, including streams, rivers, lakes, wetlands, coastal environments, and the vast expanses of the open ocean. Aquatic ecology studies a wide diversity of different organisms, ranging from tiny bacteria to large whales, facing a myriad of different processes such as biogeochemical cycles, genetic differentiation, and climate change. Fundamental research in aquatic ecology adds new discoveries almost every day. Applied research makes major contributions to biotechnology, fisheries, water management, nature conservation, and environmental policy. Reassessments and syntheses in aquatic ecology are stimulating to the discipline as a whole, as well as enormously useful to students and researchers in ecological sciences.

A series of succinct monographs and specialized evaluations in aquatic ecology has been developed. Subjects covered are topical (e.g., coastal hydrodynamics, microbial loops, alpine lakes) rather than broad and superficial.

The treatments must be comprehensive and state-of-the-art, whether the topic is at the biochemical, mathematical, population, community, or ecosystem level. The objectives are to advance the topics by the development of arguments, with documented support, that generate new insights, concepts, theories to stimulate thought, ideas, directions, controversies. The books are intended for mature as well as emerging scientists to stimulate intellectual leadership in the topics treated.

If you have questions regarding our book series or would like to submit a proposal to Aquatic Ecology, please contact the Publisher Éva Lőrinczi (eva.loerinczi@springer.com).

E. Marcelo Acha • Oscar Osvaldo Iribarne
Alberto R. Piola
Editors

The Patagonian Shelfbreak Front

Ecology, Fisheries, Wildlife Conservation

Editors
E. Marcelo Acha
Instituto de Investigaciones
Marinas y Costeras
UNMdP-CONICET
Mar del Plata, Argentina

Oscar Osvaldo Iribarne
Instituto de Investigaciones Marinas y
Costeras
UNMdP-CONICET
Mar del Plata, Argentina

Alberto R. Piola
Facultad de Ciencias Exactas y Naturales
Departamento de Ciencias de la Atmósfera
y los Océanos
Universidad de Buenos Aires
Buenos Aires, Argentina

ISSN 1573-4595 ISSN 2543-005X (electronic)
Aquatic Ecology Series
ISBN 978-3-031-71189-3 ISBN 978-3-031-71190-9 (eBook)
https://doi.org/10.1007/978-3-031-71190-9

This Springer imprint is published by the registered company Springer Nature Switzerland AG
The registered company address is: Gewerbestrasse 11, 6330 Cham, Switzerland

Preface

This book was written within the framework of the Pampa Azul Initiative, a program of the Argentine government aimed at providing a scientific basis to support the country's marine policies. Pampa Azul identified several Priority Geographic Areas, among them, the Patagonian Shelf-Break Front. Systematic scientific research of this front did not previously exist; it was hindered by its remoteness from the scientific research centers, its large spatial scale, and the considerable depths involved. These factors presented logistical and financial challenges for several decades. However, the front and its surroundings were sporadically sampled during national research cruises and, as it lies at the border of international waters, by vessels from distant countries, many of which were *en route* to Antarctica. A substantial amount of the information gathered in previous surveys of the shelf-break front remained unpublished or published in sources of limited distribution. Our main motivation for preparing this book is to compile and synthesize the available material to set a knowledge baseline for this ecosystem. Given the global significance of the Patagonia shelf-break front, in addition to the local and regional interest, we envision this book to be useful for the international community. Elucidation of frontal processes and their impact on species migrations and life cycles is crucial for resource management, conservation actions, and, in general, for understanding the ecology of the sea. Shelf-break fronts are among the largest and most productive marine fronts and are ubiquitous in the world ocean. Fronts occur at various spatial and temporal scales and are characterized by high biological activity. They are sites of interest for fisheries and wildlife conservation, and because they play a role in climate regulation through the uptake of atmospheric carbon dioxide.

Though the Southwestern Atlantic is a productive region of the world ocean, characterized by significant fisheries and rich wildlife, there is a lack of scientific information compared to Northern Hemisphere ecosystems. As marine ecosystems are interconnected, issues at large scales, such as global change, highlight the need to understand how processes occurring in distant areas may be interconnected. This book focuses on a region thought to have a wide biogeochemical impact on the South Atlantic basin, where changes in ocean circulation may have interhemispheric climate significance, aiming to fill the gap in scientific information on this region. Furthermore, this area is attracting attention from the conservation community due to its importance as a feeding ground for many charismatic species (e.g., oceanic birds, elephant seals), and also due to increasing problems with illegal, unreported, and unregulated fishing. The book includes a table containing all the species cited in the chapters, classified according to the World Register of Marine Species (WORMS). Finally, we express our gratitude to our colleagues who willingly conducted exhaustive work to recover information, analyze available data, and review existing documents to produce a comprehensive view of each theme developed in the chapters contributed.

Mar del Plata, Argentina E. Marcelo Acha

Mar del Plata, Argentina Oscar Osvaldo Iribarne

Buenos Aires, Argentina Alberto R. Piola

Acknowledgments

A special thanks to our reviewers:

Dr. Santiago Barbini. Instituto de Investigaciones Marinas y Costeras, Universidad Nacional de Mar del Plata y Consejo Nacional de Investigaciones Científicas y Técnicas, Argentina sbarbini@mdp.edu.ar

Dr. Ernesto Brugnoli. Instituto de Ecología y Ciencias Ambientales (IECA), Universidad de la República, Uruguay ebo@fcien.edu.uy

Dr. Javier Ciancio. Centro para el Estudio de los Sistemas Marinos (CESIMAR-CONICET) javier.ciancio@gmail.com

Dr. Manuel Haimovici. Instituto de Oceanografía, Universidad Federal Rio Grande (FURG), Rio Grande, Brazil manuelhaimovici@gmail.com

Dr. Federico Ibarbalz. Interacciones Clima, Ecosistemas y Biodiversidad (ICEB) Centro de Investigaciones del Mar y la Atmosfera (CIMA) Instituto Franco-Argentino sobre Estudios de Clima y sus Impactos (IFAECI) CONICET/FCEN-UBA/CNRS/IRD Buenos Aires, Argentina federico.ibarbalz@cima.fcen.uba.ar

Dr. Susanne Kortsch. University of Helsinki, Finland susanne.kortsch@abo.fi

Dr. José H. Muelbert. Instituto de Oceanografía, Universidad Federal Rio Grande (FURG), Rio Grande, Brazil jmuelbert@furg.br

Dr. Pablo Muniz Maciel. Instituto de Ecología y Ciencias Ambientales (IECA), Universidad de la República, Uruguay pmmaciel@fcien.edu.uy

Dr. Pablo Yorio. Centro para el Estudio de los Sistemas Marinos (CESIMAR-CONICET) yorio@cenpat-conicet.gob.ar

Contents

1 **Introduction** 1
E. Marcelo Acha, Oscar O. Iribarne, and Alberto R. Piola

2 **Anatomy and Dynamics of the Patagonia Shelf-Break Front** 17
Alberto R. Piola, Nicolás Bodnariuk, Vincent Combes, Bárbara C. Franco, Ricardo P. Matano, Elbio D. Palma, Silvia I. Romero, Martin Saraceno, and M. Milagro Urricariet

3 **The Phytoplankton of the Patagonian Shelf-Break Front** 49
Valeria A. Guinder, Carola Ferronato, Ana I. Dogliotti, Valeria Segura, and Vivian Lutz

4 **Zooplanktonic Crustacea and Ichthyoplankton of the Patagonian Shelf-Break Front** 73
Georgina D. Cepeda, Martín D. Ehrlich, Carla M. Derisio, Ayelén Severo, Laura Machinandiarena, Mariana Cadaveira, Paola Betti, Marina Do Souto, Carolina Pantano, and E. Marcelo Acha

5 **Nekton in the Patagonian Shelf-Break Front: Fishes and Squids** . . . 97
Daniela Alemany, Mauro Belleggia, Gabriel Blanco, Mariana Deli Antoni, Marcela Ivanovic, Nicolás Prandoni, Natalia Ruocco, María Luz Torres Alberto, and Anabela Zavatteri

6 **Benthic Assemblages and Biodiversity Patterns of the Shelf-Break Front** 137
Diego A. Giberto, Laura Schejter, María Virginia Romero, Mauro Belleggia, and C. S. Bremec

7 **Fisheries in the Patagonian Shelf-Break Front** 165
Daniela Alemany, Anabela Zavatteri, Nicolás Prandoni, and Analía Giussi

8 Seabirds in the Argentine Continental Shelf and Shelf-Break 185
Marco Favero, Juan Pablo Seco Pon, Jesica Paz,
Maximiliano Hernandez, and Sofía Copello

9 Patagonian Shelf-Break Front: The Ecosystem Services Hot-Spot of the South West Atlantic Ocean 211
Paulina Martinetto, Carolina Kahl, Daniela Alemany,
and Florencia Botto

10 Species-Dependent Conservation in a SW Atlantic Ecosystem 229
Claudio Campagna, Valeria Falabella, Pablo Filippo,
and Daniela Alemany

11 Food Web Topology Associated with the Patagonian Shelf-Break Front 261
Florencia Botto, Paulina Martinetto, Daniela Alemany,
and Clara Díaz de Astarloa

Chapter 1
Introduction

E. Marcelo Acha , Oscar O. Iribarne, and Alberto R. Piola

Acronyms and Abbreviations

AEEZ	Argentine Exclusive Economic Zone
ESS	Extended Shelf System
MC	Malvinas Current
PSBF	Patagonian Shelf-break Front
PSLME	Patagonian Shelf Large Marine Ecosystem

Supplementary Information The online version contains supplementary material available at https://doi.org/10.1007/978-3-031-71190-9_1.

E. M. Acha (✉)
Instituto de Investigaciones Marinas y Costeras, UNMdP-CONICET,
Mar del Plata, Argentina

Instituto Nacional Investigación y Desarrollo Pesquero (INIDEP), Mar del Plata, Argentina

Consejo Nacional de Investigaciones Científicas y Técnicas, Buenos Aires, Argentina
e-mail: macha@inidep.edu.ar

O. O. Iribarne
Instituto de Investigaciones Marinas y Costeras, UNMdP-CONICET,
Mar del Plata, Argentina

Consejo Nacional de Investigaciones Científicas y Técnicas, Buenos Aires, Argentina

A. R. Piola
Facultad de Ciencias Exactas y Naturales, Departamento de Ciencias de la Atmósfera y los Océanos, Universidad de Buenos Aires, Buenos Aires, Argentina

Instituto Franco-Argentino de Estudios sobre el Clima y sus Impactos, CONICET/UBA/CNRS/IRD, Buenos Aires, Argentina

E. M. Acha et al. (eds.), *The Patagonian Shelfbreak Front*, Aquatic Ecology Series 13, https://doi.org/10.1007/978-3-031-71190-9_1

This book concerns the properties and functioning of the Patagonian shelf-break front (PSBF), a narrow boundary separating waters occupying a wide continental shelf from those of the neighboring western boundary current, the Malvinas Current (MC). The PSBF front is part of the structural complexity of the Southwestern Atlantic, and it is featured by high primary production, biodiversity, and important fisheries. The book compiles scientific papers on the Patagonian shelf-break front, an integral reanalysis of this information under modern paradigms to produce new insights into its functioning, and an assessment of the risks to which it is subjected by human pressures.

1.1 Marine Fronts; Shelf-Breaks and Shelf-Break Fronts

The ocean is not homogeneous; on the contrary, it is organized into parcels of relatively uniform temperature and salinity. Within these regions, water property changes are relatively small; however, near the parcels' boundaries, salinity and/or temperature present larger changes, showing high horizontal gradients. These narrow bands are referred to as fronts. Satellite information, high-resolution field measurements, and numerical simulations show a profusion of patterns characterizing the oceans, allowing for the detection and description of fronts. Diverse forcings (e.g., winds, tides, currents, continental runoff) cause and maintain marine fronts, which separate different water masses. Fronts occur throughout the world ocean at several spatial and temporal scales and are characterized by high biological activity (Olson 2002; Acha et al. 2015). Among the several types of fronts, those related to shelf-breaks (i.e., shelf-break fronts) are among the largest and most productive, and they are ubiquitous in the world ocean.

In general, ocean bottom topography mirrors that of the adjoining land. The Argentine Continental Shelf is one of the largest and flattest in the world, mirroring the flat Argentine Pampas and the Patagonian steppe. In the flattest areas of this shelf, the bottom gradient is close to 1:10,000 (Lonardi and Ewing 1971). Depth increases gradually offshore up to a point of major change in the bottom slope: the shelf-break, located at a mean depth of about 110–165 m (Parker et al. 1997). Offshore the shelf-break, depth increases abruptly along the slope. That is, the shelf-break is the region of the seafloor where the flat continental shelf drops rapidly to form the continental slope. The continental slope has a regional northeast-southwest direction, with a width ranging between 200 and 500 km—in general, increasing southwards—an average gradient of 1:50, and a slope ranging between 2° and 5°, although there are sectors with different slopes. Around 70 deep submarine canyons, of varying widths and depths, crisscross the shelf-break (COPLA 2017). Towards deeper regions, the continental rise remarkably coincides with the 3200 m isobath and finally merges into the abyssal plains that, in the Argentine Basin, deepen to more than 5000 m (Parker et al. 1997) (Fig. 1.1).

Shelf-break fronts are distinct critical interfaces of the continental margins, which delineate the chief physiographic boundary between two major submarine

Fig. 1.1 Ocean bottom topography offshore Argentina. Note the shelf-break extension and depth. Blue arrows indicate the Malvinas Current (cold and fresher sub-Antarctic waters); red arrows indicate the Brazil Current (warmer and saltier subtropical waters)

provinces: shelf and slope. In many regions around the world, satellite imagery reveals the shelf-break as a well-defined narrow band of relatively cold surface waters, supporting high chlorophyll concentrations (Bakun 1996; Longhurst 1998). Strong horizontal gradients in water properties often develop at the shelf-break; these usually take the form of a surface-to-bottom density front with a corresponding geostrophic jet along the front (Condie 1993). In contrast with terrestrial boundaries, ocean fronts are fluctuating frontiers, but compared to other oceanographic patterns, shelf-break fronts are relatively stable due to strong dynamical control by bottom topography. Although different oceanographic processes may generate shelf-break fronts, the associated band of cold waters generally indicates the entrainment of deeper, colder, and nutrient-rich waters towards the surface (Bakun 1996; Longhurst 1998). Nutrient abundance promotes high photosynthesis rates, and the elevated primary production by phytoplankton is distributed up the trophic web. Many shelf-break regions are areas of high biological productivity and consequently support intensive commercial fisheries (Gawarkiewicz and Plueddemann 2020; Bakun 1996).

Shelf-break fronts have been recognized for a long time. A case documented nearly a century ago is that of the shelf-break front in the Middle Atlantic Bight, extending from Nova Scotia to Cape Hatteras in North America (Condie 1993).

Moreover, there are many other well-studied shelf-break fronts, including fronts observed in the eastern Bering Sea, the Celtic Sea, the southern Norwegian Sea, along southern Australia (Condie 1993 and references therein), and along the Antarctic slope (Jacobs 1991).

1.2 Importance of Shelf-Break Fronts

Shelf-break fronts are located between the shelf and the deep ocean and represent an interface in terms of bathymetric, physical, geological, and biological characteristics. In general, any boundary study necessarily focuses on understanding the regulation of flows across the heterogeneous space. In this sense, a crucial scientific issue is understanding the cross-shelf exchange of properties such as heat, salt, nutrients, and biota between the open and the coastal oceans. Shelf/slope exchange processes in many places are regulated by shelf-break fronts that usually separate lighter shelf waters from denser waters on the slope. Temporal and spatial variability of these fronts leads to the mixing and exchange between the coastal and open ocean waters (Lozier and Reed 2005; Siedlecki et al. 2011; Gawarkiewicz and Plueddemann 2020), though these fluxes tend to be an order of magnitude smaller than along-slope fluxes (Brink 2016). Several studies have demonstrated the extent to which parcels of water pass from one side to the other by a variety of mechanisms (Ashjian 1993; Sournia 1994; Piola et al. 2010). The across-frontal exchange is critical for shelf water budgets and ecosystem dynamics (Garvine et al. 1988).

High standing stocks of phytoplankton most likely occur at shelf-break fronts because of a net upward nutrient flux in that localized region. The fertilization mechanisms seem to be diverse; they could be slow and steady or stronger but intermittent and would result from the dynamic upwelling characteristics of each front (Fournier 1978). The associated biological productivity flows through the trophic webs, reaching top predators. Commercial species, and those of interest for wildlife conservation, integrate such trophic webs and take advantage of frontal productivity; as a result, interactions between fisheries and vulnerable species intensify in these regions. Frequently, intense fishing activity develops close to shelf-break fronts around the world, and fishing vessels concentrate along the shelf-breaks (e.g., Olson 2002; Acha et al. 2015). In some locations where the boundary of a country's exclusive economic zone lies near the shelf-break, domestic and foreign fishing fleets from distant countries meet near the shelf-break (e.g., Fournier 1978; Martinetto et al. 2020).

Several shelf-break fronts play a role in ocean-atmosphere CO_2 fluxes, acting as sinks, such as the Siberian Arctic shelf-break front (Humborg et al. 2017) or the Patagonian shelf-break front (Bianchi et al. 2009; Kahl et al. 2017). Others separate areas acting as sinks or sources of atmospheric CO_2, such as in the Celtic Sea (Rippeth et al. 2014), the Antarctic (Wang et al. 2021), or the Bering Sea (Chen et al. 2014). These processes are, at least partially, associated with intense biological activity concentrated around these frontal areas.

1.3 The Patagonian Shelf-Break Front

The Patagonian continental shelf is the largest in the Southern Hemisphere and one of the largest shelves in the World Ocean (e.g., Bisbal 1995). The region belongs to the Patagonian Shelf Large Marine Ecosystem (PSLME), one of the 66 world Large Marine Ecosystems (LMEs, Sherman 2005). The PSLME supports the large-scale fisheries within FAO's Area 41, which are mostly based on Argentine hake, Argentine shortfin squid, Patagonian Red Shrimp, Patagonian toothfish, and Patagonian scallop.

Extremely rich frontal patterns characterize this shelf, which contains an extensive chlorophyll hotspot (Acha et al. 2004; Belkin et al. 2009). Our area of interest, the Patagonian Shelf-break Front (PSBF), is located along the eastern boundary of the PSLME. The PSBF is clearly visible in satellite images and concentrates a significant fraction of the regional chlorophyll abundance (Romero et al. 2006; Marrari et al. 2017). An outstanding feature of the regional oceanography is the Malvinas Current, an equatorward-flowing boundary current carrying cold and nutrient-rich subantarctic waters. The PSBF occurs at the interface between the shelf waters and those of the Malvinas Current.

Due to its remoteness from research centers and large geographic extension, the scientific study of the front poses logistical and financial challenges; such circumstances have prevented the development of systematic scientific studies. However, an ambitious program devoted to understanding the PSBF's main ecological processes has recently started to unfold. The program is focused on a portion of the shelf-break front located approximately between 45 °S and 47 °S, referred to as *Agujero Azul* (Blue Hole in English, https://www.pampazul.gob.ar/). Notwithstanding, the bulk of scientific information on the PSBF continues being that generated during the past decades, taken mainly in a non-systematic way, and that remains largely dispersed in gray and/or poorly distributed literature. Such data were mostly collected during fishery research cruises focused on the shelf ecosystems, which reached different sectors of the shelf-break. These surveys, aimed at the stock assessment of commercial species, also collected data and samples of physical variables, plankton, and noncommercial benthic and nektonic species. Numerous technical reports, of limited circulation but containing high-quality scientific information, were produced. Part of the knowledge on the shelf-break and its surroundings was generated by research vessels of several countries on their way to Antarctica, which collected mostly physical oceanography and plankton observations and published the results in international scientific journals. A third source of information is satellite data. The shelf-break front is well represented in satellite images and has attracted the interest of the international research community, which has significantly contributed to improving the understanding of this system.

From a physical viewpoint, a front may be defined by a line connecting the points of maximum horizontal gradient of some physical variable, typically temperature and/or salinity. Although conceptually correct, it does not make sense for ecological studies because organisms do not live along a line but rather occur and are concentrated in a region of "frontal influence." Defining such an area of influence is not straightforward because different species experience the landscape at different scales, mainly due to their size, mobility, longevity, and life-cycle complexity. The definition of the region of influence of the shelf-break front is a point that has been determined employing different criteria by the authors, mainly due to the object of the study and the species dealt with in each chapter.

1.4 The Ecology, Fisheries, and Wildlife Conservation of the Patagonian Shelf-Break Front

Chapter 2 deals with the physical oceanography of the shelf-break ecosystem. The regional-scale characteristics of the shelf-break front are discussed based on the analysis of satellite-borne data and highlight the role of the intense advection of cold, nutrient-rich subantarctic waters of the Malvinas Current. A detailed description of the cross-shore structure of the shelf-break front reveals distinct seasonal patterns associated with the strong stratification over the continental shelf. The analyses reveal that the seasonal pycnocline provides an isopycnal cross-front connectivity, which may be a nutrient source to the outer shelf and play a significant biological role. Time-series observations at the shelf-break reveal strong variability in temperature, salinity, and velocity at a variety of timescales. These data suggest that acceleration of the along-shore flow is associated with cooling and salinification. Numerical models indicate that the nutrient supply to the outer shelf upper layer is mediated by a relatively simple process associated with the downstream spreading of the Malvinas Current due to frictional effects. The downstream divergence drives upwelling. The models indicate that the intensity of the vertical flow is strongly modulated by the intensity of the Malvinas Current, which is consistent with the observations. Both models and observations suggest that along-shore wind stress fluctuations further modulate the vertical stratification and the location of the surface expression of the shelf-break front at various timescales.

Chapter 3 shows that intense phytoplankton blooms characterize the PSBF, making it a hotspot of primary productivity and biological activity in the Southern Hemisphere. Interactions between wind stress, shelf waters, topography, and oceanic circulation cause upwelling and phytoplankton outbreaks, with maxima in spring and summer. The interplay between geomorphological and hydrographic characteristics across the shelf-to-open ocean transition, modulated by seasonal changes in light, nutrients, and grazing pressure, sets assemblages of phytoplankton functional groups and size classes at a subregional scale. Blooms of large diatoms and dinoflagellates are responsible for the high chlorophyll signals in austral spring, while blooms of pico- and nanophytoplankton (e.g., coccolithophores) occur in summer. Potentially toxic phytoplankton have been increasingly documented in the last decade, especially multispecific blooms of dinoflagellates that produce azaspiracids. The occurrence of these natural hazards raises scientific and social concern about their drivers and impacts on marine resources and human health, although until now, no impact on commercial species has been documented.

Chapter 4 is dedicated to zooplanktonic crustaceans and ichthyoplankton (fish eggs and larvae). Copepods, amphipods, and euphausiids are the main components of meso-zooplankton, regarding their abundance and relevance to the food web. These groups are a key link through which energy and matter transfer from phytoplankton to upper trophic levels, supporting economically important populations of invertebrates and fishes, as well as top predators such as marine birds and mammals. The PSBF shows moderate zooplanktonic diversity and relatively uniform patterns. In the case of copepods, species richness slightly increases northwards along the front as waters become warmer. Abundances are uniform all along the latitudinal range of the PSBF. Mean patterns show the front as a leaky boundary between the neritic and the oceanic groups, but on a synoptic scale, the front divides the low-diversity, high-abundance (neritic) shelf region from the more diverse and less abundant oceanic region. Seasonal variations in the dominance of different copepod size classes are evident and are presumably associated with phytoplankton succession along the production cycle. The occurrence of larvae of some demersal and mesopelagic fishes, typical of cold subantarctic waters, denotes reproductive activity in the PSBF and its surroundings. Ichthyoplankton can be classified, according to bottom depths, as pertaining to the shelf-edge (80–200 m), slope (200–800 m), or oceanic regions (800–3100 m). Larvae show higher abundances immediately offshore of the front, especially above the thermocline. Copepods, especially their larval stages, constitute the main feeding source for fish larvae inhabiting along the shelf-break.

Chapter 5 shows the PSBF as a hotspot of nektonic biodiversity, with more than 135 taxa reported. Twelve squid species, some of them bathypelagic, are also reported, with two species being intensively exploited: the Argentine shortfin squid (*Illex argentinus*) and the Patagonian longfin squid (*Doryteuthis gahi*). This region is characterized by a pronounced bathymetric gradient, so bottom depth has a strong influence on the demersal fish community, and three assemblages can be recognized. The outer shelf assemblage is characterized by typical continental shelf species (e.g., hakes), while the upper shelf-break and shelf-break assemblages are characterized by grenadiers (genus *Macrourus* and *Coelorinchus*), Patagonian toothfish (*Dissostichus eleginoides*), or skates. Mesopelagic fishes occur immediately offshore the front and are dominated by Myctophids. The PSBF plays an important role in the life cycles of several fishes and squids; key processes such as feeding, reproduction, and/or migration take place at the front or its surroundings. Due to their complex trophic relationships, large home ranges, and extensive feeding and spawning migrations, nektonic species connect the PSBF with other ecosystems, redistributing the matter and energy caught at the front.

Chapter 6 reveals that the benthic realm of the PSBF is a complex system of extended soft sediment areas and localized sites of hard bottoms. Sediments are mainly sandy/muddy at the slope, although shells, gravel, or rocks are also found. Deep submarine canyons break through the continental margin and slope, whose recent study resulted in some novel species and new distributional records for the region. About 150 megabenthic taxa have been reported associated with the PSBF. Biodiversity is quite constant at any latitude along the front, but a decreasing biodiversity trend with increasing depth is apparent. Several fishes, such as rays, sharks, and bony fishes, feed upon the benthic invertebrates, connecting the benthic and water column realms. The northern part of the front is characterized by dense beds of Patagonian scallop (*Zygochlamys patagonica*), which act as ecosystem engineers, providing substrate and refuge to a variety of associated organisms. Epibiosis is an important and widespread phenomenon in this region because of the scarcity of hard bottoms. Many sessile species depend on hard-body species (e.g., shells) to survive, and such species increase local richness. Southern scallop beds (40 °S to 41° 30′S) are less dense and exhibit higher species richness than the more exploited northern ones. In the southern part of the PSBF, bottoms are dominated by sponges and tunicates, looking relatively preserved from bottom trawling and could provide refuge for juvenile fishes. The deepest waters offshore the PSBF host fragile organisms like deep-water corals and sponges, known to be habitat-structuring organisms, supporting a rich and diverse associated fauna.

Chapter 7 summarizes a large amount of fisheries data related to the PSBF. The front represents an important fishing area in the Southwestern Atlantic, where fishing fleets and fishing effort concentrate. Fishing vessels from Argentina and several other distant countries (mostly China, Taiwan, South Korea, and Spain) operate along the PSBF. Argentine fleets exploit mainly the Argentine shortfin squid (*Illex argentinus*), several demersal fishes (mainly Argentine hake *Merluccius hubbsi*), and Patagonian scallop (*Zygochlamys patagonica*). The highest concentration of fishing activity along the PSBF occurs at the *Agujero Azul*, where part of the continental shelf extends outside the Argentine Exclusive Economic Zone (AEEZ). The historical development of fishing by fleets from distant countries was prompted by the interest in white meat from fish and squids, which drove the discovery of fishing grounds that began to be exploited in the mid-70s. Towards the mid-80s, Asian countries started fishing squids by introducing "jiggers," vessels with a fishing method that had not previously been used in the Southwest Atlantic. High catches and good yields highlighted the importance of the PSBF as a squid fishing ground on a global scale. Nowadays, *I. argentinus* is the most important PSBF fishing species in volume catches, and it attracts the greatest fishing effort by several fleets. Although this squid carries out almost its entire life cycle in the AEEZ, during certain months it is found along the PSBF and also occurs in international waters on its northward migration, where it is captured by jiggers from distant countries. Eastern European countries produced the highest finfish and cephalopod catches until the early 90s. Particularly, those fishing fleets targeting finfish (mainly southern blue whiting *Micromesistius australis* and long-tail hake *Macruronus magellanicus*) overexploited the stocks. Most of the PSBF fishing resources are straddling species that swim freely between the AEEZ and international waters. A wide range of fisheries regulations and management measures apply within the AEEZ, but they do not apply in international waters, immediately offshore the PSBF. To date, there are no international agreements to manage this area. Because the PSBF closely matches a political boundary, one of the greatest challenges for those fisheries is the integrated management of the resources in the AEEZ and on the high seas, where regulations, control, or monitoring of straddling marine resources do not exist.

Chapter 8 focuses on marine birds. The high levels of biological productivity at the PSBF attract an abundant and diverse marine megafauna, including seabirds and marine mammals, which use the front as a primary foraging area. Among seabirds, albatrosses and large petrels, which mostly breed in the archipelagos of the extreme south of South America and subantarctic islands, are the most abundant. Bird biodiversity at the PSBF is exposed to a variety of threats originated or enhanced by anthropogenic activities. The distribution of fishing effort is highly overlapped with foraging seabirds, showing impacts that range from food supplementation through scavenging behind vessels to resource competition and incidental mortality. Moreover, the exploration and exploitation of nonrenewable resources, activities that are expanding in this region, have the potential to alter the at-sea distribution of pelagic seabirds at microscale and mesoscale. Albatrosses and petrels are susceptible to threats operating throughout their wide distribution ranges, which extend across national boundaries and into international waters. Negative effects of seabird bycatch and other stressors affecting species on land and at sea can be exacerbated by the effects of climate change and concomitant changes in atmospheric circulation, water masses, and prey distribution. These environmental changes may generate significant shifts in the distribution of human activities and their overlap with seabirds. Addressing this issue represents one of the major environmental conservation challenges and will require well-informed management practices and the implementation of meaningful policy responses.

Chapter 9 focuses on conservation issues and presents summary information on marine mammals that visit the shelf-break front. Three pinniped species inhabit the region: the South American sea lion, *Otaria flavescens*, with an estimated population of ca. 190,000 animals, the Southern elephant seal, *Mirounga leonina*, with ca. 60,000 animals, and the South American fur seal, *Arctocephalus australis*, also with a population of ca. 60,000 individuals. Southern elephant seals breed in an extended colony only at Peninsula Valdés and nearby areas. Southern right whales, *Eubalaena australis*, reproduce in the northern gulfs of Argentine Patagonia, with a population estimated at ca. 6000 animals. Individuals of Southern elephant seals and Southern right whales, despite obvious differences in foraging behaviors and prey, show similarities in their movements, dispersing widely over the shelf, the shelf-break, and the ocean basin, with some animals making use of the PSBF as a feeding ground. Much less is known about the dispersion at sea of sea lions and fur seals, but in general, nursing females distribute on the shelf, while males may reach the shelf-break. Large home ranges and migrations characterize the megafauna that inhabit the region, so conservation issues are focused from a large spatial scale perspective, exceeding the PSBF. The PSBF is part of the Extended Shelf System (ESS), which is envisioned as a large conservation unit. Within the ESS, the PSBF and a few other frontal areas are hotspots that sustain regional biodiversity and abundance. A rich ensemble of resident and visitor species forage and migrate through the ESS, including several species of penguins, petrels, and albatrosses, as well as marine mammals such as elephant seals, sea lions, and whales. These "charismatic" species are critical for conservation actions, yet several would be threatened, with fisheries, climate change, and potential oil extraction as the main risks. The expansion of satellite tracking efforts showed that foraging individuals of many species converged over some areas of the shelf and shelf-break, often overlapping with fishing effort. Moreover, a sovereignty dispute over the Malvinas Archipelago precludes integrated management of transboundary species. The ESS lacks a general, strategic plan for wildlife conservation; yet, several Argentine National Action Plans exist, focusing on conservation and management, and the entire continental shelf is subject to fishery management. In addition, Argentina has protected two marine areas and numerous coastal-marine areas. Yet, protected spaces are not functionally integrated, habitat representation is poor, and there are gaps in the representation of biodiversity groups. It is proposed that to enhance conservation effectiveness, efforts should focus on the most endangered groups and move beyond site conservation to precautionary management. Adaptive, dynamic, and ecosystemic management could be combined with seasonally movable protected areas. Yet climate change and the sovereignty conflict over the Malvinas Archipelago pose considerable challenges in terms of wildlife conservation.

Ecosystem services provided by the PSBF are presented in Chap. 10. Primary production is linked to top-down and bottom-up processes propagating its properties onto other ecosystem levels, processes, and components. The shelf-break front shows marine consumers and biogeochemical cycles coupled to high primary production. Marine mammals, birds, fishes, and cephalopods are coupled to primary production through trophic interactions, and abundant phytoplankton photosynthesizing leads to increased CO_2 uptake from the atmosphere, through the so-called biological pump. The most clearly defined Ecosystem Services for the PSBF are: (1) climate regulation through CO_2 uptake, (2) food provision through fisheries, and (3) biodiversity as both a supporting and a cultural service. The high biological abundance as well as the elevated rates of the different ecological processes occurring in the PSBF have been suggested to support a higher provision of marine ecosystem services in comparison to adjacent areas; in this sense, the PSBF can be seen as a hotspot of ecosystem services.

A trophic web for the PSBF, containing 54 groups, is presented in Chap. 11. The mean trophic level (TL) estimated was $TL = 3$, with top predators reaching values of $TL = 4.7$. The web shows relatively low values of connectance, which could indicate intermediate sensitivity to species removal. Because of the large spatial scale of the front, the existence of weakly connected trophic subsystems cannot be discarded. Moreover, the PSBF food web (or the subsystems) would not be isolated; instead, it would be interconnected with the continental shelf and the deep ocean ecosystems. Main basal components of the trophic web are diatoms, planktonic crustaceans, planktivorous fishes, and squids, with the two latter being consumed by demersal-pelagic fishes, seabirds, and marine mammals. Due to their extended yearly migrations, shortfin squids play an important role in redistributing a high mass of organic matter and nutrients, being a crucial pathway in the exchange of resources among marine habitats. The bentho-pelagic coupling is a very important process by which organic matter produced near the surface reaches the bottom, sustaining a rich benthic fauna. The Patagonian scallop is a primary consumer, filtering phytoplankton, microzooplankton, and other small organic particles; scallops can have ecosystem-engineering effects, indirectly increasing the abundance and biomass of other benthic species. Several demersal bony fishes, sharks, and skates feed on benthic invertebrates associated with scallop beds. Many fishes show differences in prey item preferences along ontogeny, generally increasing their trophic spectra with size. Grenadiers and notothenioids occupy intermediate trophic levels, linking benthic production to higher trophic levels. Elephant seals take advantage of the PSBF, where some individuals stay there to eat, while others make longer trips to deeper waters. A large number of pelagic seabirds, mainly albatrosses and petrels, which often travel very long distances from their breeding sites, exploit the food concentrated in the PSBF. Penguins, especially those with colonies in the Malvinas Archipelago, make use of the PSBF for feeding. An important proportion of the diet of pelagic seabirds is a byproduct of fishery activity, which makes a high amount of food available for birds.

The book finishes with a table containing the classification of all the species cited in the chapters, checked with the page of the World Register of Marine Species (WORMS, https://www.marinespecies.org/). This table provides a general idea of the regional biodiversity, but it is by no means a definitive listing, as species are continuously being reported for the area and new species are being discovered, especially in the benthos and in the smallest planktonic components.

1.5 Epilogue: The Future

The PSBF is a multidimensional frontier where several important processes, both natural and anthropogenic, occur together in an intertwined way (Fig. 1.2). Solar radiation and upwelling merge along the shelf-break to promote phytoplankton growth and reproduction, making the PSBF a hotspot of primary productivity in the Southern Hemisphere. The Malvinas Current (MC) represents an anomaly in the upper ocean large-scale circulation: it is an equatorward flowing western boundary current. The MC carries cold, fresh, and nutrient-rich subantarctic waters to lower latitudes than in any other subpolar flow in the global ocean, reaching the Brazil-Malvinas Confluence near 38 °S. The MC can be thought of as the "soul" of the Patagonian shelf-break because it is the source of nutrients for phytoplankton growth and of the mechanical energy required to upwell them to the illuminated layer. The combined solar and oceanographic energies are channeled through resident (plankton, benthic invertebrates, and fishes) and visitor species (fishes, squids, birds, and mammals) to the upper trophic levels, occupied by species of interest for fisheries and wildlife conservation. Larger and more mobile animals make medium- to high-range migrations, exporting matter and energy produced at the PSBF to nearby and distant ecosystems; squids, anchovies, marine birds, and mammals are outstanding and reasonably well-known examples. These processes spread the ecological impact of the PSBF beyond its local domain. The front is also important in

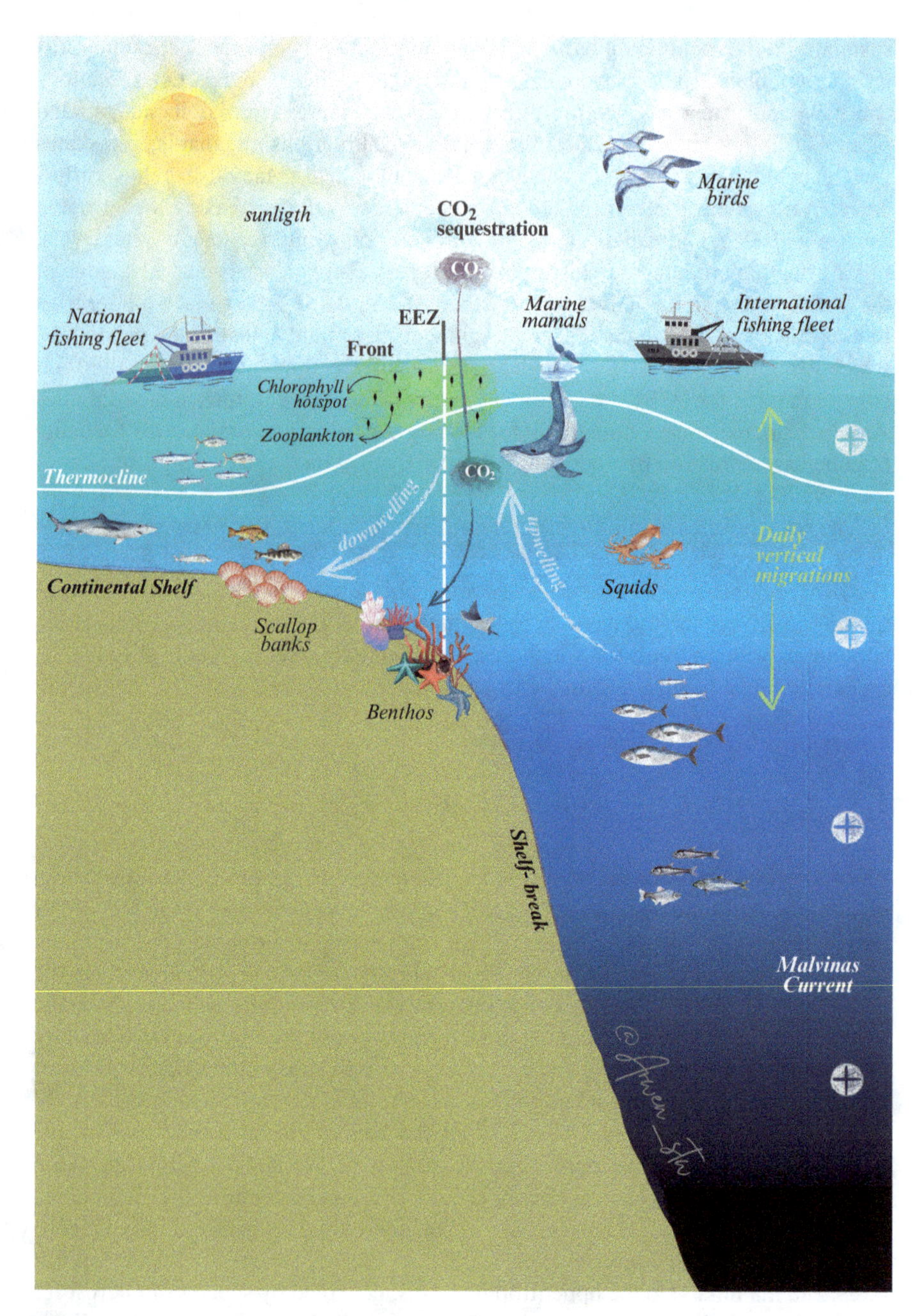

Fig. 1.2 Schematic diagram of relevant physical, biological, and biogeochemical processes in the Patagonian shelf-break front. The front is a multidimensional frontier where several natural and anthropogenic processes coexist in an intertwined way. The combination of solar and oceanographic energy (upwelling) triggers phytoplankton production and keeps the system running. ⊕ denotes the Malvinas Current flowing into the page

supporting benthic life: part of the primary production generated near the surface at the PSBF sinks and sustains an abundant and diverse benthic fauna, including concentrations of ecosystem engineers, such as scallop banks, sponge gardens, and deep-water corals, where biological interactions are presumably complex and diverse. Biogeochemical cycles are coupled to the high primary production, and high photosynthetic rates occurring along the front lead to increased CO_2 uptake from the atmosphere through the so-called biological pump.

A noteworthy scientific effort by Argentina began in recent years, which is improving our understanding of the ecological functioning of the PSBF, though several important subjects remain to be studied. The following chapters propose key scientific questions and topics that should receive priority attention in the near future:

To improve knowledge on the upwelling mechanisms that transfer nutrients to the lit upper layer, its variability at different timescales, and its impact on biological productivity through the trophic web, including commercial species and those of interest for wildlife conservation.

To identify the limiting factor(s) for photosynthesis in the Malvinas Current waters, which are defined as a High Nutrients-Low Chlorophyll ecosystem, to investigate the hypothesis that iron is the main limiting factor, and to identify putative iron sources promoting photosynthesis at the shelf-break front.

To understand benthic-pelagic coupling mechanisms. Such processes support benthic productivity and diversity and also play a key role in carbon sequestration. It will also be necessary to develop models that allow estimating CO_2 fluxes from satellite data. Understanding downwelling processes at the shelf-break and offshore the front is needed to study the role of vertical migrations of zooplankton and mesopelagic fishes in organic matter transport to the deep layers and the ocean bed.

To unveil the connectivity patterns between the PSBF and other ecosystems through larval drift and through the life cycles and migrations of nekton. This would advance the understanding of which shelf fishery resources are subsidized by the PSBF's biological production.

The PSBF lies close to the border separating the Argentine Exclusive Economic Zone from the adjacent high seas. It is, therefore, close to the limit between regulated and unregulated fishing domains. Thus, the PSBF is a boundary ecosystem from ecological and governance perspectives. In addition to the above-mentioned scientific issues, the maintenance of the ecological integrity of the PSBF will require concerted efforts by scientists, managers, and stakeholders.

Acknowledgments Many thanks to Mara Braverman and Valeria Falabella for preparing the artwork used in this chapter. This is INIDEP contribution no. 2381

References

Acha EM, Mianzan H, Guerrero R, Favero M, Bava J (2004) Marine fronts at the continental shelves of austral South America. Physical and ecological processes. J Mar Syst 44:83–105

Acha EM, Piola AR, Iribarne O, Mianzan H (2015) Ecological processes at marine fronts: oases in the ocean. In: Springer briefs in environmental science. Springer, New York

Ashjian CJ (1993) Trends in copepod species abundances across and along a gulf stream meander: evidence for entrainment and detrainment of fluid parcels from the Gulf stream. Deep Sea Research I 40(3):461–482

Bakun A (1996) Patterns in the ocean: ocean processes and marine population dynamics. Sea Grant, La Jolla

Belkin IM, Cornillon PC, Sherman K (2009) Fronts in large marine ecosystems. Prog Oceanogr 81:223–236

Bianchi AA, Ruiz Pino D, Isbert Perlender HG, Osiroff AP, Segura V, Lutz V, Clara ML, Balestrini CF, Piola AR (2009) Annual balance and seasonal variability of sea-air CO2 fluxes in the Patagonia Sea: their relationship with fronts and chlorophyll distribution. J Geophys Res 114:C03018

Bisbal GA (1995) The southeast south American shelf large marine ecosystem. Mar Policy 19(1):21–38

Brink KH (2016) Cross-shelf exchange. Annu Rev Mar Sci 8(1):59–78. https://doi.org/10.1146/annurev-marine-010814-015717

Chen L, Gao Z, Sun H, Chen B, Cai W-j (2014) Distributions and air-sea fluxes of CO2 in the summer Bering Sea. Acta Oceanol Sin 33(6):1–8. https://doi.org/10.1007/s13131-014-0483-9

Condie SA (1993) Formation and stability of shelf break fronts. J Geophys Res Oceans 98(C7):12405–12416. https://doi.org/10.1029/93JC00624

COPLA (2017) El margen continental argentino: entre los 35° S y los 55° S en el contexto del artículo 76 de la Convención de la Naciones Unidas sobre el Derecho del Mar. Edición biligûe. Ministerio de Relaciones Exteriores y Culto. Comisión Nacional del Límite Exterior de la Plataforma Continental, Ciudad Autónoma de Buenos Aires

Fournier RO (1978) Biological aspects of the Nova Scotian shelfbreak fronts. In: Oceanic fronts in coastal processes: proceedings of a workshop held at the marine sciences research center, May 25–27, 1977, 1978. Springer, p 69–77

Garvine RW, Wong K-C, Gawarkiewicz GG, McCarthy RK, Houghton RW, Aikman F III (1988) The morphology of shelfbreak eddies. J Geophys Res Oceans 93(C12):15593–15607. https://doi.org/10.1029/JC093iC12p15593

Gawarkiewicz G, Plueddemann AJ (2020) Scientific rationale and conceptual design of a process-oriented shelfbreak observatory: the OOI Pioneer Array. J Oper Oceanogr 13(1):19–36. https://doi.org/10.1080/1755876X.2019.1679609

Humborg C, Geibel MC, Anderson LG, Björk G, Mörth C-M, Sundbom M, Thornton BF, Deutsch B, Gustafsson E, Gustafsson B, Ek J, Semiletov I (2017) Sea-air exchange patterns along the central and outer East Siberian Arctic shelf as inferred from continuous CO2, stable isotope, and bulk chemistry measurements. Glob Biogeochem Cycles 31(7):1173–1191. https://doi.org/10.1002/2017GB005656

Jacobs SS (1991) On the nature and significance of the Antarctic Slope Front. Mar Chem 35(1):9–24. https://doi.org/10.1016/S0304-4203(09)90005-6

Kahl LC, Bianchi AA, Osiroff AP, Pino DR, Piola AR (2017) Distribution of sea-air CO2 fluxes in the Patagonian Sea: seasonal, biological and thermal effects. Cont Shelf Res 143:18–28. https://doi.org/10.1016/j.csr.2017.05.011

Lonardi AG, Ewing M (1971) Sediment transport and distribution in the Argentine basin. 4. Bathymetry of the continental margin, Argentine basin and other related provinces. Canyons and sources of sediemnts. In: Ahrens LH, Press F, Runcorn SK, Urey HC (eds) Physics and chemistry of the earth, vol 8. Pergamon Press, New York, p 81

Longhurst A (1998) Ecological geography of the sea. Academic, San Diego

Lozier MS, Reed MSC (2005) The influence of topography on the stability of Shelfbreak fronts. J Phys Oceanogr 35(6):1023–1036. https://doi.org/10.1175/JPO2717.1

Marrari M, Piola A, Valla D (2017) Variability and 20-year trends in satellite-derived surface chlorophyll concentrations in large marine ecosystems around south and Western Central America. Front Marine Science 4(372):1–17. https://doi.org/10.3389/fmars.2017.00372

Martinetto P, Alemany D, Botto F, Mastrángelo M, Falabella V, Acha EM, Antón G, Bianchi A, Campagna C, Cañete G (2020) Linking the scientific knowledge on marine frontal systems with ecosystem services. Ambio 49:541–556

Olson DB (2002) Biophysical dynamics of ocean fronts. In: Robinson AR, McCarthy JJ, Rothschild BJ (eds) The sea. Biological-physical interactions in the sea, vol 12. Wiley, New York, pp 187–218

Parker G, Paterlini MC, Violante RA (1997) El fondo marino. In: Boschi EE (ed) El Mar Argentino y sus recursos pesqueros, vol Tomo 1: Antecedentes históricos de la exploraciones en el mar y las características ambientales. Instituto Nacional de Investigación y Desarrollo Pesquero, Mar del Plata, Argentina, p 65–87

Piola AR, Martínez Avellaneda N, Guerrero RA, Jardón FP, Palma ED, Romero SI (2010) Malvinas-slope water intrusions on the northern Patagonia continental shelf. Ocean Sci 6:345–359

Rippeth TP, Lincoln BJ, Kennedy HA, Palmer MR, Sharples J, Williams CAJ (2014) Impact of vertical mixing on sea surface pCO2 in temperate seasonally stratified shelf seas. J Geophys Res Oceans 119(6):3868–3882. https://doi.org/10.1002/2014JC010089

Romero SI, Piola AR, Charo M, Eiras Garcia CA (2006) Chlorophyll a variability off Patagonia based on SeaWiFS data. J Geophys Res 111:C05021

Sherman K (2005) The large marine ecosystem approach for assessment and management of ocean coastal waters. In: Hennessey TM, Sutinen JG (eds) Sustaining large marine ecosystems: the human dimension. Elsevier, Amsterdam, pp 3–16

Siedlecki SA, Archer DE, Mahadevan A (2011) Nutrient exchange and ventilation of benthic gases across the continental shelf break. J Geophys Res Oceans 116(C6). https://doi.org/10.1029/2010JC006365

Sournia A (1994) Pelagic biogeography and fronts. Prog Oceanogr 34:109–120

Wang Y, Qi D, Wu Y, Gao Z, Sun H, Lin H, Pan J, Han Z, Gao L, Zhang Y, Chen L (2021) Biological and physical controls of pCO2 and air-sea CO2 fluxes in the austral summer of 2015 in Prydz Bay. East Antarct Marine Chem 228:103897. https://doi.org/10.1016/j.marchem.2020.103897

Chapter 2
Anatomy and Dynamics of the Patagonia Shelf-Break Front

Alberto R. Piola, Nicolás Bodnariuk, Vincent Combes, Bárbara C. Franco, Ricardo P. Matano, Elbio D. Palma, Silvia I. Romero, Martin Saraceno, and M. Milagro Urricariet

Abstract The Patagonia shelf-break front presents sharp offshore changes in surface temperature, salinity, chlorophyll, and horizontal velocity. In summer, the cross-shore temperature and salinity changes are not uniform, suggesting the existence of multiple fronts. In winter, the offshore changes are fairly uniform, displaying a single thermohaline front located just offshore from the shelf-break. Cross-front temperature and salinity present significant seasonal variations associated with intense vertical stratification over the shelf during summer. The thermocline provides a density interval for cross-front isopycnal exchange, which may fertilize the outer shelf waters. The salinity front extends from the surface to the bottom and is observed year-round. Frontal displacements occur throughout the water column. The high surface chlorophyll along the front suggests a sustained nutrient flux to the shelf-break upper layer. Numerical experiments indicate intense frontal upwelling

Supplementary Information The online version contains supplementary material available at https://doi.org/10.1007/978-3-031-71190-9_2.

A. R. Piola (✉)
Facultad de Ciencias Exactas y Naturales, Departamento de Ciencias de la Atmósfera y los Océanos, Universidad de Buenos Aires, Buenos Aires, Argentina

Instituto Franco-Argentino de Estudios sobre el Clima y sus Impactos, CONICET/UBA/CNRS/IRD, Buenos Aires, Argentina

N. Bodnariuk · M. Saraceno
Facultad de Ciencias Exactas y Naturales, Departamento de Ciencias de la Atmósfera y los Océanos, Universidad de Buenos Aires, Buenos Aires, Argentina

Instituto Franco-Argentino de Estudios sobre el Clima y sus Impactos, CONICET/UBA/CNRS/IRD, Buenos Aires, Argentina

Centro de Investigaciones del Mar y la Atmósfera, CONICET/UBA, Buenos Aires, Argentina

V. Combes
Institut Mediterrani d'Estudis Avançats (IMEDEA), Esporles, Spain

Departament de Física, Universitat de les Illes Balears, Palma de Mallorca, Spain

E. M. Acha et al. (eds.), *The Patagonian Shelfbreak Front*, Aquatic Ecology Series 13, https://doi.org/10.1007/978-3-031-71190-9_2

mediated by the interaction of the Malvinas Current with the bottom topography and suggest that upwelling in upstream portions of the shelf-break, advected northward along the shelf edge, may further modulate the nutrient fluxes required to sustain frontal productivity. A southward displacement of the northernmost extension of the front observed during the past decades may have biological and biogeochemical impacts.

Acronyms and Abbreviations

ADCP	Acoustic Doppler current profiler
DUCAS	Data unification and altimeter combination system
MIS	Malvinas Islands Shelf
MODIS	Moderate-resolution imaging spectroradiometer
PSBF	Patagonia Shelf-Break Front
SMAP	Soil moisture active passive
SMOS	Soil moisture ocean salinity
SSS	Sea surface salinity
SST	Sea surface temperature
SWOT	Surface water and ocean topography

B. C. Franco
Instituto Franco-Argentino de Estudios sobre el Clima y sus Impactos, CONICET/UBA/CNRS/IRD, Buenos Aires, Argentina

Centro de Investigaciones del Mar y la Atmósfera, CONICET/UBA, Buenos Aires, Argentina

R. P. Matano
Oregon State University, Corvallis, OR, USA

E. D. Palma
Departamento de Física, Universidad Nacional del Sur, Bahía Blanca, Argentina

Instituto Argentino de Oceanografía (IADO), CONICET, Bahía Blanca, Argentina

S. I. Romero
Facultad de Ciencias Exactas y Naturales, Departamento de Ciencias de la Atmósfera y los Océanos, Universidad de Buenos Aires, Buenos Aires, Argentina

Instituto Franco-Argentino de Estudios sobre el Clima y sus Impactos, CONICET/UBA/CNRS/IRD, Buenos Aires, Argentina

Servicio de Hidrografía Naval, Buenos Aires, Argentina

Universidad de la Defensa Nacional, Buenos Aires, Argentina

M. M. Urricariet
Facultad de Ciencias Exactas y Naturales, Departamento de Ciencias de la Atmósfera y los Océanos, Universidad de Buenos Aires, Buenos Aires, Argentina

Servicio de Hidrografía Naval, Buenos Aires, Argentina

2.1 Introduction

Shelf-break fronts mark the transition between distinct continental shelf and deep ocean waters. Shelf-break fronts are primarily driven by local wind forcing, western boundary currents, and, in some regions (e.g., the Mid-Atlantic Bight, the eastern Bering Sea, and the Celtic Sea), by the offshore flux of fresher shelf waters (Gawarkiewicz and Chapman 1992). The shelf-break circulation is largely dominated by along-shore flows, but given the relatively large cross-front gradients of momentum and properties, even weak cross-front and associated vertical fluxes can play a significant role in shelf dynamics, biology, and biogeochemistry (Brink 2016). Thus, shelf-break regimes have a profound impact throughout the marine food web (Simpson and Sharples 2012).

Classical shelf-break fronts were associated with a density contrast induced by the continental discharge of freshwater, which lowers the salinity and density of the shelf waters (Mann and Lazier 2005). Enhanced vertical circulations around shelf-break fronts were thought to be the main drivers of frontal productivity. The observation of low surface temperatures in the vicinity of high chlorophyll-*a* concentration patches has been interpreted as indicative of upwelling along the shelf-break. Throughout the remainder of this chapter, chlorophyll refers specifically to chlorophyll-*a*. The proposed upwelling mechanism has been related to the downslope buoyancy flux associated with the presence of a shelf current (i.e., Gawarkiewicz and Chapman 1992) or tidally driven internal waves (Pingree and Mardell 1981). In addition, other processes, such as internal tides generated at the shelf-break, have been postulated as modulators of the depth of the surface mixed layer leading to nutrient entrainment (e.g., Longhurst 2007; Xie et al. 2018 and references therein). Frontal instabilities were thought to further enhance upward nutrient fluxes.

The Patagonia shelf-break front (PSBF) is physically and biologically unique in several aspects: it is bounded inshore by a wide shelf and offshore by an equatorward flowing western boundary current (the Malvinas Current) carrying cold subantarctic waters as far as 38 °S (Fig. 2.1). The latitudinal span and width of the shelf, along with the presence of the Malvinas Current, conspire to generate a shelf-break front extending for more than 1500 km and located several hundred kilometers from the coast. Shelf-break upwelling is not driven by the shelf circulation nor tidal forcing but by the lateral spreading of the Malvinas Current (Matano and Palma 2008). Frontal productivity at the PSBF, however, may be further enhanced by turbulent wind-driven mixing, tidal forcing, and topographic features observed along the shelf-break.

The PSBF upwelling fuels a primary productivity hotspot that spreads throughout the marine food web (Acha et al. 2004; Martinetto et al. 2020, and other chapters published in this book). Along the PSBF, there is an extensive and continuous band of high chlorophyll, which is apparent in both in-situ (Brandhorst and Castello 1971; Carreto et al. 1995; Garcia et al. 2008; Lutz et al. 2010) and satellite-based observations (Brown and Podestá 1997; Acha et al. 2004; Saraceno et al. 2005; Rivas et al. 2006; Romero et al. 2006; Saraceno et al. 2006). Model estimates of

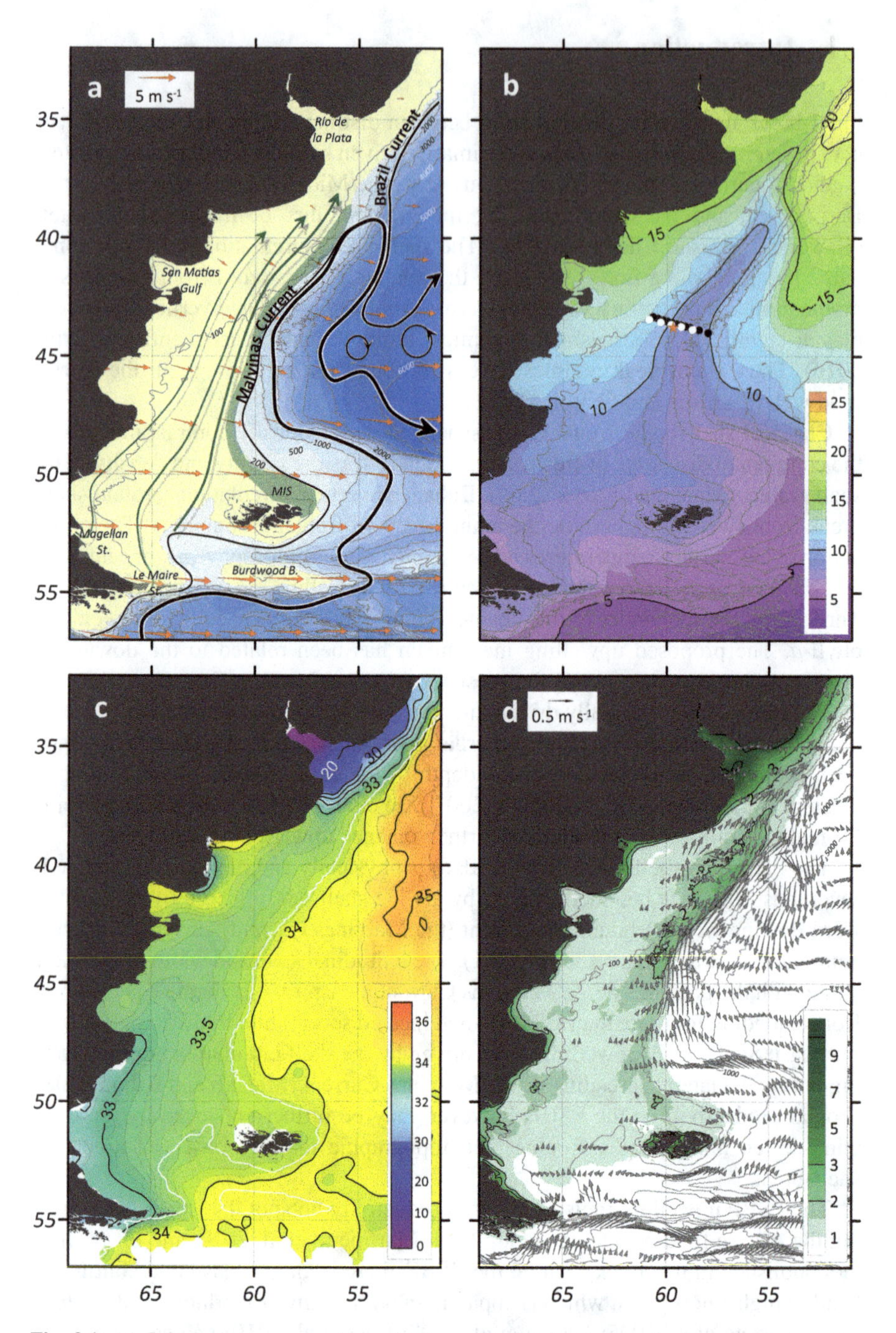

Fig. 2.1 (**a**) Schematic mean circulation in the southwest South Atlantic. The shelf circulation (green lines) is based on Palma et al. (2008) and Combes and Matano (2018). The offshore flow (heavy black lines) is based on the CNES-CLS22 mean dynamic topography. The background

upwelling velocities are comparable to those of eastern boundary current systems (Matano et al. 2010). The northern extension of the high-chlorophyll band along the shelf-break, which matches the northernmost extension of the Malvinas Current, and the abrupt drop in chlorophyll concentration east of the shelf-break suggest that the current and bottom topography play a key role in the upwelling dynamics, as indicated by numerical models.

This chapter focuses on the physical and dynamical characteristics of the PSBF based on satellite and in-situ observations, and numerical simulations. The chapter is organized as follows: Sect. 2.2 presents the regional setting; Sect. 2.3 describes the cross-shelf characteristics at the sea surface at selected latitudes; Sect. 2.4 presents the subsurface characteristics of the shelf-break front; Sect. 2.5 discusses the proposed upwelling mechanisms; and Sect. 2.6 briefly describes the front's long-term variability and trends. Finally, Sect. 2.7 summarizes and presents challenges and perspectives for future research.

2.2 Regional Setting

South of the Río de la Plata estuary (~35 °S), the continental shelf of eastern South America widens from ~180 km at 38 °S to 430 km at 45 °S and exceeds 800 km at 50.5 °S (Fig. 2.1a). Excluding the relatively deep north-Patagonian gulfs, the bottom of the shelf is relatively flat, with a gentle offshore slope (~1:1000) and depths mostly shallower than 180 m. The offshore edge of the continental shelf is characterized by an abrupt increase in the bottom slope, reaching about 1:50, and it is broadly defined by the location of the 200 m isobath (Parker et al. 1997).

Figure 2.1b presents the climatological average of sea surface temperature (SST) prepared using monthly 4 km resolution nighttime temperature data from the Moderate-resolution Imaging Spectroradiometer (MODIS) on the Aqua satellite. Daily data are available at NASA's Physical Oceanography Distributed Active Data Center (NASA OBPG 2020; Kilpatrick et al. 2015). We use data from January 2003 to December 2022. The most prominent feature in Fig. 2.1b is the wedge of cold waters extending along the upper continental slope, approximately following the 1000–2000 m isobaths, and reaching 38 °S. The cold waters are primarily advected by the Malvinas Current, a northward-flowing western boundary current observed along the western flank of the Argentine Basin (e.g., Piola and Gordon 1989; Peterson and Whitworth 1989; Saunders and King 1995; Vivier and Provost 1999;

Fig. 2.1 (continued) shading indicates the bottom topography (in m) based on GEBCO and the green shaded area indicates the shelf-break front. (**b-d**) Time-averaged surface distributions of temperature (**b**, °C, 2003–2022), salinity (**c**), and chlorophyll-*a* concentration (**d**, mg m^{-3}, 2003–2022). Mean winds are shown by orange arrows in (**a**) (m s^{-1}). Mean current vectors are shown by gray arrows in (**d**). Black and white circles in (**b**) indicate the locations of stations used to prepare Fig. 2.3. The orange star in (**b**) marks the position of the shelf-break mooring. The 100, 200, 1000, 2000, 3000, 4000, 5000, and 6000 m isobaths are shown in light gray lines in panels (**a**), (**b**), and (**c**), and the thin white line in panel (**c**) is the 200 m isobath

Paniagua et al. 2021; Frey et al. 2021). Analyses of satellite-derived SST fronts based on edge detection algorithms indicate that the shelf-break is associated with a thermal front (Saraceno et al. 2004; Rivas and Pisoni 2010). These studies show that north of 45 °S, the front is located roughly over the 300 m isobath and is relatively stable, but its intensity presents substantial seasonal variability, being more intense during the austral summer and fall. Franco et al. (2008) described multiple thermal fronts north of 44 °S close to the shelf-break, with the main front being closely located over the 200 m isobath, and two additional fronts located farther onshore and offshore. The fronts emerged at the onshore side of cold filaments, which are more frequently observed during the austral summer and fall. Numerical simulations suggest that these multiple fronts reflect the westward drift of eddies shed at the Malvinas/Brazil Confluence (Combes and Matano 2014).

The relatively fresh subantarctic waters flowing over the Patagonian shelf are diluted by low-salinity inflows from the Le Maire and Magellan Straits. The freshening of these waters is driven by an excess of precipitation and continental runoff along the southern Chilean coast (Dávila et al. 2002). A surface salinity distribution, prepared from 29,710 historical hydrographic stations, is presented in Fig. 2.1c. The data were obtained from holdings at the Instituto Nacional de Investigación y Desarrollo Pesquero (Mar del Plata, Argentina, https://www.argentina.gob.ar/inidep), the Centro Argentino de Datos Oceanográficos (http://www.hidro.gov.ar/ceado/ceado.asp), and the World Ocean Database (https://www.ncei.noaa.gov/products/world-ocean-database). Numerical models indicate additional inflows to the continental shelf of relatively cold, salty subantarctic waters east of Isla de los Estados (Combes and Matano 2018; Guihou et al. 2020, Palma et al. 2021). As a result, salinities over the continental shelf are typically lower than 33.8, except for local near-coastal maxima associated with excess evaporation over precipitation, such as in San Matías Gulf (Fig. 2.1c, Brun et al. 2020 and references therein). The 33.8–33.9 isohalines lie close to the 200 m isobath at the shelf edge.

Observations reveal a substantial increase in sea surface chlorophyll close to the PSBF, attributed to a sustained upwelling of nutrient-rich Malvinas Current waters (Brandhorst and Castello 1971; Matano and Palma 2008). In-situ observations, for example, show surface chlorophyll concentrations exceeding 2 mg m^{-3} near 39 °S (Carreto et al. 1995) and values exceeding 10 mg m^{-3} near 44 °S (García et al. 2008; Lutz et al. 2010; Carreto et al. 2016). High chlorophyll concentrations are associated with similarly high nitrate concentrations, frequently exceeding 20 μM. There is a nearly linear relationship between temperature and nitrate concentration, associating cold waters with nutrient-rich waters (Carreto et al. 1995). Satellite observations indicate that, excluding the region under the influence of the Río de la Plata, which is characterized by sediment-laden, optically complex waters, the PSBF presents the highest chlorophyll concentration in the southwest South Atlantic (Saraceno et al. 2005), characterized by a band of high chlorophyll concentration that is particularly notable during the austral summer (e.g., Romero et al. 2006; Rivas et al. 2006; Rivas 2006; Delgado et al. 2023), when nutrients have been mostly exhausted after the spring bloom in the mid-shelf region (Carreto et al. 1995). The long-term annual mean of chlorophyll concentration presents a maximum exceeding 2 mg m^{-3}

along the shelf-break north of 45 °S, and 1.7 mg m^{-3} farther south (Fig. 2.1d), while in austral summer it exceeds 3 mg m^{-3} (Romero et al. 2006). The chlorophyll concentration drops abruptly east of the shelf-break, which is indicative of the impact of the shelf-break on the abundance of phytoplankton. Field observations (Bowie et al. 2002) and numerical simulations (Tagliabue et al. 2009, Matano et al. 2019) suggest that shelf-break upwelling may also contribute to the injection of sediment-borne micronutrients into the photic layer, required to sustain the growth of phytoplankton (e.g., Garcia et al. 2008). However, other models suggest that micronutrients required to sustain phytoplankton production at the shelf-break are mostly advected into the region from the southeast Pacific (Song et al. 2016). In addition, aeolian sources may also contribute to the supply of micronutrients at a regional scale, but their potential impact on phytoplankton growth is controversial (Cosentino et al. 2020; Simonella et al. 2022). The scarcity of micronutrient observations poses a serious limitation to the validation of numerical models and to advancing the understanding of processes leading to their input into the euphotic layer.

The high primary productivity at the shelf-break sustains the growth of numerous species throughout the food web, reaching top predators, including sea birds and commercially relevant pelagic and benthic species (Acha et al. 2004; Martinetto et al. 2020, and references therein). In addition, many migratory species are associated with the shelf-break at some stage of their life cycle (e.g., Arkhipkin et al. 2012; Martinetto et al. 2020; Torres Alberto et al. 2021). Up-to-date descriptions of the ecological impact of the PSBF are provided in other chapters of this book.

2.3 The Surface Expression of the Shelf-Break Front

The cross-shelf characteristics at the surface of the PSBF and their variability in time and space are described based on satellite data. The cross-front structure of surface temperature, salinity, chlorophyll concentration, and meridional velocity during the peak months of austral summer (February) and winter (August) at 45 °S is displayed in Fig. 2.2. The zonal profiles span from the mid-shelf (63 °W) to the region beyond the Malvinas Current, where the bottom depth exceeds 4500 m (57 °W). At 45 °S, the shelf-break is located close to the 170 m isobath. In these profiles, the cross-front distance is displayed relative to the location of the shelf-break.

2.3.1 Sea Surface Temperature

To analyze the cross-shore structure of SST, we used the same satellite product described in Sect. 2.2 (Kilpatrick et al. 2015). Mean patterns at 45 °S are representative of those observed at other latitudes (not shown). In February, SST decreases from 17 °C in the mid-shelf to a minimum of 12 °C close to the core of the Malvinas Current, which is located approximately 50–80 km from the shelf-break (Fig. 2.2a). The cross-shore SST gradient is not uniform; it displays well-defined minima exceeding −3 °C/100 km (f1-f3, Fig. 2.2b). The negative sign denotes the eastward

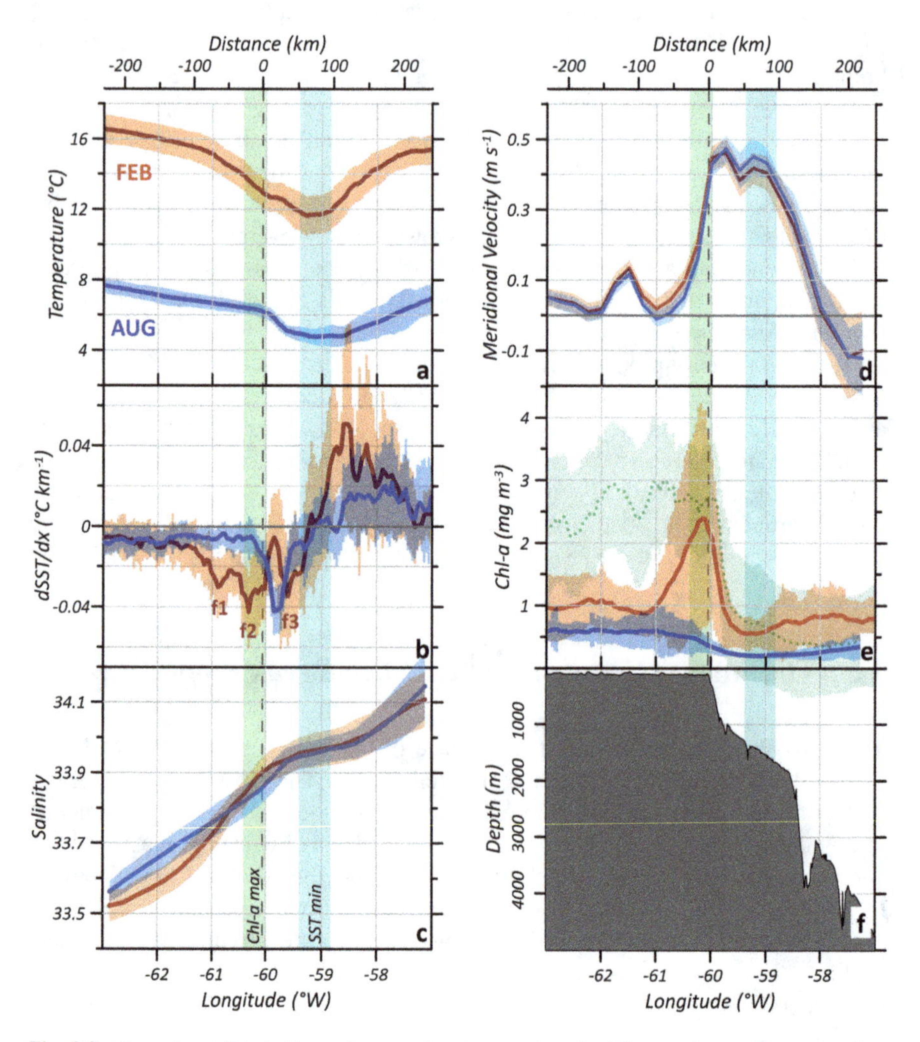

Fig. 2.2 Cross-front distributions of sea surface temperature (**a**, °C), zonal temperature gradient (**b**, °C/km), sea surface salinity (**c**), along-shelf geostrophic velocity (**d**, m s^{-1}), chlorophyll-*a* concentration (**e**, mg m^{-3}), and bottom depth (**f**, m). All sections correspond to a nominal latitude of 45 °S. Means and standard deviations in February (red, summer) and August (blue, winter) are indicated by lines and shadings in panels (**a–e**), respectively. The mean and standard deviation of chlorophyll-*a* distribution in October (spring) are also shown in panel **e** (green dotted line and shading, respectively). f1–f3 in panel **b** indicate the three cross-shore SST fronts observed in summer. The vertical light green and cyan shadings indicate the position of the chlorophyll maximum and sea surface temperature minimum in summer, respectively, and the vertical dashed gray line marks the location of the shelf-break. Distance (km) relative to the shelf-break is shown on the upper axes

temperature decrease ($\partial T/\partial x < 0$). A broad onshore SST front is located over the outer continental shelf; it is approximately 100 km wide and includes two distinct gradient minima (f1 and f2, Fig. 2.2b). The offshore front is located on the western edge of the wedge of cold surface waters, about 40 km east of the shelf-break (f3, Fig. 2.2b). In agreement with previous studies, the intensity of the cross-front temperature gradient intensifies at lower latitudes (Saraceno et al. 2004; Franco et al. 2008; Rivas and Pisoni 2010). For example, at 43 °S, the cross-shelf gradient exceeds −6 °C/100 km (not shown).

In August, the peak of the austral winter, the SST over the continental shelf decreases uniformly from 8 to 5 °C in the offshore direction and displays a smaller standard deviation than in summer, indicating a similarly smaller interannual variability (Fig. 2.2a). At the shelf-break, the SST drops abruptly, forming a well-defined surface temperature front. In contrast with summer, during winter there is only one intense SST front located about 20 km east of the shelf-break, close to the midpoint between the summer fronts (Fig. 2.2b). This is also observed at 43 °S and 47 °S (not shown). The front in winter is as intense as in summer, reaching −4 °C/100 km. A similar pattern is also observed farther south, though the thermal fronts are displaced offshore of the shelf-break (not shown).

2.3.2 *Sea Surface Salinity*

The sea surface salinity (SSS) analysis (Fig. 2.2c) is based on the global Level-4 analyses Multi Observation Global Ocean Sea Surface Salinity and Sea Surface Density, obtained through a multivariate optimal interpolation algorithm that combines multi-platform satellite observations (NASA's Soil Moisture Active Passive, SMAP, and ESA's Soil Moisture Ocean Salinity, SMOS) with in-situ salinity measurements developed by the Italian Consiglio Nazionale delle Ricerche. We use monthly 0.25° × 0.25° data downloaded from https://data.marine.copernicus.eu/product/MULTIOBS_GLO_PHY_S_SURFACE_MYNRT_015_013/description. The analysis includes data from January 1993 to December 2021. As expected from the long-term mean distribution, SSS increases in the offshore direction (Fig. 2.2c). The summer offshore increase in SSS is not spatially uniform; it increases sharply on the outer shelf and decreases off the continental shelf, while in winter, SSS increases uniformly over the shelf, displaying a sharp increase just east of the shelf-break, close to the location of the largest winter SST gradient. Thus, the cross-shore structures of SST and SSS present similar patterns: both show an increased zonal gradient (of opposite sign) on the outer shelf in summer and a uniform gradient over the mid and outer shelf in winter (Fig. 2.2c). Though the available resolution of satellite-derived salinity (several tens of km) is too low to quantify the cross-front salinity gradients and their time variability, the offshore salinity increase combined with the offshore cooling leads to an offshore increase in surface density, which maximizes at the shelf-break. The cross-shelf density pattern will be described in Sect. 2.4 based on hydrographic observations.

2.3.3 *Sea Surface Velocity*

The meridional component of geostrophic velocity derived from the altimetry-based absolute dynamic topography is employed to unveil the mean cross-slope structure of the along-slope flow (Fig. 2.2d). To estimate the meridional component of the geostrophic velocity, we combined daily sea-level anomalies with the mean dynamic topography. Sea-level anomalies are available at 0.25° × 025° resolution derived from the Data Unification and Altimeter Combination System (DUCAS) multi-mission altimeter data processing system (https://data.marine.copernicus.eu/product/SEALEVEL_GLO_PHY_L4_MY_008_047/description) for the period from January 1993 to December 2022. The mean dynamic topography CNES-CLS-2022 is the recent global estimate of the mean sea surface height above the geoid with a 0.125° × 0.125° resolution over the period 1993–2012. The data are available at the Copernicus Marine data portal (https://doi.org/10.48670/moi-00150). The CNES-CLS-2022 was constructed following similar procedures to the previous versions (e.g., MDT CNES-CLS18). In addition to the improved spatial resolution, it incorporates extended gravimeter, CTD, Argo, and altimetry data (Mulet et al. 2023). Though compared with the previous CNES-CLS18, the CNES-CLS22 represents a moderate improvement in global RMS differences between the geostrophic velocities and independent surface drifter observations, the improvement seems to be better in the western South Atlantic (Mulet et al. 2023). To estimate the absolute dynamic topography and the associated meridional component of absolute geostrophic velocity, the mean dynamic topography was subsampled and interpolated to the sea-level anomaly grid. The resulting geostrophic velocities are presented in Fig. 2.2d.

The meridional velocity data at 45 °S display a sharp increase at the shelf-break, with a broad northward maximum exceeding 0.45 m s^{-1} associated with the intense Malvinas Current (Fig. 2.2d). In contrast to the temperature and salinity distributions, which present different patterns in austral summer and winter, the velocity estimates in February (red line in Fig. 2.2d) are similar to the August velocities (dark blue line in Fig. 2.2d). In addition, both velocity profiles show a relative maximum (~0.12 ms^{-1}) over the shelf, located about 100 km onshore from the shelf-break. Previous analyses of altimetry-derived circulation over the shelf close to 45 °S suggest an along-shore velocity increase in March compared to September and a sharp increase approaching the shelf-break (Ruiz Etcheverry et al. 2016). Two velocity maxima are observed offshore from the shelf-break; the most intense is located 10–20 km east of the shelf-break, and the other is observed about 40 km further offshore. The above observations therefore suggest that the dual jet structure of the Malvinas Current is a permanent feature at around 45 °S. It should be noted that the two-jet configuration only emerges when the altimetry sea-level anomalies are combined with the recently developed high-resolution mean dynamic topography CNES-CLS22.

Ship-borne acoustic Doppler profiler observations, mostly collected in austral summer near 45 °S, indicate that the flow is organized into two well-defined high-velocity branches (e.g., Piola et al. 2013). Similarly, acoustic Doppler current profiler (ADCP) data collected at other locations suggest that the multiple-core structure of the Malvinas Current is ubiquitous (e.g., Painter et al. 2010; Krechik 2020; Frey et al. 2021; Salyuk et al. 2022), with these cores converging towards the Brazil/Malvinas Confluence (Frey et al. 2023). Near 45 °S, ADCP data consistently display an onshore high-velocity core located within 10–20 km offshore from the shelf-break; this flow exceeds 0.4 m s^{-1}, is about 20 km wide, and extends close to the bottom (200–500 m, Piola et al. 2013; Frey et al. 2021; Salyuk et al. 2022). These data also show another high-velocity core over the 1400 m isobath. A comparison of the ADCP observations near 45 °S suggests that the offshore current branch is more variable in width and intensity (Painter et al. 2010; Piola et al. 2013; Frey et al. 2021; Salyuk et al. 2022). The offshore branch extends deeper into the water column, accounting for over 80% of the total northward transport of the Malvinas Current (Frey et al. 2021).

The existence of two jets near 45 °S is supported by numerical experiments. Based on the results of a global, eddy-permitting model, Fetter and Matano (2008) reported that after leaving the Drake Passage, the northernmost portion of the Antarctic Circumpolar Current—the source of the Malvinas Current—splits into two branches, which flow along the Malvinas Plateau and merge into a single jet north of 45 °S. This pattern is similar to that depicted by altimeter observations (e.g., Fig. 2.1d). High-resolution experiments, furthermore, indicate that although a multiple jet structure of the Malvinas Current is frequently observed in snapshots of cross-slope velocity sections, north of 45 °S the jets are highly intermittent and are not apparent in the time-averaged circulation (Combes and Matano 2014). Instead, they are associated with the transient signal generated by the westward drift of mesoscale eddies detached from the Brazil/Malvinas Confluence. Recent Lagrangian analyses of drifters and altimetry data suggest that topographic waves propagating along the slope could modulate the location of the jets and associated chlorophyll maxima at synoptic timescales (Saraceno et al. 2024; Bodnariuk et al. 2024). Northward propagation of sea-level anomalies along the slope modulates the eastern portion of the Malvinas Current at this latitude (Paniagua et al. 2021). Altimetry-derived sea-level anomaly shows high eddy kinetic energy associated with the mesoscale activity in the Brazil/Malvinas Confluence and a sharp decay close to the core of the Malvinas Current (e.g., Artana et al. 2018). This is consistent with the increased standard deviations in the monthly mean meridional velocities observed east of 58 °W (Fig. 2.2d).

2.3.4 Sea Surface Chlorophyll

To describe the long-term average distribution of the cross-shelf structure of surface chlorophyll, we use NASA's MODIS Aqua Level 3 monthly mean observations of 4 km resolution from January 2002 to December 2022 (Hu et al. 2012). Surface chlorophyll concentration is not normally distributed (Campbell 1995); therefore, long-term averages are computed over the log-transformed (base 10) chlorophyll. The chlorophyll concentration over the continental shelf undergoes a strong seasonal cycle characterized by a spring bloom and a minimum in winter (e.g., Romero et al. 2006). For reference, the October concentration representative of the spring bloom is also displayed in Fig. 2.2e (green dotted line). During austral summer, the chlorophyll concentration over the mid-shelf region drops to significantly lower values as stratification develops and nutrients are consumed in the upper layer (Carreto et al. 1995). Thus, the cross-shelf distribution of surface chlorophyll in summer at 45 °S presents relatively low concentrations over the continental shelf (~1 mg m^{-3}) and a relatively sharp peak in the outer shelf and upper slope on average reaching 2.5 mg m^{-3} (Fig. 2.2e). In agreement with the regional distribution (Fig. 2.1d), the surface chlorophyll sharply decreases further offshore, displaying the low values characteristic of the open ocean subantarctic waters (0.5 mg m^{-3}). In winter (August), the chlorophyll concentrations are much lower throughout the region, presenting values close to 0.5 mg m^{-3} over the continental shelf and decreasing to ~0.2 mg m^{-3} in the open ocean. The surface chlorophyll distributions present substantial interannual variability, as displayed by the large standard deviations. This will be discussed in Sects. 2.5.2 and 2.6.

2.4 The Subsurface Characteristics

2.4.1 Cross-Shelf Distributions

Hydrographic sections occupied across the shelf-break near 44 °S illustrate late austral summer and winter patterns in vertical stratification across the front (Fig. 2.3). Although some of the features described below are present in several cross-shelf-break occupations, it should be noted that the following descriptions are based on synoptic observations. The summer observations were collected on 12 March 1994 with an MKIII Neil Brown Instruments Systems CTD profiler. The winter data were obtained on 9 September 2006 using a 911+ SeaBird Scientific CTD (Charo and Piola 2014). The summer and winter data are available at NOAA's National Centers for Environmental Information World Ocean Database under accession numbers 0038589 and 0110317, respectively.

In summer, the outer shelf presents a shallow mixed layer (~15 °C, 20 m), a sharp seasonal thermocline, with Brunt-Väisälä frequencies exceeding 15 cycles per hour (Bianchi et al. 2009; Valla and Piola 2015; Zhang et al. 2018), and a cold-fresh

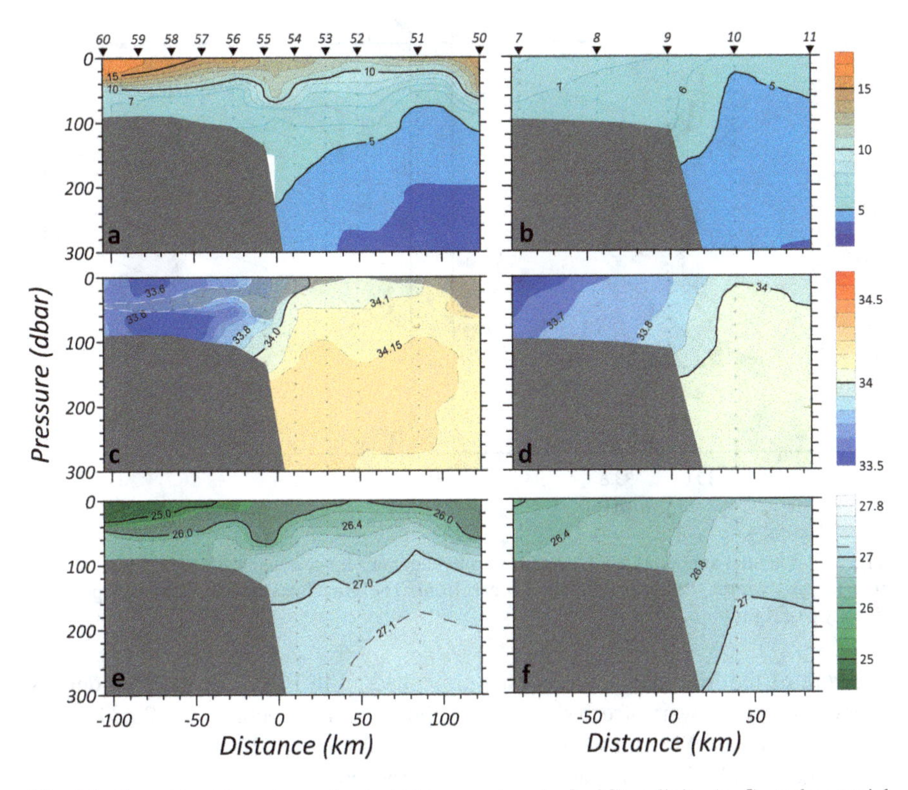

Fig. 2.3 Cross frontal sections of potential temperature (**a**, **b**, °C), salinity (**c**, **d**), and potential density anomaly (**e**, **f**, σ_θ, kg m^{-3}) as observed on 12 March 1994 (austral summer, left panels) and 9 September 2007 (late austral winter, right panels). The gray shadings in panels (**c**) and (**e**) indicate the 25.4–26 σ_θ range. Distance (km) is shown relative to the shelf-break. Station locations are shown in Fig. 2.1b

bottom layer (<7 °C, 33.6). The upper layer temperature decreases offshore, reaching 11 °C about 50 km east of the shelf-break (station 52, Fig. 2.3a). Salinity increases in the offshore direction from 33.6–33.7 on the outer shelf to 34.1 within the core of cold waters, characteristic of pure subantarctic waters (Fig. 2.3c). At the shelf-break, isohalines slope downward in the onshore direction, forming a relatively intense cross-slope salinity gradient near the bottom. This is a permanent feature of the PSBF (e.g., Martos and Piccolo 1988; Carreto et al. 1995; Romero et al. 2006; Painter et al. 2010; Piola et al. 2013; Carreto et al. 2016; Severo et al. 2024). The vertical stratification is mostly controlled by temperature, displaying a strong pycnocline over the continental shelf at about 40 m depth, with shallower isopycnals (σ_θ ~25–25.6 kg m^{-3}) outcropping near the shelf-break (Fig. 2.3e). Note that, in the shelf-break region, the salinity gradient below 50 m induces a well-defined cross-shelf density gradient, which is indicative of limited cross-shelf exchange (Piola et al. 2010). Consequently, a relatively well-defined range of densities across the shelf-break provides an isopycnal connection between the shelf and

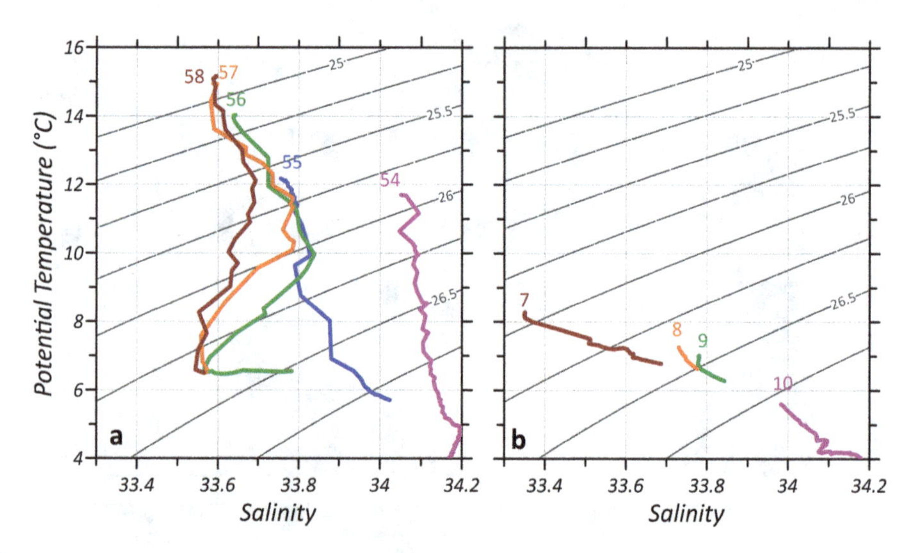

Fig. 2.4 Potential temperature (°C) versus salinity diagrams for selected stations occupied in summer (**a**) and winter (**b**) across the shelf-break front. The gray lines indicate constant potential density (σ_θ, Kg m^{-3})

the core of the Malvinas Current (see gray shading in Fig. 2.3c, e). This density range is set by the warm-fresh shelf waters above the thermocline ($\sigma_\theta < 25.8$ kg m^{-3}) and the cold-salty Malvinas Current waters ($\sigma_\theta > 26.4$ kg m^{-3}).

Cross-frontal mixing and advection impact the temperature and salinity distributions. Note, for example, the high-salinity intrusion in the outer shelf at 40–60 m depth (stations 55–57 in Fig. 2.3c). Enhanced mixing along this density range is apparent from the frequent observation of similar high-salinity intrusions at the thermocline level in the outer shelf during austral summer (e.g., station 56 in Fig. 2.3b, see also Severo et al. 2024, their Fig. 2.2a). Further evidence of this exchange is provided by T–S distributions (Fig. 2.4). Diapycnal mixing with low-salinity shelf waters progressively erodes the salinity anomaly in the onshore direction (station 58, Fig. 2.4). At station 56, the intrusion is 25 m thick, and salinity at the core reaches 33.84, but as a result of diapycnal mixing, at station 57, the salty intrusion widens vertically to 40 m, and salinity at its core decreases to 33.73.

The intrusion of salty waters at mid-thermocline level in the outer shelf enhances the vertical stratification at the top of the intrusion, where both the vertical temperature and salinity gradients lead to a downward density increase. In contrast, at the base of the intrusion, the downward salinity decrease acts against the temperature gradient and weakens the vertical stratification. This effect leads to the nearly orthogonal intersection of the T–S profiles and the isopycnals at the top of the intrusion, and a tendency to become parallel to the isopycnals at its base (Fig. 2.4). Due to vertical mixing, as the distance from the front increases, the intrusion becomes less salty and warmer (compare stations 55–58, Fig. 2.4a). The injection of salty, nutrient-rich subantarctic waters into the thermocline in the outer shelf fertilizes the otherwise nutrient-depleted shelf waters in summer.

The potential temperature-salinity distribution highlights the thermohaline contrast across the PSBF (Fig. 2.4). The T–S diagram displays the sharp salinity drop throughout the water column observed between stations 54 and 55, and the surface warming further onshore (stations 56–58, Fig. 2.4a). The T–S diagram also illustrates the sharp thermohaline contrast in the lower layer across the shelf-break front. The cold waters observed below the seasonal pycnocline in the outer shelf (e.g., stations 56–58) are too fresh (< 33.6, stations 57 and 58, orange and brown lines in Fig. 2.4a) to be a result of mixing with the substantially saltier slope waters (>33.9, station 55, blue line in Fig. 2.4a). Evidence of cross-front mixing in the bottom layer is only apparent at station 56, occupied close to the shelf-break (green line in Fig. 2.4a). As we shall show in the following section, this is consistent with time series observations of the thermohaline variability below the seasonal thermocline. The cold-fresh waters (6.5 °C, 33.56) observed in the lower layer onshore of station 56 most likely originate over the shelf farther south.

During austral winter, the shelf is vertically homogeneous or very weakly stratified (Rivas and Piola 2002; Bianchi et al. 2009). The winter section occupied in early September already displays some thermal stratification over the continental shelf, with temperatures close to those observed in the deep layer during the summer (~6–8 °C, Fig. 2.3b). Note that the station separation in the winter section (40 km) is about twice as large as in the summer section, thus providing a much coarser resolution of the cross-shelf patterns. The offshore region presents temperatures lower than 6 °C. The cross-slope salinity gradient in winter appears to be less intense than in summer, but the isohalines present a similar orientation: deepening in the onshore direction and intersecting the bottom close to the shelf-break. Consequently, in contrast with the summer situation, in winter the PSBF displays a moderate density contrast, with $\sigma_\theta < 26.6$ kg m^{-3} over the shelf and $\sigma_\theta > 26.8$ kg m^{-3} farther offshore (Fig. 2.4b), thus suggesting reduced isopycnal mixing across the shelf-break. In addition to the low light intensity, the weak winter stratification limits the development of phytoplankton close to the shelf-break.

2.4.2 *Time Variability and Evidence of Shelf-Break Upwelling*

There are scant observations available to characterize the subsurface variability of the PSBF. Thermohaline observations collected from mooring deployments at the shelf-break are described in Valla and Piola (2015), Carranza et al. (2017), Paniagua et al. (2018, 2021), and Zhang et al. (2018). These moorings, located at about 180 m depth, included meteorological sensors installed in a surface buoy and a downward-looking ADCP. In addition, some deployments had temperature and salinity sensors attached at six levels in the upper 100 m (see Valla and Piola 2015). Hourly observations collected close to 44 °S from mid-October to early December 2005 show the onset of the vertical stratification during the austral spring (Fig. 2.5). In October, the

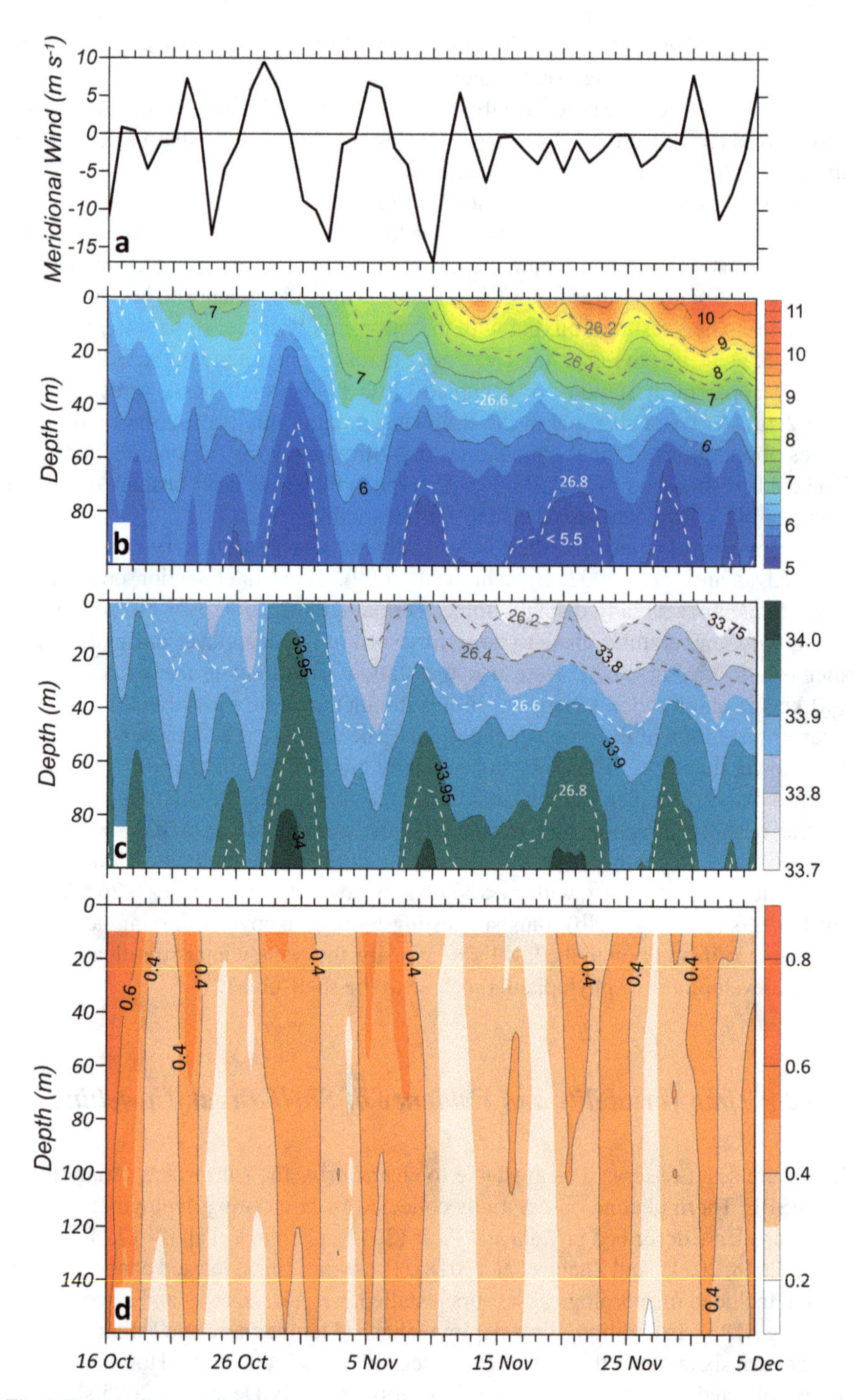

Fig. 2.5 Time series of meridional wind velocity (**a**, m s^{-1}) and Hovmoller diagrams (depth versus time) of temperature (**b**, °C), salinity (**c**), and meridional velocity (**d**, m s^{-1}) collected from October to December 2005. The white dashed lines in (**b**) and (**c**) are isopycnals (σ_θ, Kg m^{-3}). See Fig. 2.1b for mooring location

upper 100 m present very weak vertical stratification, with a near-surface temperature and salinity of ~6 °C and 33.9, and about 5.5 °C and 33.97 at 100 m. The slight salinity increase with increasing depth at the shelf-break is consistent with the cross-shelf sections (Fig. 2.3c, d). By early December, the near-surface temperature increased to 11 °C, while at 100 m, it remained close to the 5.5 °C observed during the early part of the record. During spring, the near-surface salinity decreases, reaching 33.7 by early December. This freshening suggests an offshore displacement of the outer shelf waters, and possibly of the surface expression of the shelf-break front. This observation could be associated with the weak northerly winds that prevailed after mid-November 2005 (Fig. 2.5a).

The near-surface temperature data display substantial variability at diurnal, semi-diurnal, synoptic, and intraseasonal time scales, while the variance decreases with increasing depth (Valla and Piola 2015; Zhang et al. 2018). Temperature and salinity fluctuations in the upper 20 m are very weakly correlated, suggesting differences in the variability of its drivers, i.e., its heat and freshwater sources. In contrast, temperature and salinity variations at 100 m are highly coherent (> 0.8) and 180° out of phase over a wide range of frequencies (see Fig. 2.6 and Valla and Piola 2015). These observations, which are well below the depth of the seasonal thermocline (~40–60 m, see Fig. 2.3a), represent the deep layer characteristics at the shelf-break and indicate transitions between cold-salty slope waters and warm-fresh shelf

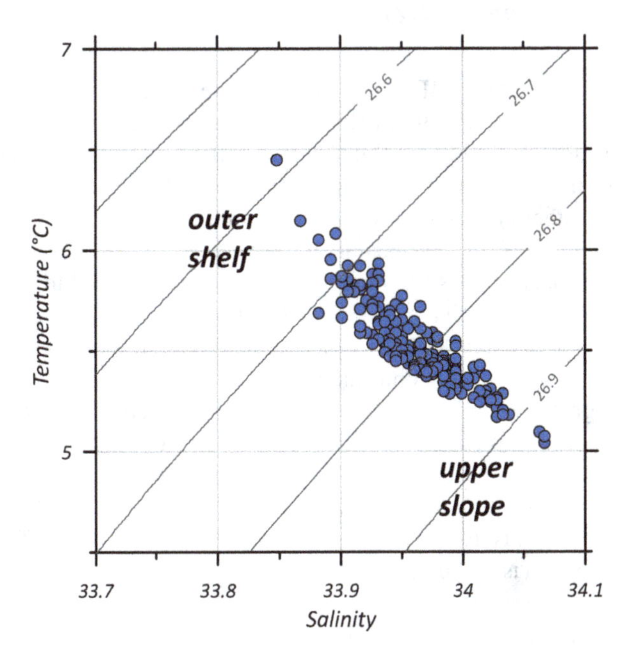

Fig. 2.6 Potential temperature (°C) versus salinity diagram at 100 m depth. The data were collected at the shelf-break near 43 °S (see Fig. 2.1b for location) from 16 October to 5 December 2005. The gray lines indicate constant potential density (σ_θ, Kg m^{-3})

waters. The data also display variations of several tens of meters in the depth of isotherms and isohalines.

Mid-summer observations from another shelf-break mooring deployment at 41 °S display an intense upper ocean cooling event (~ −3 °C) lasting nearly 10 days. Concurrent ADCP observations show that the event was associated with unusually high along-shore velocities exceeding 0.45 m s^{-1} (Valla and Piola 2015). An evaluation of the temperature conservation equation suggests that this intense cooling event could not have been driven by surface heat fluxes or horizontal heat advection. Valla and Piola (2015) concluded that the most plausible explanation for the observed cooling was an intense upwelling of cold subsurface waters, with vertical velocities exceeding 10 m day^{-1}. A combined analysis of satellite-derived SST and chlorophyll concentration collected during 2002–2013 shows similarly intense and extended cooling events near the shelf-break, most of which were associated with anomalously high chlorophyll patches, further supporting the upwelling hypothesis (Valla and Piola 2015). These findings suggest that the cold-salty waters (< 5.5 °C, > 34) observed at 100 m depth (Fig. 2.6) may also be associated with the upwelling of subantarctic waters from the Malvinas Current on the outer shelf.

2.5 Dynamics of the Patagonian Shelf-Break Upwelling

2.5.1 The Malvinas Current

The main driver of the PSBF productivity is the northward-flowing Malvinas Current, which brings cold, nutrient-rich subantarctic waters to both the subsurface and surface portions of the water column at the shelf-break. These waters are locally upwelled as well as advected by the Malvinas Current from other regional upwelling centers of southern Patagonia, generating a continuous band of high chlorophyll extending from the Malvinas Islands Shelf (MIS) to 38 °S (Fig. 2.1d).

The physical mechanism sustaining local upwelling is related to the interaction of the Malvinas Current with the seafloor (Matano and Palma 2008). As the current flows along the continental slope, frictional effects (bottom friction and/or horizontal turbulent mixing) spread it onto the neighboring shelf, thus generating along-shelf pressure gradients and a divergent cross-shelf circulation pattern that leads to shelf-break upwelling. A simplified analytical model shows that the upwelling rate increases with the transport of the Malvinas Current and with the ratio between the inclination of the the continental slope and shelf (Miller et al. 2011). Upwelling rates estimated by this model for the PSBF are on the order of 10^{-4} m s^{-1} (~10 m day^{-1}), which is of the same order of magnitude as typical upwelling velocities in eastern boundary regions.

Numerical experiments using a 3D baroclinic model show that the lateral spreading of the slope current and the subsequent local upwelling is a robust physical phenomenon because its occurrence does not depend on a single physical process:

bottom topography, shelf currents, stratification, bottom friction, and turbulent mixing (Matano and Palma 2008). Although the core of the upwelling cell in the model is located below the surface, enhanced vertical mixing by local winds and tides might ultimately pump the nutrients to the surface layer. It is important to note that, in contrast to previous studies of shelf-break dynamics (e.g., Gawarkiewicz and Chapman 1992), the PSBF upwelling is not controlled by the downslope buoyancy flux associated with the presence of a shelf current but by the along-shelf pressure gradient associated with the presence of a slope current. The significant correlation between the transport of the inshore branch of the Malvinas Current and the vertical velocities derived from a realistic numerical model and integrated over an extended portion of the shelf-break north of 48 °S agrees with results from the analytical and simplified numerical experiments (Combes and Matano 2014). Despite the highly variable intensity of the Malvinas Current transport (e.g., Artana et al. 2018; Paniagua et al. 2021), the proposed mechanism would drive quasi-permanent shelf-break upwelling.

Figure 2.7 illustrates the regional scale of the shelf-break upwelling as represented in a numerical experiment. The model is described in detail by Combes and Matano (2014, 2018), but a brief description is provided here. The model is based on the Regional Ocean Modeling System (Shchepetkin and McWilliams 2005), implemented for the ocean around South America. The model uses a 1/12° horizontal resolution and 40 vertical levels and is nested within a parent model of the southern hemisphere oceans. At the surface, the model is forced by momentum, heat, and freshwater fluxes from the ERA-Interim reanalysis (Dee et al. 2011). A passive tracer set to a concentration of 1 was released below 200 m and between the 200 and 1000 m isobaths (e.g., within the Malvinas Current, green-outlined region in

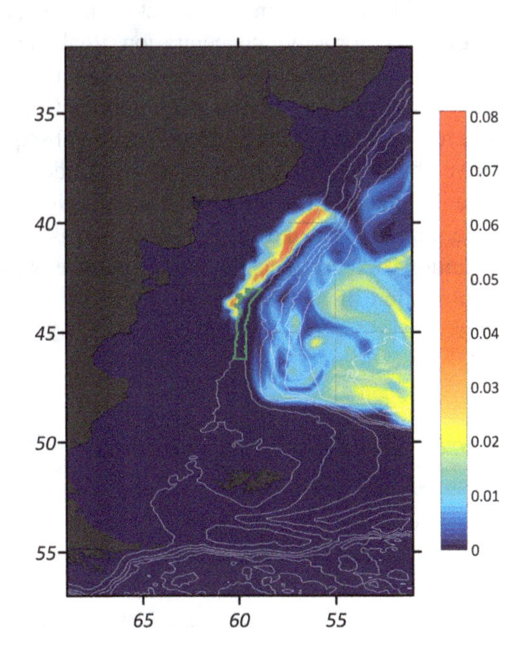

Fig. 2.7 Snapshot of tracer concentration (unitless) at the sea surface 180 days after the initial release in a numerical model (see text). The tracer was originally released below 200 m depth between the 200 and 1000 m isobaths in the region outlined in green. Selected model isobaths are indicated in white (200, 1000, 2000, 3000, 4000, 5000, and 6000 m)

Fig. 2.7). Figure 2.7 displays a snapshot of tracer concentration at the sea surface 180 days after the initial release. At that time, a substantial amount of tracer has been transferred to the deep ocean and is apparent within the rich eddy field associated with the Brazil/Malvinas Confluence, though with relatively low concentrations. The highest concentrations, however, are observed along the outer continental shelf northeast of the release region. Thus, in the model, the tracer was advected north-northeastward from the upper slope to the outer shelf and upwelled 200 m. Note that the region of high tracer concentration resembles the region of high surface chlorophyll (Fig. 2.1d). A similar spatial pattern, though with higher tracer concentrations, is observed 360 days after the initial release. This suggests that upwelling from the release region consistently reaches the outer shelf where the tracer concentrates and is consistent with similar tracer release experiments discussed by Combes and Matano (2014).

It is physically difficult to discriminate between local nutrient supply and those advected northwards by the Malvinas Current from remote southern locations. Recent numerical studies have noted that other upwelling centers, e.g., the Burdwood Bank and the MIS, might also contribute to the fertilization of the PSBF. Chlorophyll blooms are not regularly observed over the Burdwood Bank, most likely due to light limitations, but realistic models have shown that the bank is an active upwelling center of nutrient-rich subantarctic waters, which are subsequently advected to the PSBF (Matano et al. 2019). These nutrient-enriched waters possibly contribute to the maintenance of the high primary productivity observed farther north, where higher stratification and solar radiation might allow the development of intense chlorophyll blooms, particularly on the offshore side of the shelf-break around the MIS.

Satellite observations show that the northward portion of the MIS is also a region with high chlorophyll concentration. Realistic numerical simulations using passive tracers initially located in the deep ocean indicate that the place where the northernmost branch of the Antarctic Circumpolar Current collides with the MIS is the main site for the outcropping of deep waters. Subsequent upwelling on the shelf is largely driven by the synergistic interaction of tides and wind forcing. In agreement with satellite images of chlorophyll concentration, in early summer, the core of the surface plume is located on the lee of the islands and confined to the shelf region. Additional experiments employing passive particles show that there is an important contribution of MIS waters to the PSBF, particularly its inshore portion (Palma et al. 2021).

2.5.2 *Wind-Induced Variability*

The satellite-derived chlorophyll over the shelf-break reveals substantial interannual variability, with spring-summer maxima ranging between ~4 and 10 mg m^{-3} (Saraceno et al. 2005; Romero et al. 2006; Delgado et al. 2023). A combined analysis of surface wind and chlorophyll variability at interannual scales during

1998–2003 suggests that shelf-break spring blooms are more intense and occur earlier under the influence of northerly winds (Saraceno et al. 2005). This study argues that offshore Ekman transport of shelf waters over denser Malvinas Current waters leads to increased stratification, thus favoring the retention of phytoplankton in the photic layer. At synoptic time scales, however, the chlorophyll concentration at the shelf-break appears to increase in response to southerly winds (Carranza et al. 2017). The observations show a moderate correlation between temperature and along-front winds at synoptic time scales (less than 10 days, Carranza et al. 2017). Southerly wind events lasting more than a day led to the cooling of the upper 100 m observed 2–4 days later. These observations are consistent with the vertically coherent temperature and salinity oscillations below 20 m depth apparent in Fig. 2.5 and with the modulation of vertical stratification by along-shore winds, with southerly winds weakening the vertical stratification. As expected from Ekman dynamics, under northerly wind conditions, relatively warmer-fresher shelf waters are observed in the upper 30 m of the water column, while under southerly winds this pattern is reversed (Carranza et al. 2017). Changes in vertical stratification due to northerly winds produce the slumping of isopycnals, thus enhancing vertical stability (Siedlecki et al. 2011; Carranza et al. 2017). The observations are also consistent with the onshore displacement of the surface chlorophyll maximum under southerly winds. A similar phytoplankton response to wind variability has been reported elsewhere (e.g., Hosegood et al. 2008). Thus, local wind variability over the shelf-break front can further modulate frontal upwelling.

Under uniform meridional wind, the relatively large surface velocity observed close to the shelf-break will induce a nonzero curl in the effective surface stress, which can be parameterized as a quadratic function of the difference between the surface wind and current velocities (e.g., Dewar and Flierl 1987; Pacanowski 1987). Over the high northward velocity core near the shelf-break, the stress produced by a southerly wind is smaller than that produced by a northerly wind. Regardless of the sign of the meridional wind component, however, the cross-shore pattern of surface stress creates a band of negative wind stress curl and divergent Ekman transport on the onshore side of the Malvinas Current, and convergent Ekman transport offshore. This pattern emerges from the combination of high-resolution winds and numerical models (Combes and Matano 2018). Assuming uniform along-shore conditions, the divergence will be compensated by upwelling onshore of the current jet and downwelling offshore of the jet. If the mass divergence in the Ekman layer is distributed over a narrow along-shore band (~10 km), under relatively strong meridional winds (~10 m s^{-1}), the associated vertical velocities are of the order of 1×10^{-5} m s^{-1} (~1 m day^{-1}), or an order of magnitude smaller than caused by other processes, as suggested by numerical models (Combes and Matano 2014) and derived from observations (Valla and Piola 2015). This agrees with high-resolution numerical simulations that include the current velocity to estimate the effective wind stress (Combes and Matano 2018). The model shows that shelf-break upwelling integrated over the upper slope between 41.5 °S and 46 °S is moderately correlated with the wind stress curl at high frequencies and that the correlation decreases at longer time scales. This is also in agreement with numerical experiments forced with different monthly mean climatological winds, which show similar shelf-break upwelling intensities, suggesting that local wind variations at monthly time scales do not significantly impact the upwelling dynamics (Combes and Matano 2014).

2.5.3 *Mesoscale and Submesoscale Variability*

Observations and biophysical models have shown that eddies may contribute to restratification near large-scale fronts even before the development of seasonal upper layer warming in early spring (Mahadevan et al. 2012). Although in altimetry-derived eddy kinetic energy distributions, the Malvinas Current is characterized as a low-energy domain, despite its energetic mean flow (e.g., Saraceno and Provost 2012), high-resolution true color and chlorophyll images display complex small-scale structures at the shelf-break resembling small mesoscale and submesoscale features observed throughout the region (e.g., Glorioso et al. 2005; Capet et al. 2008; Matano et al. 2019; Becker et al. 2023). Direct long-term current meter observations from the upper continental slope (bottom depth ~ 1320 m) close to 45 °S present an energetic northward mean flow (0.37 m s^{-1}) and substantial variability, with frequent peaks exceeding 0.5 m s^{-1} at 300 m depth (Paniagua et al. 2021). During summer, temperature and density sections frequently display vertical displacements of several tens of meters at the shelf-break. For example, the deepening of the thermocline and pycnocline observed at the shelf-break (station 55, Fig. 2.3a, e) suggests a small-scale (20–40 km) anticyclonic eddy, which appears to extend vertically only to the thermocline level (20–80 m depth). Although the hydrographic stations are not close enough to properly resolve this feature, note that the high-salinity intrathermocline intrusion of slope waters into the outer shelf region appears to emerge from this feature (see Sect. 2.4.1 and Fig. 2.3c). Similar small-scale thermocline perturbations combined with patterns of change in mixed layer depth and optical properties observed at the shelf-break near 45 °S have been interpreted as indicative of upwelling and downwelling (Salyuk et al. 2022). Combined analysis of phytoplankton distributions and surface transport suggests very weak turbulent stirring at the shelf-break (D'Ovidio et al. 2010). The stability of the Malvinas Current at the space and time scales resolved by the available data also emerges from the analysis of Lagrangian Coherent Structures based on satellite altimetry fields and historical surface drifter trajectories, suggesting the region presents weak flows across the PSBF (Beron-Vera et al. 2020). These conclusions, however, might be biased by the lack of sufficient spatial and temporal resolution of the presently available altimetric and drifter data.

2.6 Long-Term Variability and Trend

Satellite observations show substantial variability in sea surface temperature and chlorophyll concentration at interannual time scales over the entire Patagonian region (e.g., Romero et al. 2006; Rivas 2010; Bodnariuk et al. 2021; Risaro et al. 2022). The drivers of this variability, however, are unclear. Analyses of regional models (e.g., Palma et al. 2008; Combes and Matano 2018) and ocean reanalysis (Guihou et al. 2020; Bodnariuk et al. 2021) highlight the impact of wind variability on the shelf circulation from seasonal to interannual time scales. The studies point out the influence of the Southern Annular Mode[1] on interannual variations in the shelf circulation. It is unclear, however, how changes in the shelf circulation impact the variability of the shelf-break front. It seems reasonable to surmise, however, that these properties would be highly susceptible to large transport variations of the Malvinas Current (e.g., Artana et al. 2018, and references therein) on account of its dominant influence over the outer shelf region.

Analyses of satellite-derived surface chlorophyll concentration indicate that the Patagonia continental shelf presents a positive trend of about 0.25 mg m^{-3} during the period between 2003 and 2017 (Marrari et al. 2017), while north of 44 °S, the mid-shelf and shelf-break regions present an increase exceeding 1 mg m^{-3} during 1998–2021 (Delgado et al. 2023). Likewise, analysis of satellite-derived SST data during 2008–2017 reveals a moderate positive trend over the continental shelf north of about 49 °S (0.15 °C $decade^{-1}$), and moderate cooling farther south (−0.1 °C $decade^{-1}$, Risaro et al. 2022). In contrast, no significant trend in SST was detected in the core of the Malvinas Current south of 40 °S.

Several studies have reported a significant poleward shift and associated surface warming in global western boundary currents (Wu et al. 2012; Yang et al. 2016, 2020). The estimated poleward shift of the Brazil Current (0.11° $decade^{-1}$, Yang et al. 2020) agrees with the southward shift of the PSBF suggested by changes in sea level anomaly (Ruiz Etcheverry et al. 2016) and meridional gradients of SST (Franco et al. 2022). The latter study confirms the positive trend in sea surface chlorophyll over the shelf-break front (0.2–0.5 mg m^{-3} $decade^{-1}$). Warming over the continental shelf (0.5 °C $decade^{-1}$) and cooling over the cold wedge of the Malvinas Current (<−0.1 °C $decade^{-1}$) during the past decades have led to an increase in the cross-shelf temperature gradient over the shelf-break front (Franco et al. 2020). It is important to note that Malvinas Current transport estimates based on satellite altimetry and direct current observations at 41 °S reveal no significant long-term trend over the period 1993–2016 (Artana et al. 2018).

Coupled Model Intercomparison Project Phase 6 models under the high emission Shared Socio-economic Pathway scenario referred to as SSP5–8.5 predict that subtropical gyres and the southern hemisphere western boundary currents, including the

[1] The Southern Annular Mode is the leading mode of atmospheric variability in the extratropical Southern Hemisphere. It is characterized by meridional fluctuations of the westerly winds associated with hemispheric-scale fluctuations in the meridional pressure gradient (Kidson 1999; Mo 2000).

Brazil Current, will intensify in the twenty-first century in response to changes in wind forcing (IPCC 2021). The poleward expansion of the Brazil Current is likely to lead to a similar poleward displacement of the northernmost extension of the Malvinas Current as observed during the past few decades (Franco et al. 2022). The possible biological and biogeochemical impacts of these changes at the northern extension of the shelf-break front are unknown. However, given the impact of the Malvinas Current on the nutrient supply to the PSBF, its southward retraction is likely to have a substantial impact on primary producers. Likewise, since productivity over the shelf is the main driver of intense uptake of atmospheric carbon dioxide (Kahl et al. 2017, also see Chap. 9), changes in the distribution of phytoplankton will likely impact the distribution of the exchange of CO_2 between the ocean and the atmosphere. There is evidence that the reproductive success of squid *Illex argentinus* is modulated by the export of larvae from the continental shelf to the deep ocean at the northernmost extent of the Malvinas Current (Torres Alberto et al. 2021), which may be altered by a meridional displacement of the export route. Meridional fluctuations of the northernmost extension of the PSBF may also impact the success of benthic species that depend on the productivity of the water column (e.g., Mauna et al. 2008).

2.7 Summary and Perspective

This chapter described the characteristics of the Patagonia shelf-break front based on satellite and in-situ observations. The front is characterized by a relatively sharp change in sea surface temperature and salinity that extends along the outer shelf and is most notable north of about 50 °S. The temperature and salinity fronts are formed between the relatively warm-fresh shelf waters and cold-salty subantarctic waters advected northwards by the Malvinas Current along the upper slope of the western Argentine Basin, reaching approximately 38 °S. The salinity front is due to the freshening of the shelf waters due to inflows of diluted waters from the southeast Pacific. The Malvinas Current mean velocities generally exceed 0.45 m s^{-1}, in sharp contrast with the shelf, which is characterized by a comparatively sluggish mean flow of a few cm per second. Thus, the PSBF is also characterized by a sharp velocity shear. Satellite and in-situ observations show that the PSBF is associated with an extensive band of high chlorophyll concentration that persists from early spring to late fall. This band extends from about 50 °S to 38 °S, closely following the onshore flank of the Malvinas Current.

Numerical models indicate that the upwelling of dense, nutrient-rich subantarctic waters to the upper layer is mediated by the interaction of the Malvinas Current with the bottom. The models suggest that the intensity of the vertical velocity is modulated by the intensity of the Malvinas Current. Observations and models suggest that the vertical stratification at the PSBF and the upwelling of nutrient-rich waters are modulated by local wind variations at a variety of time scales. Since the Malvinas Current waters are denser than the shelf waters, they tend to reach the shelf-break near the bottom. However, during summer, the intense seasonal thermocline that develops over the shelf provides isopycnal connections that facilitate the exchange between the shelf and nutrient-rich slope waters at the intrathermocline level, which may further promote the growth of phytoplankton. The combined effect of these processes, together with other biological factors, may explain the large variability observed in chlorophyll distributions at a wide range of time scales. Though subantarctic waters are nutrient-rich, models show that further nutrient enrichment can be derived from intense upwelling regions around Burdwood Bank and the southern portions of the continental shelf, which are advected northward along the upper slope.

Observations show that while the intensity of the Malvinas Current does not present a trend during the past decades, the surface waters in its northern portion have moderately cooled, while the northern shelf waters have warmed. There is also evidence of a southward displacement of the Brazil Current and of the northernmost extent of the Malvinas Current. The impact of these changes on the PSBF and the associated ecosystem is yet to be investigated. Given the primary role of the Malvinas Current on the PSBF, in-situ transport monitoring is required to improve the understating of its dynamics, and to better quantify its temporal variability and impact.

There is a significant gap in the understanding of the dynamical implications of small-scale features apparent in high-resolution satellite images. These features resemble patterns emerging from frontal instabilities at scales that are not well resolved by most numerical simulations currently available. Likewise, no observations are available on the scale of kinetic energy dissipation due to turbulent mixing processes. These processes may play a key role in promoting vertical nutrient fluxes and the fertilization of the shelf-break. Although very high-resolution numerical models for the Patagonia shelf-break front will likely be developed in the next few years, given the fundamental role of bottom topography in shaping the shelf-break circulation and dynamics, similarly high-resolution (multibeam) bathymetric data are necessary to allow the models to realistically represent the processes at play. High-resolution altimetry from the Surface Water and Ocean Topography (SWOT) mission recently began to provide snapshots of sea surface height anomaly at unprecedented resolution (~1 km). These data will be capable of capturing submesoscale structures, though the revisit time will be much longer than required to observe their time evolution. However, SWOT observations will serve as a base to evaluate how well submesoscale features are represented in very high-resolution numerical simulations. Satellite data, including infrared, optical, and synthetic aperture radar, will continue to provide fundamental information on the regional patterns at the sea surface and their time variability. Likewise, Argo floats provide valuable subsurface information in the deep ocean. However, very few long-term subsurface observations are available over the continental shelf. Such observations are urgently needed to better understand the dynamics of the shelf-break front, to detect future changes, and to more rigorously test the performance of numerical models and assimilate them into ocean reanalyses. Likewise, there are scant micronutrient observations in the western South Atlantic, and none are available over the shelf and in the PSBF. These observations are essential to determine their sources and to better understand the mechanisms that sustain the large primary productivity at the shelf-break.

Acknowledgments Among others, this chapter summarizes findings from project VOCES, grant CRN3070 from the Inter-American Institute for Global Change Research, and US NSF grant GEO-1128040. ARP, EDP, MS, and MU acknowledge support from CONICET (Argentina) grant PIP 2021-2023 GI 11220200103112CO and IAI/CONICET RD3347. NB is supported by a post-doctoral fellowship from CONICET, Argentina. VC and RPM acknowledge the support of NSF grant 2149292 and NASA grant 80NSSC21K0559. MS and ARP acknowledge projects EUMETSAT/CNES OSTST CASSIS and SABIO. EDP and ARP acknowledge support from Agencia Nacional de Promoción Científica y Tecnológica (grant PICT2020-02024) and Universidad Nacional del Sur (grant 24F079), Argentina. MU is supported by a doctoral fellowship from CONICET, Argentina.

References

Acha EM, Mianzan HW, Guerrero RA et al (2004) Marine fronts at the continental shelves of austral South America: physical and ecological processes. J Mar Syst 44(1–2):83–105. https://doi.org/10.1016/j.jmarsys.2003.09.005

Arkhipkin A, Brickle P, Laptikhovsky V et al (2012) Dining hall at sea: feeding migrations of nektonic predators to the eastern Patagonian Shelf. J Fish Biol 81(2):882–902. https://doi.org/10.1111/j.1095-8649.2012.03359.x

Artana C, Ferrari R, Koenig Z et al (2018) Malvinas current volume transport at 41°S: a 24 year-long time series consistent with mooring data from 3 decades and satellite altimetry. J Geophys Res Oceans 123:378–398. https://doi.org/10.1002/2017JC013600

Becker F, Romero SI, Pisoni JP (2023) Detection and characterization of submesoscale eddies from optical images: a case study in the Argentine continental shelf. Int J Remote Sens 44(10):3146–3159. https://doi.org/10.1080/01431161.2023.2216853

Beron-Vera FJ, Bodnariuk N, Saraceno M et al (2020) Stability of the Malvinas current. Chaos 30(1). https://doi.org/10.1063/1.5129441

Bianchi AA, Ruiz-Pino D, Isbert Perlender H et al (2009) Annual balance and seasonal variability of sea-air CO2 fluxes in the Patagonia Sea: their relationship with fronts and chlorophyll distribution. J Geophys Res 114:C03018. https://doi.org/10.1029/2008JC004854

Bodnariuk N, Simionato CG, Saraceno M (2021) SAM-driven variability of the southwestern Atlantic shelf sea circulation. Cont Shelf Res 212:104313. https://doi.org/10.1016/j.csr.2020.104313

Bodnariuk N, Saraceno M, Ruiz-Etcheverry LA et al (2024) Multiple Lagrangian jet-core structures in the Malvinas Current. J Geophys Res Oceans 129:e2023JC020446. https://doi.org/10.1029/2023JC020446

Bowie AR, Whitworth DJ, Achterberg EP et al (2002) Biogeochemistry of Fe and other trace elements (Al, Co, Ni) in the upper Atlantic Ocean. Deep-Sea Res I Oceanogr Res Pap 49(4):605–636. https://doi.org/10.1016/S0967-0637(01)00061-9

Brandhorst W, Castello JP (1971) Evaluación de los recursos de anchoíta (Engraulis anchoita) frente a la Argentina y Uruguay. I. Las condiciones oceanográficas, sinopsis del conocimiento actual sobre la anchoíta y el plan para su evaluación. Proy Des Pesq FAO 29: p 63. Mar del Plata, Argentina

Brink KH (2016) Cross-shelf exchange. Annu Rev Mar Sci 8:59–78. https://doi.org/10.1146/annurev-marine-010814-015717

Brown CW, Podestá GP (1997) Remote sensing of coccolithophore blooms in the western South Atlantic Ocean. Remote Sens Environ 60(1):83–91. https://doi.org/10.1016/S0034-4257(96)00140-X

Brun AA, Ramírez N, Pizarro O et al (2020) The role of the Magellan Strait on the southwest South Atlantic shelf. Estuar Coast Shelf Sci 237:106661. https://doi.org/10.1016/j.ecss.2020.106661

Campbell JW (1995) The lognormal distribution as a model for bio-optical variability in the sea. J Geophys Res 100:13237–13254. https://doi.org/10.1029/95JC00458

Capet X, Campos EJ, Paiva AM (2008) Submesoscale activity over the Argentinian shelf. Geophys Res Lett 35:L15605. https://doi.org/10.1029/2008GL034736

Carranza MM, Gille ST, Piola AR et al (2017) Wind modulation of upwelling at the shelf-break front off Patagonia: observational evidence. J Geophys Res Oceans 122:2401–2421. https://doi.org/10.1002/2016JC012059

Carreto JI, Lutz VA, Carignan MO et al (1995) Hydrography and chlorophyll *a* in a transect from the coast to the shelf-break in the Argentinian Sea. Cont Shelf Res 15(2–3):315–336. https://doi.org/10.1016/0278-4343(94)E0001-3

Carreto JI, Montoya NG, Carignan MO et al (2016) Environmental and biological factors controlling the spring phytoplankton bloom at the Patagonian shelf-break front–degraded fucoxanthin pigments and the importance of microzooplankton grazing. Prog Oceanogr 146:1–21. https://doi.org/10.1016/j.pocean.2016.05.002

Charo M, Piola AR (2014) Hydrographic data from the GEF Patagonia cruises. Earth Syst Sci Data 6:265–271. https://doi.org/10.5194/essd-6-265-2014

Combes V, Matano RP (2014) A two-way nested simulation of the oceanic circulation in the Southwestern Atlantic. J Geophys Res Oceans 119:731–756. https://doi.org/10.1002/2013JC009498

Combes V, Matano RP (2018) The Patagonian shelf circulation: drivers and variability. Prog Oceanogr 167:24–43. https://doi.org/10.1016/j.pocean.2018.07.003

Cosentino NJ, Ruiz-Etcheverry LA, Bia GL et al (2020) Does satellite chlorophyll-a respond to southernmost Patagonian dust? A multi-year, event-based approach. J Geophys Res Biogeo 125:e2020JG006073. https://doi.org/10.1029/2020JG006073

d'Ovidio F, De Monte S, Alvain S et al (2010) Fluid dynamical niches of phytoplankton types. Proc Natl Acad Sci 107(43):18366–18370. https://doi.org/10.1073/pnas.1004620107

Dávila PM, Figueroa D, Müller E (2002) Freshwater input into the coastal ocean and its relation with the salinity distribution off austral Chile (35–55 S). Cont Shelf Res 22(3):521–534. https://doi.org/10.1016/S0278-4343(01)00072-3

Dee DP, Uppala SM, Simmons AJ et al (2011) The ERA-interim reanalysis: configuration and performance of the data assimilation system. QJR Meteorol Soc 137:553–597. https://doi.org/10.1002/qj.828

Delgado AL, Hernández-Carrasco I, Combes V et al (2023) Patterns and trends in chlorophyll-a concentration and phytoplankton phenology in the biogeographical regions of southwestern Atlantic. J Geophys Res Oceans 128:e2023JC019865. https://doi.org/10.1029/2023JC019865

Dewar WK, Flierl GR (1987) Some effects of the wind on rings. J Phys Oceanogr 17(10):1653–1667. https://doi.org/10.1175/1520-0485(1987)017%3C1653:SEOTWO%3E2.0.CO;2

Fetter AF, Matano RP (2008) On the origins of the variability of the Malvinas current in a global, eddy-permitting numerical simulation. J Geophys Res Oceans 113(C11):C11018. https://doi.org/10.1029/2008JC004875

Franco BC, Piola AR, Rivas AL et al (2008) Multiple thermal fronts near the Patagonian shelf break. Geophys Res Lett 35:L02607. https://doi.org/10.1029/2007GL032066

Franco BC, Defeo O, Piola AR et al (2020) Climate change impacts on the atmospheric circulation, ocean, and fisheries in the southwest South Atlantic Ocean: a review. Clim Chang 162:2359–2377. https://doi.org/10.1007/s10584-020-02783-6

Franco BC, Ruiz-Etcheverry LA, Marrari M et al (2022) Climate change impacts on the Patagonian shelf break front. Geophys Res Lett 49:e2021GL096513. https://doi.org/10.1029/2021GL096513

Frey DI, Piola AR, Krechik VA et al (2021) Direct measurements of the Malvinas current velocity structure. J Geophys Res Oceans 126(4):e2020JC016727. https://doi.org/10.1029/2020JC016727

Frey DI, Piola AR, Morozov EG (2023) Convergence of the Malvinas current branches near 44°S. Deep-Sea Res I 196:104023. https://doi.org/10.1016/j.dsr.2023.104023

Garcia VM, Garcia CA, Mata MM et al (2008) Environmental factors controlling the phytoplankton blooms at the Patagonia shelf-break in spring. Deep-Sea Res I 55(9):1150–1166. https://doi.org/10.1016/j.dsr.2008.04.011

Gawarkiewicz G, Chapman DC (1992) The role of stratification in the formation and maintenance of shelf-break fronts. J Phys Oceanogr 22(7):753–772. https://doi.org/10.1175/1520-0485(1992)022%3C0753:TROSIT%3E2.0.CO;2

Glorioso PD, Piola AR, Leben RR (2005) Mesoscale eddies in the Subantarctic front – Southwest Atlantic. Sci Mar 69(Suppl 2):7–15. https://doi.org/10.3989/scimar.2005.69s27

Guihou K, Piola AR, Palma ED et al (2020) Dynamical connections between large marine ecosystems of austral South America based on numerical simulations. Ocean Sci 16(2):271–290. https://doi.org/10.5194/os-16-271-2020

Hosegood PJ, Gregg MC, Alford MH (2008) Restratification of the surface mixed layer with submesoscale lateral density gradients: diagnosing the importance of the horizontal dimension. J Phys Oceanogr 38:2438–2460. https://doi.org/10.1175/2008JPO3843.1

Hu C, Lee Z, Franz B (2012) Chlorophyll *a* algorithms for oligotrophic oceans: a novel approach based on three-band reflectance difference. J Geophys Res Oceans 117(C1). https://doi.org/10.1029/2011JC007395

IPCC (2021) Climate change 2021: the physical science basis. In: Masson-Delmotte V, Zhai P, Pirani A et al (eds) Contribution of working group I to the sixth assessment report of the intergovernmental panel on climate change. Cambridge University Press, Cambridge and New York, p 2391. https://doi.org/10.1017/9781009157896

Kahl LC, Bianchi AA, Osiroff AP et al (2017) Distribution of sea-air CO_2 fluxes in the Patagonian Sea: seasonal, biological and thermal effects. Cont Shelf Res 143:18–28. https://doi.org/10.1016/j.csr.2017.05.011

Kidson JW (1999) Principal modes of southern hemisphere low frequency variability obtained from NCEP-NCAR reanalyses. J Clim 12:2808–2830. https://doi.org/10.1175/1520-0442(1999)012%3C2808:PMOSHL%3E2.0.CO;2

Kilpatrick KA, Podestá G, Walsh S et al (2015) A decade of sea surface temperature from MODIS. Remote Sens Environ 165:27–41. https://doi.org/10.1016/j.rse.2015.04.023

Krechik VA (2020) The upper layer of the Malvinas/Falkland current: structure, and transport near 46 S in January 2020. Russ J Earth Sci 20(5):3. https://doi.org/10.2205/2020ES000715

Longhurst A (2007) Ecological geography of the sea, 2nd edn. Elsevier, Burlington. https://doi.org/10.1016/B978-0-12-455521-1.X5000-1

Lutz VA, Segura V, Dogliotti AI et al (2010) Primary production in the Argentine sea during spring estimated by field and satellite models. J Plankton Res 32(2):181–195. https://doi.org/10.1093/plankt/fbp117

Mahadevan A, D'Asaro E, Lee C et al (2012) Eddy-driven stratification initiates North Atlantic spring phytoplankton blooms. Science 336(6090):54–58. https://doi.org/10.1126/science.1218740

Mann KH, Lazier JR (2005) Dynamics of marine ecosystems: biological-physical interactions in the oceans, 3rd edn. Blackwell, Oxford. https://doi.org/10.1002/9781118687901

Marrari M, Piola AR, Valla D (2017) Variability and 20-year trends in satellite-derived surface chlorophyll concentrations in large marine ecosystems around South and Western Central America. Front Mar Sci 4:372. https://doi.org/10.3389/fmars.2017.00372

Martinetto P, Alemany D, Botto F et al (2020) Linking the scientific knowledge on marine frontal systems with ecosystem services. Ambio 49:541–556. https://doi.org/10.1007/s13280-019-01222-w

Martos P, Piccolo MC (1988) Hydrography of the Argentine continental shelf between 38 and 42°S. Cont Shelf Res 8(9):1043–1056. https://doi.org/10.1016/0278-4343(88)90038-6

Matano RP, Palma ED (2008) On the upwelling of down welling currents. J Phys Oceanogr 38(11):2482–2500. https://doi.org/10.1175/2008JPO3783.1

Matano RP, Palma ED, Piola AR (2010) The influence of the Brazil and Malvinas currents on the southwestern Atlantic shelf circulation. Ocean Sci 6:983–995. https://doi.org/10.5194/os-6-983-2010

Matano RP, Palma ED, Combes V (2019) The Burdwood Bank circulation. J Geophys Res Oceans 124:6904–6926. https://doi.org/10.1029/2019JC015001

Mauna AC, Franco BC, Baldoni A et al (2008) Cross-front variations in adult abundance and recruitment of Patagonian scallop (*Zygochlamys patagonica*) at the SW Atlantic Shelf Break Front. ICES J Mar Sci 65(7):1184–1190. https://doi.org/10.1093/icesjms/fsn098

Miller RN, Matano RP, Palma ED (2011) Shelfbreak upwelling induced by alongshore currents: analytical and numerical results. J Fluid Mech 686:239–249. https://doi.org/10.1017/jfm.2011.326

Mo KC (2000) Relationships between low-frequency variability in the Southern Hemisphere and sea surface temperature anomalies. J Clim 13:3599–3620. https://doi.org/10.1175/1520-0442(2000)013%3C3599:RBLFVI%3E2.0.CO;2

Mulet S, Jousset S, Pujol M-I et al (2023) Sea level TAC – mean dynamic topography (MDT) products, Issue 3.0. Copernicus Marine Service. https://catalogue.marine.copernicus.eu/documents/QUID/CMEMS-SL-QUID-008-063-066-067.pdf

NASA OBPG (2020) MODIS Aqua Global Level 3 Mapped SST. Ver. 2019.0. PO.DAAC, CA, USA. https://doi.org/10.5067/MODAM-1D4N9. Accessed 12 Dec 2023

Pacanowski RC (1987) Effect of equatorial currents on surface stress. J Phys Oceanogr 17(6):833–838. https://doi.org/10.1175/1520-0485(1987)017%3C0833:EOECOS%3E2.0.CO;2

Painter SC, Poulton AJ, Allen JT et al (2010) The COPAS'08 expedition to the Patagonian shelf: physical and environmental conditions during the 2008 coccolithophore bloom. Cont Shelf Res 30(18):1907–1923. https://doi.org/10.1016/j.csr.2010.08.013

Palma ED, Matano RP, Piola AR (2008) A numerical study of the Southwestern Atlantic shelf circulation: Stratified Ocean response to local and offshore forcing. J Geophys Res Oceans 113:C11010. https://doi.org/10.1029/2007JC004720

Palma ED, Matano RP, Combes V (2021) Circulation and cross-shelf exchanges in the Malvinas Islands Shelf region. Prog Oceanogr 198:102666. https://doi.org/10.1016/j.pocean.2021.102666

Paniagua GF, Saraceno M, Piola AR et al (2018) Dynamics of the Malvinas current at 41°S: first assessment of temperature and salinity temporal variability. J Geophys Res Oceans 123:5323–5340. https://doi.org/10.1029/2017JC013666

Paniagua GF, Saraceno M, Piola AR et al (2021) Malvinas current at 44.7°S: first assessment of velocity temporal variability from in situ data. Prog Oceanogr 195:102592. https://doi.org/10.1016/j.pocean.2021.102592

Parker G, Paterlini CM, Violante RA (1997) El Fondo Marino. In: Boschi E (ed) El Mar Argentino y sus Recursos Pesqueros, vol 1. Inst Nac de Investigación y Desarrollo Pesquero, Mar del Plata, pp 65–87

Peterson RG, Whitworth T III (1989) The Subantarctic and Polar Fronts in relation to deep water masses through the southwestern Atlantic. J Geophys Res Oceans 94(C8):10817–10838. https://doi.org/10.1029/JC094iC08p10817

Pingree RD, Mardell GT (1981) Slope turbulence, internal waves and phytoplankton growth at the Celtic Sea shelf-break. Philos Trans R Soc London Ser A, Math Phys Sci 302(1472):663–682. https://doi.org/10.1098/rsta.1981.0191

Piola AR, Gordon AL (1989) Intermediate waters in the southwest South Atlantic. Deep Sea Res Part A Oceanogr Res Papers 36(1):1–16. https://doi.org/10.1016/0198-0149(89)90015-0

Piola AR, Martínez Avellaneda N, Guerrero RA et al (2010) Malvinas-slope water intrusions on the northern Patagonia continental shelf. Ocean Sci 6(1):345–359. https://doi.org/10.5194/os-6-345-2010

Piola AR, Franco BC, Palma ED et al (2013) Multiple jets in the Malvinas current. J Geophys Res Oceans 118:2107–2117. https://doi.org/10.1002/jgrc.20170

Risaro DB, Chidichimo MP, Piola AR (2022) Interannual variability and trends of sea surface temperature around southern South America. Front Mar Sci 9:8291440. https://doi.org/10.3389/fmars.2022.829144

Rivas AL (2006) Quantitative estimation of the influence of surface thermal fronts over chlorophyll concentration at the Patagonian shelf. J Mar Syst 63(3–4):183–190. https://doi.org/10.1016/j.jmarsys.2006.07.002

Rivas AL (2010) Spatial and temporal variability of satellite-derived sea surface temperature in the southwestern Atlantic Ocean. Cont Shelf Res 30(7):752–760. https://doi.org/10.1016/j.csr.2010.01.009

Rivas AL, Piola AR (2002) Vertical stratification on the shelf off northern Patagonia. Cont Shelf Res 22:1549–1558. https://doi.org/10.1016/S0278-4343(02)00011-0

Rivas AL, Pisoni JP (2010) Identification, characteristics and seasonal evolution of surface thermal fronts in the Argentinean Continental Shelf. J Mar Syst 79(1–2):134–143. https://doi.org/10.1016/j.jmarsys.2009.07.008

Rivas AL, Dogliotti AI, Gagliardini DA (2006) Seasonal variability in satellite-measured surface chlorophyll in the Patagonian Shelf. Cont Shelf Res 26(6):703–720. https://doi.org/10.1016/j.csr.2006.01.013

Romero SI, Piola AR, Charo M et al (2006) Chlorophyll-a variability off Patagonia based on SeaWiFS data. J Geophys Res 111:C05021. https://doi.org/10.1029/2005JC003244

Ruiz Etcheverry LA, Saraceno M, Piola AR et al (2016) Sea level anomaly on the Patagonian continental shelf: trends, annual patterns and geostrophic flows. J Geophys Res Oceans 121:2733–2754. https://doi.org/10.1002/2015JC011265

Salyuk PA, Mosharov SA, Frey DI et al (2022) Physical and biological features of the waters in the outer Patagonian shelf and the Malvinas current. Water 14(23):3879. https://doi.org/10.3390/w14233879

Saraceno M, Provost C (2012) On eddy polarity distribution in the southwestern Atlantic. Deep-Sea Res I Oceanogr Res Pap 69:62–69. https://doi.org/10.1016/j.dsr.2012.07.005

Saraceno M, Provost C, Piola AR et al (2004) The Brazil Malvinas frontal system as seen from nine years of AVHRR data. J Geophys Res 109(C5):C05027. https://doi.org/10.1029/2003JC002127

Saraceno M, Provost C, Piola AR (2005) On the relationship of satellite retrieved surface temperature fronts and chlorophyll-a in the Western South Atlantic. J Geophys Res 110:C11016. https://doi.org/10.1029/2004JC002736

Saraceno M, Provost C, Lebbah M (2006) Biophysical regions identification using an artificial neuronal network: a case study in the South Western Atlantic. Adv Space Res 37(4):793–805. https://doi.org/10.1016/j.asr.2005.11.005

Saraceno M, Bodnariuk N, Ruiz-Etcheverry LA et al (2024) Lagrangian characterization of the southwestern Atlantic from a dense surface drifter deployment, Deep Sea Res Part I 208. https://doi.org/10.1016/j.dsr.2024.104319

Saunders PM, King BA (1995) Oceanic fluxes on the WOCE A11 section. J Phys Oceanogr 25(9):1942–1958. https://doi.org/10.1175/1520-0485(1995)025%3C1942:OFOTWA%3E2.0.CO;2

Severo A, Cepeda GD, Acha EM (2024) The effects of the Patagonian shelf-break front on copepod abundance, biodiversity, and assemblages. J Mar Syst 241:103921. https://doi.org/10.1016/j.jmarsys.2023.103921

Shchepetkin A, McWilliams JC (2005) The regional oceanic modeling system (ROMS): a split explicit, free-surface, topography-following-coordinate oceanic model. Ocean Model 9:347–404. https://doi.org/10.1016/j.ocemod.2004.08.002

Siedlecki SA, Archer DE, Mahadevan A (2011) Nutrient exchange and ventilation of benthic gases across the continental shelf break. J Geophys Res Oceans 116:C06023. https://doi.org/10.1029/2010JC006365

Simonella LE, Cosentino NJ, Montes ML et al (2022) Low source-inherited iron solubility limits fertilization potential of South American dust. Geochim Cosmochim Acta 335:272–283. https://doi.org/10.1016/j.gca.2022.06.032

Simpson JH, Sharples J (2012) Introduction to the physical and biological oceanography of shelf seas. Cambridge University Press. https://doi.org/10.1017/CBO9781139034098

Song H, Marshall J, Follows MJ et al (2016) Source waters for the highly productive Patagonian shelf in the southwestern Atlantic. J Mar Syst 158:120–128. https://doi.org/10.1016/j.jmarsys.2016.02.009

Tagliabue A, Bopp L, Aumont O (2009) Evaluating the importance of atmospheric and sedimentary iron sources to Southern Ocean biogeochemistry. Geophys Res Lett 36:L13601. https://doi.org/10.1029/2009GL038914

Torres Alberto ML, Bodnariuk N, Ivanovic M et al (2021) Dynamics of the confluence of Malvinas and Brazil currents, and a southern Patagonian spawning ground, explain recruitment fluctuations of the main stock of *Illex argentinus*. Fish Oceanogr 30(2):127–141. https://doi.org/10.1111/fog.12507

Valla D, Piola AR (2015) Evidence of upwelling events at the northern Patagonian shelf break. J Geophys Res Oceans 120:7635–7656. https://doi.org/10.1002/2015JC011002

Vivier F, Provost C (1999) Direct velocity measurements in the Malvinas current. J Geophys Res Oceans 104(C9):21083–21103. https://doi.org/10.1029/1999JC900163

Wu L, Cai W, Zhang L et al (2012) Enhanced warming over the global subtropical western boundary currents. Nat Clim Chang 2(3):161–166. https://doi.org/10.1038/nclimate1353

Xie X, Liu Q, Zhao Z et al (2018) Deep sea currents driven by breaking internal tides on the continental slope. Geophys Res Lett 45(12):6160–6166. https://doi.org/10.1029/2018GL078372

Yang H, Lohmann G, Wei W et al (2016) Intensification and poleward shift of subtropical western boundary currents in a warming climate. J Geophys Res Oceans 121(7):4928–4945. https://doi.org/10.1002/2015JC011513

Yang H, Lohmann G, Krebs-Kanzow U et al (2020) Poleward shift of the major ocean gyres detected in a warming climate. Geophys Res Lett 47(5):e2019GL085868. https://doi.org/10.1029/2019GL085868

Zhang T, Yankovsky AE, Piola AR et al (2018) Observations of semidiurnal internal tides on the Patagonian shelf. Cont Shelf Res 167:46–54. https://doi.org/10.1016/j.csr.2018.08.004

Chapter 3
The Phytoplankton of the Patagonian Shelf-Break Front

Valeria A. Guinder, Carola Ferronato, Ana I. Dogliotti, Valeria Segura, and Vivian Lutz

Abstract Intense phytoplankton blooms in the Patagonian Shelf-Break Front (PSBF) in the southwest Atlantic Ocean (~35 °S–55 °S) are responsible for making this large marine ecosystem one of the most productive and rich in resources and biodiversity of the Global Seas. This systematic review presents up-to-date knowledge of the phytoplankton community composition, biomass and primary productivity along the permanent PSBF from temperate to subpolar latitudes. The interaction between wind stress, shelf waters, the steep slope and western energetic edge currents (i.e. Malvinas and Brazil Currents) originates upwelling areas and phytoplankton outbreaks, with maxima in spring and summer. On a sub-regional scale, the structure of the phytoplankton community is driven by the interplay between the complex geomorphological and hydrographical characteristics across the shelf-to-open ocean transition, also modulated by seasonal changes in light and

Supplementary Information The online version contains supplementary material available at https://doi.org/10.1007/978-3-031-71190-9_3.

V. A. Guinder (✉) · C. Ferronato
Instituto Argentino de Oceanografía (IADO), Consejo Nacional de Investigaciones Científicas y Técnicas (CONICET), Universidad Nacional del Sur (UNS), Bahía Blanca, Argentina
e-mail: vguinder@iado-conicet.gob.ar

A. I. Dogliotti
Facultad de Ciencias Exactas y Naturales (CONICET -UBA), Instituto de Astronomía y Física del Espacio (IAFE), Universidad de Buenos Aires, Buenos Aires, Argentina

Instituto Franco-Argentino para el Estudio del Clima y sus Impactos (UMI IFAECI/CNRS-CONICET-UBA), Buenos Aires, Argentina

V. Segura
Instituto Nacional de Investigación y Desarrollo Pesquero (INIDEP), Mar del Plata, Argentina

V. Lutz
Instituto Nacional de Investigación y Desarrollo Pesquero (INIDEP), Mar del Plata, Argentina

Instituto de Investigaciones Marinas y Costeras (IIMyC), Consejo Nacional de Investigaciones Científicas y Técnicas–Universidad Nacional de Mar del Plata (CONICET – UNMdP), Mar del Plata, Argentina

E. M. Acha et al. (eds.), *The Patagonian Shelfbreak Front*, Aquatic Ecology Series 13, https://doi.org/10.1007/978-3-031-71190-9_3

nutrients. Hence, different assemblages of phytoplankton functional groups and size classes are observed along contrasting areas of the PSBF. Phytoplankton blooms have been studied from field observations in oceanographic cruises, satellite imagery and modelling. As a general outcome, blooms of micro-phytoplankton, e.g. large diatoms and dinoflagellates, are responsible for the high chlorophyll levels in austral spring, while blooms of pico- and nanophytoplankton occur in summer, e.g. blooms of calcified and non-calcified haptophytes such as the coccolithophore *Emiliania huxleyi* and *Phaeocystis antarctica*. Potentially toxic phytoplankton and associated phycotoxins have been increasingly documented in the last decade, especially multispecific blooms of dinoflagellates of the genera *Azadinium* and *Amphidoma*, producers of azaspiracids (AZAs). The occurrence of these natural hazards raises scientific and social concern about their drivers and impacts on marine resources and human health. This review motivates further discussion on the taxonomical and functional biodiversity of phytoplankton and their biogeochemical roles in the PSBF, which will help to elucidate possible shifts at the base of the food web driven by Global Change that might have repercussions on ecosystem services.

Acronyms and Abbreviations

ACC	Antarctic Circumpolar Current
ASP	Amnesic shellfish poisoning
AZAs	Azaspiracids
AZP	Azaspiracid shellfish poisoning
BB	Burdwood Bank
BC	Brazil Current
BMC	Brazil–Malvinas Confluence
chl-a	chlorophyll *a* concentration
CIs	Cycloimines
DA	Domoic acid
DSP	Diarrhetic shellfish poisoning
HAB	Harmful algal bloom
HNLC	High nutrient-low chlorophyll
MC	Malvinas Current
MIS	Malvinas Islands Shelf
PBPT	Photosynthetic and bio-optic phytoplankton type
PIC	Particulate inorganic carbon
PP	Primary production
PSBF	Patagonian Shelf-Break Front
PSP	Paralytic shellfish poisoning
PTXs	Pectenotoxins
SSP	Spiroimine shellfish poisoning
YTXs	Yessotoxins

3.1 Introduction

3.1.1 The Permanent Patagonian Shelf-Break Front

The Argentine Sea in the southwestern South Atlantic Ocean is one of the largest and most productive ecosystems in the World Ocean (reviewed in Martinetto et al. 2020). The part attached to the South American continent covers 1.2 million km^2 and extends from the La Plata River (at ~35 °S) to the Beagle Channel (at ~55 °S) and ~200 km to ~900 km offshore to the shelf-break. The Argentine Shelf sustains high primary productivity associated with several fronts (Schloss et al. 2007; García et al. 2008; Lutz et al. 2010; Dogliotti et al. 2014), including the large Patagonian Shelf-Break Front (PSBF), which spans 1500 km and is one of the most extensive chlorophyll hotspots globally, clearly observed by ocean colour sensors (Romero et al. 2006; Dogliotti et al. 2014; Marrari et al. 2017) (Fig. 3.1). This permanent front is characterized by a pronounced thermal gradient shaped by the interplay between

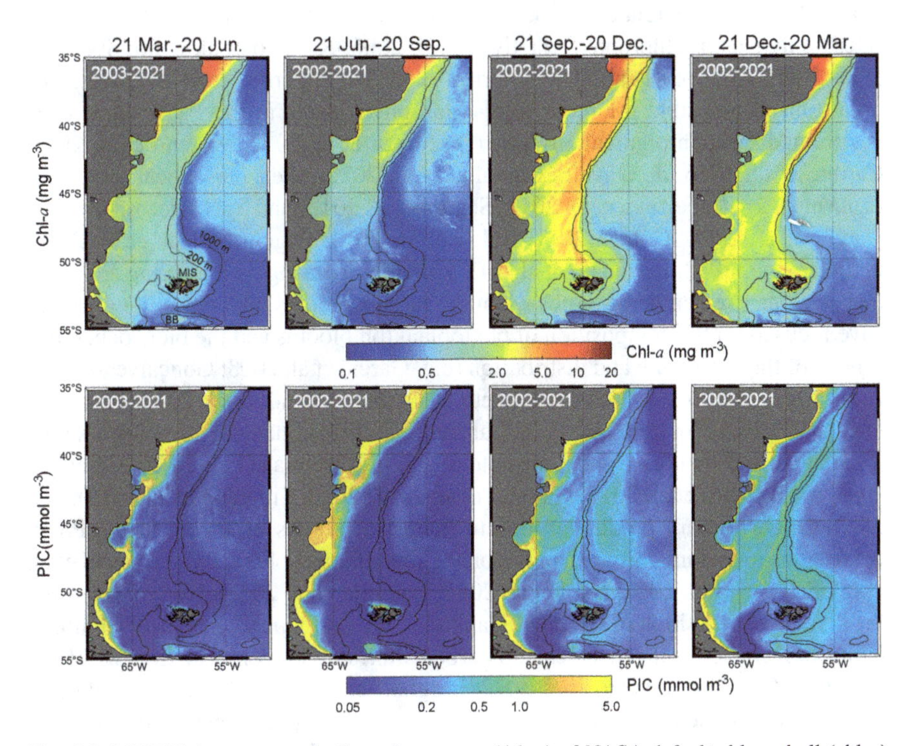

Fig. 3.1 MODIS-Aqua seasonal climatology maps (4 km) of NASA default chlorophyll (chl-*a*) and particulate inorganic carbon (PIC) products in autumn (21 March to 20 June 2003–2021), winter (21 June to 20 September 2002–2021), spring (21 September to 20 December 2002–2021) and summer (21 December to 20 March 2002–2021) along the Patagonian Shelf and Shelf-Break Front. Note: the high PIC values on the coastal strip are probably produced by high suspended fine sediments. The 200 and 1000 m isobaths, Malvinas Islands Shelf (MIS) and Burdwood Bank (BB) are indicated

different water masses in the shelf, the sharp slope and the strength, variability and pathways of the Brazil and Malvinas Currents, as well as strong winds (Palma et al. 2008; Matano et al. 2010, see Chap. 2 in this Book). The conspicuously high phytoplankton biomass along the PSBF in spring and summer provides suitable feeding and spawning ground conditions for several species of commercial and ecological importance (see Chaps. 5, 6, 7, 8, and 9 in this Book). The high phytoplankton yield supports vast fisheries, which include the exploitation of large fish and squid stocks (Brunetti et al. 1998; Martinetto et al. 2020) and benthic beds of the Patagonian scallop *Zygochlamys patagonica* (Bogazzi et al. 2005; Campodónico et al. 2019).

3.1.2 Studies Available on Phytoplankton in the PSBF

The occurrence of phytoplankton blooms in the PSBF has been studied through field estimations (e.g. Lutz and Carreto 1991; Gayoso and Podestá 1996; Carreto et al. 2003, 2008; García et al. 2008; Segura et al. 2013; Ferronato et al. 2023) and satellite images of chlorophyll *a* (Rivas et al. 2006; Romero et al. 2006; Signorini et al. 2006; Dogliotti et al. 2014; Marrari et al. 2017; Ferronato et al. 2023). The organisms responsible for these high-chlorophyll levels in the PSBF have been determined through pigment analysis (De Souza et al. 2012; Moreno et al. 2012; Carreto et al. 2016; de Oliveira Carvalho et al. 2022), their bio-optical properties (Lutz et al. 2010; Ferreira et al. 2013; Segura et al. 2013) and light microscopy (e.g. García et al. 2008; Gómez 2011; Balch et al. 2014; Olguín Salinas et al. 2015; Guinder et al. 2018; Ferronato et al. 2023), mainly focusing on the spring and summer seasons (September–March) when the largest blooms take place. Few studies have focused on the composition of phytoplankton blooms and the biogeochemical drivers of their structure and distribution (e.g. Garcia et al. 2008; Gonçalves-Araujo et al. 2012; Poulton et al. 2013; Carreto et al. 2016; Smith et al. 2017; Ferronato et al. 2023) and ecophysiology (e.g. Valiadi et al. 2014), including toxin-producing microalgae and their associated phycotoxins (e.g. Akselman et al. 2015; Almandoz et al. 2017; Fabro et al. 2017; Guinder et al. 2018, 2020; Tillmann et al. 2019).

The high primary production due to intensive blooms occurring in the PSBF contributes to making this area an important sink of atmospheric CO_2 (Bianchi et al. 2009; Schloss et al. 2007; Kahl et al. 2017; Laurelle et al. 2018). Notwithstanding, the diverse composition of the phytoplankton community plays a key role in modulating biogeochemical processes and carbon fluxes along contrasting areas of the PSBF (de Oliveira Carvalho et al. 2022; Berghoff et al. 2023). The ecological function, especially the different potential to sequester CO_2 (i.e. transferring carbon to deep water layers, where it remains for long periods), has not been investigated in detail (Schloss et al. 2007). A recent study on a cross-shelf transect in the northern Argentinian shelf found that picophytoplankton, e.g. *Synechococcus*, were the main contributors to CO_2 fixation in summer in the outer shelf (~39 °S) (Berghoff et al. 2023). While diatom blooms are well recognized for their important contribution to the biological carbon pump (Tréguer et al. 2018), coccolithophores—calcifying haptophytes—have also become of large interest in this regard (Rivero-Calle et al. 2015). In the PSBF, extensive and massive blooms of the coccolithophore *Emiliania huxleyi* occur during early summer (Dec-Jan). These blooms have been detected by remote sensing algorithms (Brown and Podestá 1997; Signorini et al. 2006), field studies (Poulton et al. 2013; Balch et al. 2014; Smith et al. 2017) and by both approaches combined (García et al. 2011). They could be an important carbon contribution to the deep seafloor by biomineralizing nanoplankton.

Despite all the multivariate data available, significant gaps remain in our understanding of the spatiotemporal dynamics of phytoplankton blooms from a holistic ecosystem approach. We revise the main hypotheses that have been postulated about the physical and biogeochemical drivers producing the extensive blooms in the PSBF. For instance, the dynamics of the front lead to optimal conditions for phytoplankton growth on the outer shelf in spring and summer, where the stability of the upper layer keeps the cells in the illuminated zone while receiving a continuous supply of nutrients from the Malvinas Current, including upwelling and the enriched air-borne iron supply. Other, less-studied factors, more related to biological interactions such as grazing control and species competition and succession, are discussed as well. Special emphasis is put into attempting to elucidate factors contributing to the growth of different phytoplankton groups in contrasting areas along the PSBF, such as near the shelf of the Malvinas Islands, over transversal submarine canyon systems and in the Brazil–Malvinas Confluence. Overall, this revision synthesizes and narrows down the main factors controlling phytoplankton in the PSBF but equally important highlights voids of information and questions that need to be addressed—such as the effects of global warming on microbial communities' composition and abundance—to enhance our knowledge of the role of primary producers in the area.

3.2 Drivers of Phytoplankton Blooms in the PSBF

Large phytoplankton biomass along the PSBF is revealed as a persistent band of high satellite chlorophyll *a* concentration (chl-*a*) (Fig. 3.1) (Saraceno et al. 2005; Rivas et al. 2006; Romero et al. 2006), located at or inshore of the surface temperature minimum and density maximum, which are associated with high nitrate concentrations in relation to the surrounding shelf and offshore surface waters (Carreto et al. 2016; Balch et al. 2014). The western boundary current system shapes the phytoplankton dynamics throughout the year, with the upwelling of cold, nutrient-rich waters carried northwards by the Malvinas Current (MC) being the most important driver of blooms along the PSBF (García et al. 2008). The MC originates as a branch of the Antarctic Circumpolar Current (ACC) around 55 °S (Matano et al. 2019) and converges around 38 °S with the warm, salty and oligotrophic waters of the Brazil Current (BC), which runs southwards (de Oliveira Carvalho et al. 2022). At the Brazil–Malvinas Confluence (BMC), the mixed waters are advected eastwards by the BC recirculation gyre along the western South Atlantic Ocean (Fig. 3.2a).

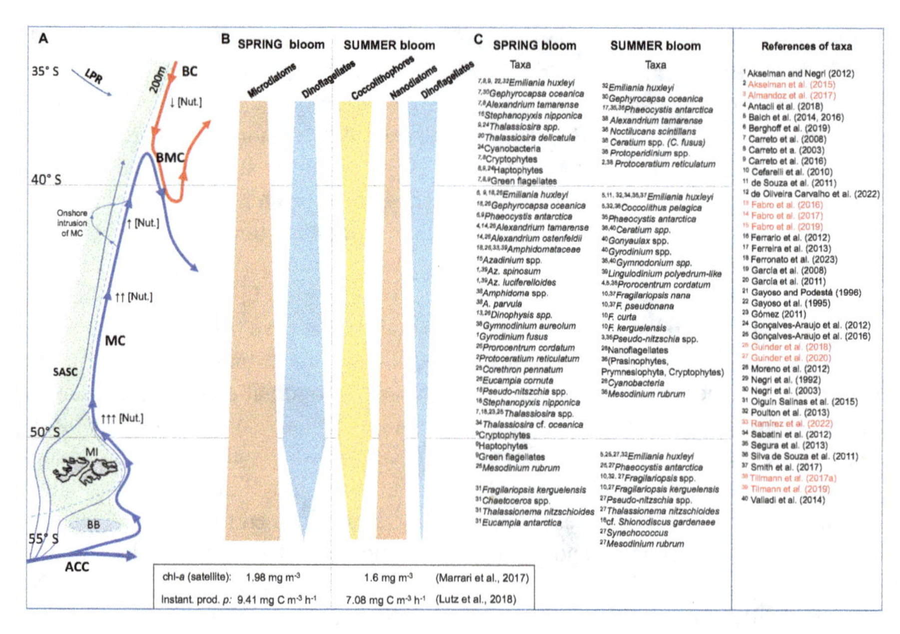

Fig. 3.2 Schematic representation of the main physical drivers of the Patagonian Shelf-Break Front (PSBF) and the phytoplankton composition in spring and summer blooms along different areas of the PSBF: 35–40 °S, 40–50 °S and 50–55 °S, after a systematic review of the literature. (**a**) Nutrient-rich waters ↑↑↑ [Nut.] of the Malvinas Current (MC) running northward (in blue), nutrient-poor waters ↓[Nut.] of the Brazil Current (BC) running southward (in red), and the Brazil–Malvinas Confluence (BMC) around 38 °S. *LPR* La Plata River, *MI* Malvinas Islands, *BB* Burdwood Bank, *ACC* Antarctic Circumpolar Current, *SASC* Sub-Antarctic Shelf Currents, and onshore intrusions of the Malvinas Current around 42 °S. (**b**) Distribution of the dominant phytoplankton groups in spring and summer along the PSBF. Microdiatoms: diatom cells >20 μm; Nanodiatoms: diatom cells <20 μm, dinoflagellates and coccolithophores. Note: data on nanoflagellates and picophytoplankton in the literature were not enough to assess their distribution with the same level of confidence as the other phytoplankton groups. The mean chl-*a* (satellite concentration) and primary productivity (field estimations) in spring and summer are shown at the bottom, extracted from Marrari et al. (2017) and Lutz et al. (2018), respectively. (**c**) Common phytoplankton taxa in spring and summer in the different areas of the PSBF, according to the corresponding references (superscript numbers) listed on the right. The references in red include observations of toxin-producing species and associated phycotoxins

3.2.1 *Upwelling*

High biological primary production at the PSBF results from a variety of mechanisms that act at different time and spatial scales. One important factor is the upwelling, which emerges as a compensation for the horizontal diverging velocities of the MC caused by frictional effects (Matano and Palma 2008). The magnitude of the upwelling has been found to correlate with the transport of the inshore portion of the MC, to have low seasonal variability, and to be heterogeneous along the shelf-break due to changes in the bottom slope and the presence of transversal submarine canyons (Combes and Matano 2018). The inshore branch and the main branch of the

MC converge near 44 °S, narrowing the PSBF to the north (Frey et al. 2023). Around 41 °S, the middle shelf is affected by water intrusions from the MC (Lucas et al. 2005; Piola et al. 2013), in addition to the diluted subantarctic waters that fuel nutrients into the shelf (Fig. 3.2a). The MC intrusions play a significant role in the development of the Mid Shelf Front (between 37.5 °S and 41 °S) during autumn and spring (Romero et al. 2006) through fertilization and stimulation of phytoplankton outbursts (Díaz et al. 2018; Ferronato et al. 2021). Together with the nutrient supply from the MC upwelling, phytoplankton growth in the onshore portion of the PSBF is stimulated by strong vertical stability that maintains the cells in the well-illuminated surface layers (García et al. 2008, Signorini et al. 2006; Ferronato et al. 2023). The seasonal thermal cycle promotes water column stratification and marks the timing of the spring bloom initiation (Rivas et al. 2006): north of ~45 °S, the bloom starts in early spring (September and October), while in the south, it starts in late spring to early summer (December–January) and continues until autumn (March) (Rivas et al. 2006; Romero et al. 2006).

Observational evidence showed that at sub-seasonal timescales, upwelling can also be induced by southerly winds, which erode the stratification of the water column through Ekman transport and allow the input of nutrients into the euphotic zone. This mechanism is particularly important in summer when the mixed layer is stronger, and nutrients at the surface are depleted after the intense spring bloom (Carranza et al. 2017). In fact, reduced nutrient availability and grazing pressure by microzooplankton seem to regulate the spring bloom decay (Gonçalves-Araujo et al. 2012; Carreto et al. 2016). Towards summer, phytoplankton growth is restricted to species that are adapted to highly stratified and relatively poor nutrient conditions—especially low silicates (García et al. 2011; Moreno et al. 2012; Balch et al. 2014; Gonçalves-Araujo et al. 2016; Smith et al. 2017). Upwelling has intraseasonal variability at the PSBF, and continental-trapped waves have been suggested as the forcing mechanism (Saraceno et al. 2005). Changes in the direction of meridional winds also trigger chlorophyll variability on an interannual scale; for instance, northerly (southerly) winds promote the initiation of the spring bloom in the first days of September (October). Moreover, stronger northerly winds, which might induce Ekman transport of shelf waters to the Malvinas domain, resulting in higher vertical stability that maintains phytoplankton in the well-illuminated surface, have been correlated with enhanced satellite-derived chl-*a* (Saraceno et al. 2005).

3.2.2 *Other Mechanisms Responsible of High Productivity in the PSBF*

A study of satellite images covering 20 years of surface chl-*a* records revealed an annual increase of 2% in the middle shelf and frontal areas of the Patagonian Large Marine Ecosystem (Marrari et al. 2017). The drivers of this rising chl-*a* are still not well understood, and more in situ data are required to understand why this is occurring and whether it is related to long-term natural variability or to anthropogenic effects. In a more recent assessment, Delgado et al. (2023) characterized different bioregions in the SW Atlantic Ocean based on 24-year satellite-derived chl-*a* to study the phenology of the blooms and potential drivers. They report that the bioregions along the PSBF, especially those in the northern Patagonian Shelf, show a positive trend in chl-*a* concentration, probably related to an increase in sea surface temperature and the shoaling of the mixed layer depth. Different studies of the underlying hydrological drivers seem to agree on the changes in the interaction between the Malvinas Current and the shelf-break that may be affecting the supply of nutrients onto the shelf (Matano and Palma 2008). On the one hand, the poleward displacement of the BMC implies a southward shift in the northernmost penetration of the MC (Franco et al. 2022), and a model analysis suggests that upwelling at the PSBF has been weakening since 1999–2000 due to reduced Malvinas transport (Combes and Matano 2018). On the other hand, satellite-derived MC transport at 41 °S showed no significant trend in spite of large interannual variability. Changes in surface wind patterns have displayed a positive trend in southerly winds affecting the Patagonian shelf over the past decade (Risaro et al. 2022) and probably enhancing upwelling at the PSBF (Carranza et al. 2017). Another mechanism that may additionally contribute to the fertilization of the northern part of the PSBF (north of ~52 °S), which has been derived from observational and modelled data (Frey et al. 2021; Palma et al. 2021), involves the advection of nutrient-rich waters from the Malvinas Islands Shelf (MIS) (Fig. 3.1), with higher inputs between the 100 and 200 m isobaths during spring and summer. The MIS waters have their origin mainly in the northern portion of the Drake Passage. At this location, upwelling of the bottom layer is promoted by the collision of the northern branch of the Antarctic Circumpolar Current (ACC) with the MIS. It has been suggested that waters from the Drake Passage have enough macronutrient levels (particularly silicates) to maintain primary production (e.g. nitrates, 12–18 μM; phosphates, 1.1–1.6 μM; silicates, 5–15 μM; Paparazzo and Esteves 2018). South of the Malvinas Islands, the circulation around the Burdwood Bank, located between 53.5 °S and 55 °S (Fig. 3.1), seems to explain the initiation of the summer bloom at those latitudes. Two branches of the nutrient-rich ACC contour the bank along its eastern and western sides and merge to the north, running into the Malvinas Current along the shelf-break. A recent high-resolution circulation model (Matano et al. 2019) showed that mixing and upwelling driven by tides and the ACC interacting with the complex topography of the bank may supply nutrients from deep waters to fertilize phytoplankton growth in the surface layers. This is in agreement with satellite images of chl-*a* in early summer (December), where a dense phytoplankton signal is visible around the Burdwood Bank and is distributed northward into the jet of the Malvinas Current (Fig. 20 in Matano et al. 2019). The scarce data on phytoplankton abundance and composition (Valiadi et al. 2014; Charalampopoulou et al. 2016; Guinder et al. 2020) and nutrient levels (Paparazzo and Esteves 2018) along the PSBF south of the Malvinas Islands call for further investigation into the biogeochemical processes in this area.

Further south (>55 °S), some regions of the Drake Passage and the Southern Ocean have been characterized as high-nutrient low-chlorophyll (HNLC) areas in which co-limitation of light and micronutrients, such as iron, may offer the proper conditions for the prevalence of smaller-size phytoplankton, which outcompete large cells of different phytoplankton groups (Jickells et al. 2005; Treguer et al. 2018). Although iron limitation has been mentioned as a possible driver affecting the growth of phytoplankton in the Southwest Atlantic, so far the evidence is scarce. Iron measurements at sea are not frequent in the global ocean, less so in the Southwest Atlantic, and non-existent in the waters of the Argentinian continental shelf and shelf-break. Geochemical studies on iron limitation are laborious, demanding the use of ultra-clean sophisticated techniques and the identification of its bioavailable forms (e.g. Thuroczy et al. 2011; Birchill et al. 2017). Furthermore, studies in the Southern Oceans performed during the GEOTRACES programme (https://www.geotraces.org/) showed that there is always a combination of factors (light, physical mixing, concentration of different macro- and other micro-nutrients) rather than iron alone limiting the growth of different types of phytoplankton (Viljoen et al. 2019). A recent work on the influence of dust deposition on the Southwest Atlantic, near Tierra del Fuego ~55 °S, showed no clear effect of a possible airborne iron fertilization on a phytoplankton increase in the area using satellite chl-*a* (Cosentino et al. 2020).

As final remarks of this section, we highlight that contrasting geomorphological and hydrographical settings along the extensive PSBF define the ensemble of conservative and nonconservative properties of water masses that lead to phytoplankton blooms at regional and local scales. Along the wide latitudinal range, different time-scale processes such as water circulation on the shelf, edge currents, winds, tides and frontal nutrient injections underpin the development of the large phytoplankton blooms. Hence, although the PSBF is a distinctive hotspot of productivity in the Southwestern South Atlantic, it may be considered as an ensemble of sub-ecosystems with distinct hydrological and biological dynamics. In fact, as we discussed here, studies using different approaches have shown seasonal, semi-annual and interannual variability in physical and biogeochemical processes, which altogether converge to formulate complementary explanations for the large-scale productivity of the PSBF.

3.3 Primary Productivity and Phytoplankton Composition in Spring and Summer Blooms

3.3.1 Primary Production

Historical field measurements of primary production (PP) in different seasons in the Southwestern Atlantic Shelf between 23 °S and 55 °S indicate that the most productive area within the shelf is found south of 37.5 °S (Lutz et al. 2018). Unfortunately, field measurements of PP in the PSBF are scarce, and the last published data date from 2005 to 2006 (Segura et al. 2013; Dogliotti et al. 2014). This constrains the assessment of potential long-term changes in ecosystem productivity and biodiversity related to phenological shifts at the base of the food webs. Production measured in the field along the shelf-break in spring 2005 was highly variable, ranging from 511.66 to 5477.40 mg m^{-2} d^{-1} with a mean value of 1706.95 mg m^{-2} d^{-1}. Maxima values were associated with a bloom of the nano-diatom Thalassiosira cf. oceanica (Lutz et al. 2010; Sabatini et al. 2012; Segura et al. 2013), and up to 7800 mg C m^{-2} d^{-1} related to blooms of large centric diatoms in 2004 (Garcia et al. 2008). In addition to these maxima of PP during diatom blooms, other high values of PP (3201 mg $m^{-2}d^{-1}$) were related to a diverse photosynthetic and bio-optic phytoplankton type (PBPT 8) during spring in the PSBF, composed mainly of large phytoplankton cells (Segura et al. 2013). In summer, field-measured PP in the Patagonian Shelf south of 37.5 °S was considerably lower than in spring (Lutz et al. 2018). In the PSBF during the summer of 2006, the PP varied between 71.78 and 416.36 mg C m^{-2} d^{-1} with a mean of 253.30 mg m^{-2} d^{-1}, and relatively low instantaneous production associated with different PBPT composed of ultra- and nanophytoplankton (Segura et al. 2013). Satellite estimation of PP displayed low values in fall and winter (Lutz et al. 2018). For instance, in late winter 2006, mean PP in the PSBF was 179.92 mg m^{-2} d^{-1}, with higher values north of ~47 °S, likely indicating the beginning of spring (Dogliotti et al. 2014).

3.3.2 *Distribution of Phytoplankton Species in the PSBF*

Following, we present phytoplankton taxa in spring and summer, revised and compiled from 40 publications corresponding to a total of 37 oceanographic cruises in the PSBF, in order to assess the distribution of the main phytoplankton groups and species assemblages in relation to the main hydrological drivers of each area (Fig. 3.2b). These *snapshot* field surveys of phytoplankton biodiversity were based on surface water samples collected with Niskin bottles and phytoplankton net tows. Traditional optical microscopy—complemented with scanning electron microscopy—and pigment analysis were the most used techniques to characterize the phytoplankton abundance and composition, sometimes combined with analysis of satellite data on chl-*a* and particulate inorganic carbon (PIC) concentration.

3.3.2.1 Species Composition in Spring Blooms

The knowledge gained so far indicates that diatoms are widespread along the PSBF and are usually major components of spring blooms—especially in the middle (40–50 °S) and southern (50 °S–55 °S) sectors of the PSBF (Fig. 3.2c)—although high abundances of dinoflagellates and flagellates (including coccolithophores) have been reported in mixed blooms in this season and area. A shift in the structure and composition of the phytoplankton blooms is observed around 40 °S, associated with the particularly complex circulation of the Brazil–Malvinas Confluence (BMC). Overall, the phytoplankton assemblage of the northern (35 °S–40 °S) PSBF resembles a transition from subtropical to subantarctic waters, shifting from higher SST and salinity and low phytoplankton abundances (Brazil Current domain) to lower SST and salinity values, together with higher diatom biomass (Malvinas Current domain) (de Oliveira Carvalho et al. 2022). For instance, in 1989, an early spring bloom (September) composed of the diatom *Thalassiosira delicatula* (up to 5.5×10^5 cells L^{-1}) took place at the BMC (Gayoso and Podestá 1996). In the same year, but at a later stage of the bloom (November–December), the coccolithophore *Emiliania huxleyi* reached densities between 4.3 and 6.1×10^5 cells L^{-1}, related to strong mixing and nutrient renovation of the BC waters (Gayoso 1995). South of the BMC, the dinoflagellate *Gyrodinium* cf. *aureolum* (nowadays renamed *Gymnodinium aureolum*) reached up to 6.8×10^5 cells L^{-1} during an early spring bloom in 1988, favoured by a shallow pycnocline and high nitrate (15–20 μM) and phosphate (1.5–2.0 μM) from subantarctic waters of the Malvinas Current (Negri et al. 1992). This bloom was accompanied by diatom species such as *Thalassiosira mendiolana*, *T. anguste-lineata*, *Nitzschia* cf. *delicatissima*, *N. longissima*, *Corethron criophilum*, and dinoflagellates including *Alexandrium excavatum*, *Gyrodinium fusus*, *Torodinium robustum* (Negri et al. 1992), *Ceratium lineatum*, *Protoperidinium* spp. and *Gonyaulax* spp. (Gayoso 1995). Akselman and Negri (2012) also described two consecutive spring blooms of dinoflagellates of the *Amphidomataceae* group in the outer shelf off Mar del Plata (around 38 °S–40 °S) related to subantarctic waters, dominated by *Azadinium luciferelloides* (Tillmann and Akselman 2016). These dinoflagellates reached 3×10^6 cells L^{-1} at the middle/outer shelf and were replaced by diatoms such as *Corethron criophilum*, *Thalassiosira anguste-lineata*, *T. allenii* and *T. antarctica* at the oceanic sector east of the shelf-break. Similarly, in spring 2015, an abundant multispecific bloom of *Amphidomataceae* took place in the same area at 41 °S (Guinder et al. 2018), composed of *Azadinium* (Tillmann et al. 2019) and *Amphidoma* (Tillmann et al. 2017a) species.

The above-mentioned studies were performed using only traditional optical microscopy and might have underestimated the small-sized fraction of the community (e.g. pico- and nanoflagellates). More recent studies using chemotaxonomic pigments approach pointed out that chlorophyceans (revealed by high amounts of chl-*b*) were particularly relevant at the chl-*a* maximum of the PSBF (36–38 °S, Carreto et al. 2008). Other pigments, such as Hex-fucoxanthin, indicated the presence of haptophytes (e.g. *Phaeocystis* sp., *Chrysochromulina* sp., *Emiliania huxleyi* and *Gephyrocapsa oceanica*) (Carreto et al. 2003, 2008) as an important component of the spring bloom decay. Fuco-pigments corresponding to nanoplanktonic diatoms are also abundant at the shelf-break (Carreto et al. 2016), with species such as *Thalassiosira bioculata* and *Chaetoceros* spp. (8.5–12 μm) distributed over a large latitudinal range (Segura et al. 2013; Carreto et al. 2016).

Blooms of *Pseudo-nitzschia* species are a common feature in spring and summer in the inner and outer shelf, some of which produce the phycotoxin domoic acid (Almandoz et al. 2017; Guinder et al. 2018, and see section on toxin-producing phytoplankton). Detailed morphological studies reported *Pseudo-nitzschia fraudulenta* and *P. pungens* as cosmopolitan diatoms that occurred from 38° to 55 °S (Olguín Salinas et al. 2015; Almandoz et al. 2017). The *P. pseudodelicatissima* complex is common north of 41 °S (Almandoz et al. 2017), although Sabatini et al. (2012) also found it at 47 °S. High densities of the diatom *Eucampia cornuta* (8×10^4 cells L^{-1}) are commonly found in spring at the outer shelf near the PSBF around 41 °S (Guinder et al. 2018). The large diatom *Stephanopyxis turris* appears restricted to subantarctic waters (south of 50 °S, Olguín Salinas et al. 2015), but it was also found in intermediate temperature waters (9.3–11.4 °C, from 41° to 45 °S, Ferrario et al. 2012) and at 47 °S (5–7 °C) as a companion species in a spring bloom of *Thalassiosira* cf. *oceanica* (1.3×10^6 cells L^{-1} and a chlorophyll value of 7.7 mg m^{-3}, Sabatini et al. 2012). The potentially toxic nanodinoflagellate *Prorocentrum cordatum* also displays a cosmopolitan distribution in spring and summer (e.g. Sabatini et al. 2012; Antacli et al. 2018), being prominent at the shelf-break at ~38 °S (Gonçalves-Araujo et al. 2012), ~47 °S (Antacli et al. 2018) and ~ 50 °S (Gómez 2011; Gonçalves-Araujo et al. 2016), as well as in nearshore waters at 51 °S, reaching blooming densities of up to 10×10^6 cells L^{-1} (Sabatini et al. 2012).

Consistent with the paradigm that high-latitude diatoms usually dominate spring phytoplankton blooms, *Pseudo-nitzschia* spp. (Almandoz et al. 2008) and *Fragilariopsis* spp. (Cefarelli et al. 2010) are common in the southern PSBF and the Drake Passage. Other microplanktonic diatoms reported in the literature as cold-water species in spring and summer are *Pseudo-nitzschia turgidula* and *P. australis* (44 °S–46 °S, Almandoz et al. 2017), *Eucampia antarctica* (49°–55 °S, Olguín Salinas et al. 2015; Guinder et al. 2020), *Stephanopyxis nipponica* (52°–56 °S, Ferrario et al. 2012), *Corethron pennatum* (~50 °S, Gonçalves-Araujo et al. 2016), *Fragilariopsis kerguelensis* (47°–55 °S, Olguín Salinas et al. 2015; Smith et al. 2017; Guinder et al. 2020) and *Thalassionema nitzschioides* (47°–55 °S, Smith et al. 2017; Guinder et al. 2020). This high diversity and predominance of diatoms between 40 °S and 55 °S, associated with cold, nutrient-rich Malvinas waters, play a central role in the modulation of sea–air CO_2 exchanges and highlight this sector as an important CO_2 sink (de Oliveira Carvalho et al. 2022).

3.3.2.2 Species Composition in Summer Blooms

The relatively lower biomass (chl-*a*, Fig. 3.1) observed during this season compared to spring is attributed to the predominance of smaller-size phytoplankton (e.g. De Souza et al. 2012; Balch et al. 2014; Antacli et al. 2018; de Oliveira Carvalho et al. 2022), likely related to nutrient depletion after the blooms of large diatoms and dinoflagellates and grazing by microzooplankton (Fig. 3.2b, c). The haptophytes *Phaeocystis antarctica* and coccolithophores are common features in early summer (December–January), along with cyanobacteria (*Trichodesmium* and *Synechococcus* in the northern part, de Oliveira Carvalho et al. 2022, and *Synechococcus* over the Burdwood Bank, Guinder et al. 2020), nanoflagellates (prasinophytes, cryptophytes), nanodiatoms and dinoflagellates (De Souza et al. 2012). Abundance of dinoflagellates (e.g. *Noctiluca*, *Alexandrium*, *Prorocentrum*, *Protoperidinium*, *Ceratium* and *Gonyaulax*) decreases southwards along the PSBF and is absent below the Malvinas Islands, where diatoms prevail in the cold, nutrient-rich and mixed waters (Valiadi et al. 2014).

The coccolithophorid *Emiliania huxleyi* is an important component of early (December–January) summer blooms, commonly co-occurring with nanodiatoms such as *Fragilariopsis pseudonana*, *F. nana*, the microdiatom *Pseudo-nitzschia* and the nanodinoflagellate *Prorocentrum* (Poulton et al. 2013; Balch et al. 2014; Smith et al. 2017) (Fig. 3.2b, c). In general, blooms of the calcified *E. huxleyi* are underestimated by satellite chlorophyll products due to their weak chl-*a* signal (<0.5 mg m^{-3}). Instead, summer blooms of coccolithophores are detected by high calcite (6 to 10 mmol m^{-3}) and thus by satellite PIC (Signorini et al. 2006). In the Patagonian Shelf, *Gephyrocapsa oceanica* is the other common coccolithophore, although it is more abundant inshore (Negri et al. 2003; Guinder et al. 2018; Ferronato et al. 2021; Delgado et al. 2019). The large-sized coccolithophore *Coccolithus pelagicus* is also a frequent species in summer in the PSBF, but in considerably lower abundance (0.05×10^6 cells L^{-1}) than the blooming *E. huxleyi* (maxima abundances recorded in the PSBF: 11×10^6 cells L^{-1} at ~49 °S in January 2008, García et al. 2011), and 6.3×10^6 cells L^{-1} in January 2019 at ~39 °S, Berghoff et al. 2023). Blooms of *E. huxleyi* in the PSBF are seen from space (Fig. 3.1) and are one of the most prominent blooms of coccolithophores in the Global Seas (Tyrrell and Merico 2004). They are also in part responsible for the elevated PIC feature occurring seasonally alongside high chlorophyll in spring and summer in the Southern Ocean, known as the Great Calcite Belt (Smith et al. 2017). *E. huxleyi* also appears in variable abundances near the Burdwood Bank (Guinder et al. 2020) and across the Drake Passage, related to particular sea-surface temperature, irradiance and calcite saturation conditions (Charalampopoulou et al. 2016). Future studies of the ecophysiological responses of coccolithophores to changing environmental conditions (e.g. carbonate chemistry, resource availability and microbial interactions) in the PSBF will help to elucidate the drivers of their massive blooms and their role in carbon fluxes between the atmosphere and the deep sea.

3.4 Toxin Producing Species and Phycotoxins in the PSBF

3.4.1 Historical Records of Harmful Algal Blooms (HABs)

A recent comprehensive assessment (Ramírez et al. 2022) of historical (1980–2018) records of toxin-producing phytoplankton and associated toxins in water samples from the Argentine continental shelf unveiled an increasing trend in the number of potentially toxigenic species. The assessment by Ramírez et al. (2022) disclosed that the rising trend over 40 years of Harmful Algal Bloom (HAB) studies in the Argentine Sea has resulted from the intensification of sampling efforts and the expansion of the monitored area from coastal ecosystems towards the shelf-break

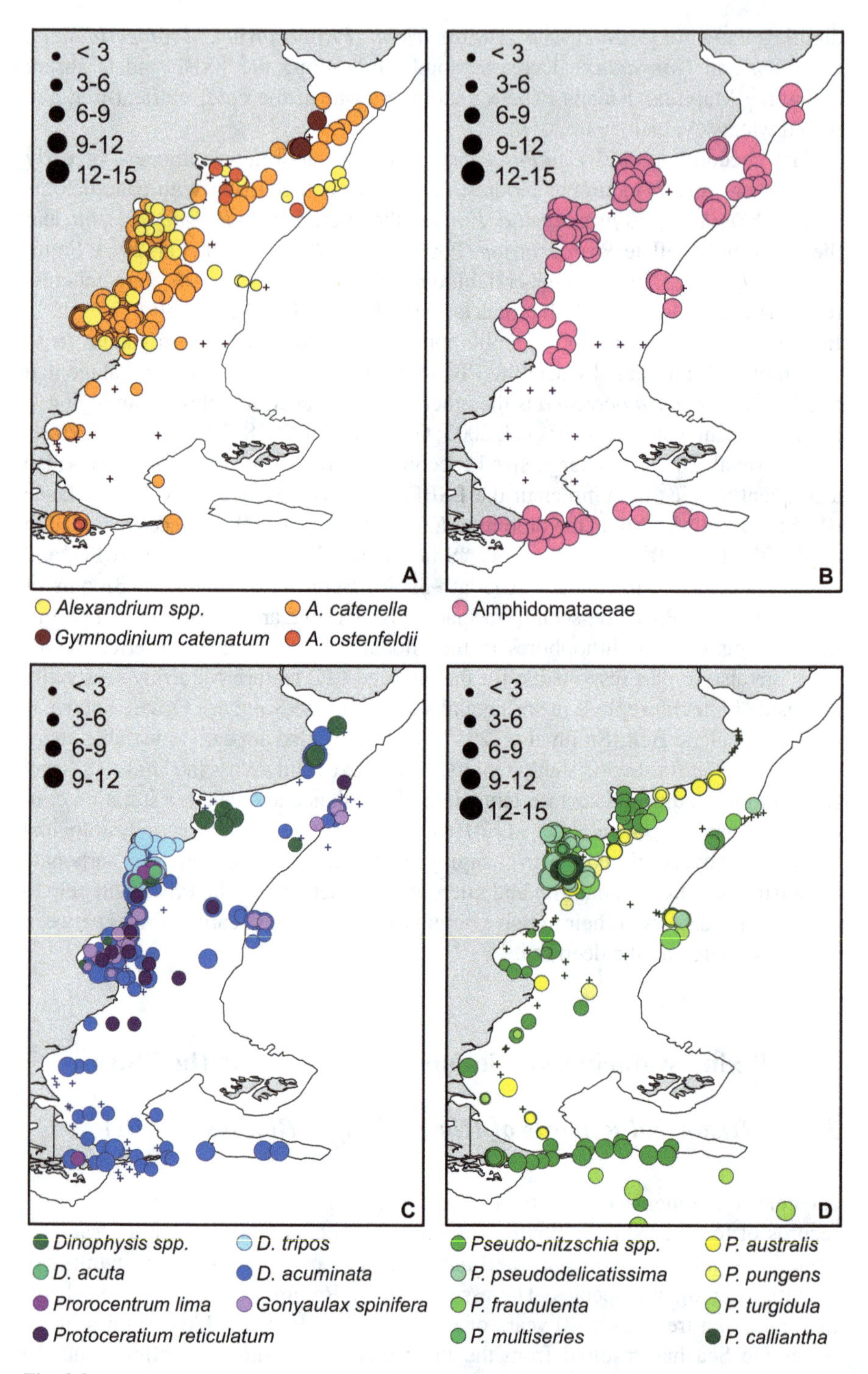

Fig. 3.3 Geographic distribution and abundance (log-transformed cells L^{-1}) of toxigenic species over the period 1980–2018. The species are grouped according to the phycotoxins they produce. (**a**) *Alexandrium* species and *Gymnodinium catenatum* (PSTs producers). (**b**) *Amphidomataceae*

(Figs. 3.3 and 3.4). This has led to the detection of numerous HAB species in the last decade along the PSBF (Akselman et al. 2015; Fabro et al. 2016; Guinder et al. 2018; Tillmann et al. 2019). Initially, long-term monitoring systems of HABs were carried out alongshore on a sustained basis, focusing on paralytic shellfish toxins (PSTs) produced by *Alexandrium catenella* in the northern Patagonian gulfs and the Beagle Channel, settled as early alarm systems related to aquaculture activities (e.g. Gayoso and Fulco 2006; Sastre et al. 2018; Almandoz et al. 2019).

More recently, growing social interest in monitoring these natural hazards and their ecological impacts has extended to productive areas offshore, with discrete sampling through sporadic oceanographic cruises (Almandoz et al. 2007; Akselman and Negri 2012). Scientific interest has focused on understanding the geographical distribution, prevalence and triggering factors of toxic phytoplankton in the PSBF, and oceanographic expeditions specifically planned for scanning HABs have been carried out since the early 2000s (Akselman et al. 2015; Fabro et al. 2016, 2017; Guinder et al. 2018, 2019; Tillmann et al. 2019). In addition, the implementation of sophisticated techniques for the taxonomic identification of species (isolation of strains, electron microscopy and DNA analyses) and the chemical characterization of marine biotoxins (liquid chromatography coupled to tandem mass spectrometry, LC-MS/MS) has resulted in the detection of numerous toxic species and toxins widespread across the large latitudinal gradient of the PSBF, some of which were previously undiscovered in the region (Akselman et al. 2015; Guinder et al. 2018; Almandoz et al. 2017; Tillmann et al. 2019).

3.4.2 Toxin-Producing Microalgae, Associated Toxins and Syndromes

Potentially toxic microalgae have been documented in general studies of plankton assemblages in the late 1990s and the early 2000s during scientific expeditions in the outer shelf (e.g. Carreto et al. 2003; Gayoso and Podestá 1996; Akselman and Negri 2012; Antacli et al. 2018). Nevertheless, the detection and characterization of phycotoxins in offshore waters started in 2012, and numerous phytoplankton species responsible for causing different human syndromes have been documented so far (Figs. 3.3 and 3.4). For instance: (1) diatoms of the genera *Pseudo-nitzschia* producers of domoic acid (DA) (Amnesic Shellfish Poisoning, ASP) are widely dispersed along the PSBF (Almandoz et al. 2017; Guinder et al. 2018), and the dinoflagellates (2) *Alexandrium catenella* and *A. ostenfeldii* and their associated toxins (Paralytic and Spiroimine Shellfish Poisoning, PSP and SSP) (Fabro et al.

Fig. 3.3 (continued) (AZAs producers). (**c**) *Dinophysis* species, *Prorocentrum lima*, *Gonyaulax spinifera* and *Protoceratium reticulatum* (LSTs producers). (**d**) *Pseudo-nitzschia* species (DA producers). Crosses indicate the presence of the species when no data of abundance were available. (Figure originally published in Ramírez et al. 2022, Harmful Algae 118: 102317, Copyright Elsevier)

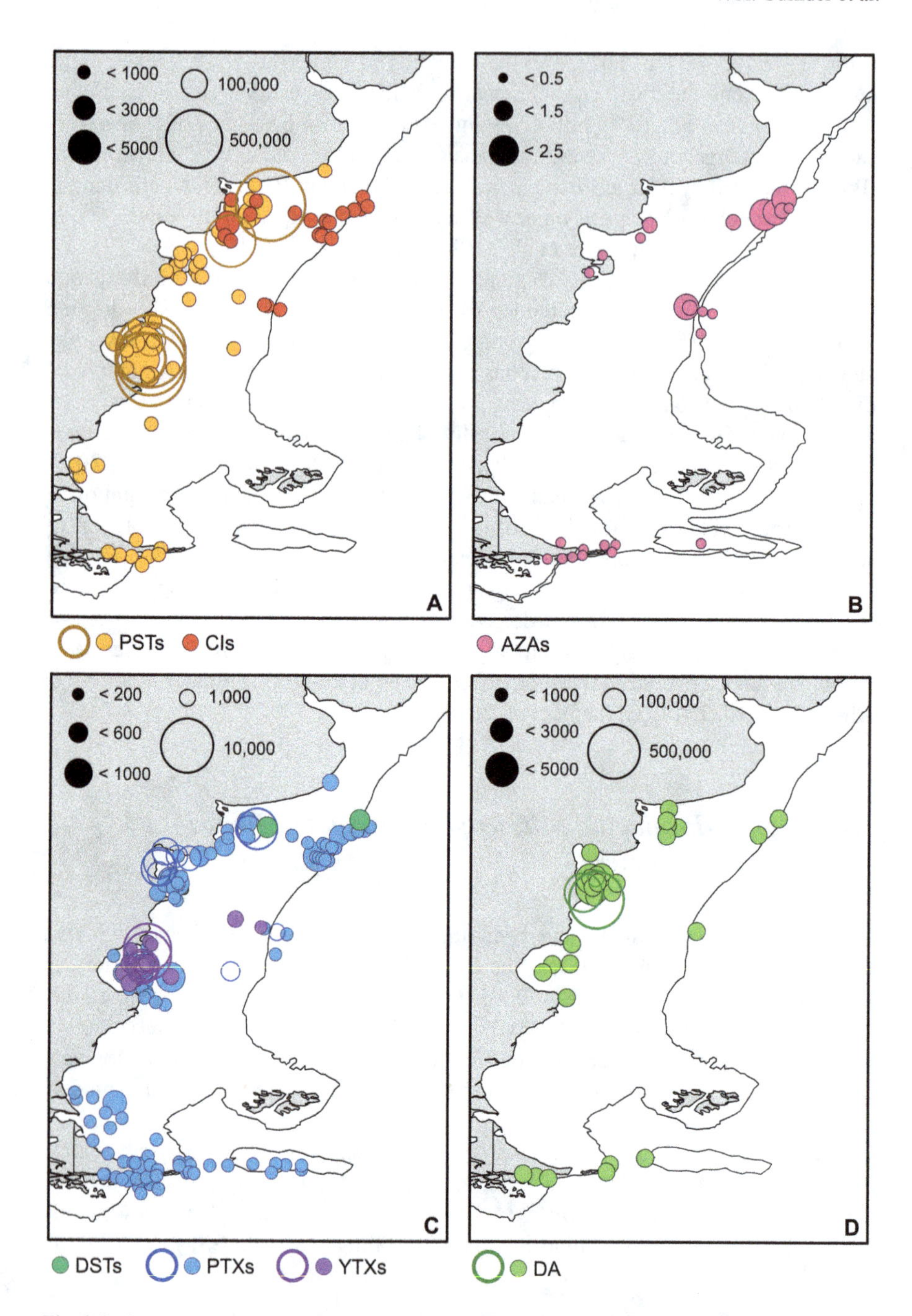

Fig. 3.4 Geographic distribution and abundance (in ng net tow^{-1}; AZAs in ng L^{-1}) of phycotoxins registered in water samples over the period 1980–2018. (**a**) Paralytic shellfish toxins (PSTs) and Cycloimines (CIs). (**b**) Azaspiracids (AZAs), represented only by AZA-2. (**c**) Lipophilic toxins: Diarrhetic shellfish toxins (DSTs), pectenotoxins (PTXs) and yessotoxins (YTXs). (**d**) Domoic acid (DA). (Figure originally published in Ramírez et al. 2022, Harmful Algae 118: 102317, Copyright Elsevier)

2017; Guinder et al. 2018), (3) *Dinophysis* spp. (Diarrhetic Shellfish Poisoning, DSP) (Fabro et al. 2016; Guinder et al. 2018) including high levels of pectenotoxins (PTXs, non-diarrheagenic) over the Burdwood Bank (Guinder et al. 2020), (4) *Protoceratium reticulatum* producer of yessotoxins (YTXs) (Akselman et al. 2015), and (5) the abundant multispecific spring and summer blooms of small dinoflagellates of the group *Amphidomataceae* (Akselman and Negri 2012; Tillmann and Akselman 2016), producers of Azaspiracids (AZAs) (Azaspiracid Shellfish Poisoning, AZP) (Fabro et al. 2019; Tillmann et al. 2019; Guinder et al. 2020).

Blooms of *Amphidomataceae* have been detected only in the PSBF between 35 °S and 45 °S, with no records in coastal or inner shelf areas (Figs. 3.3 and 3.4; Ramírez et al. 2022). These findings reinforce that frontal zones may act as 'pelagic seed banks' for toxic dinoflagellates (Smayda 2002), where they may bloom, enhanced by upwelling and the resuspension of cysts from benthic banks. In marine environments, dinoflagellates are the most common group of toxin-producing phytoplankton, likely attributed to their advantageous ecological traits such as motility, cyst production and mixotrophy, which confer them ecological plasticity to adapt to changing systems (e.g. availability of nutrients, light and prey) and outcompete other phytoplankton groups (Litchman et al. 2007; Mitra and Flynn 2021). Examples in the PSBF are the abundant viable cysts of *P. reticulatum* in plankton and benthic reservoirs in frontal areas, with a high cell quota of YTXs (Akselman et al. 2015); recurrent blooms of Amphidomataceans in different regions of the PSBF (Ramírez et al. 2022), which in the springs of 1990 and 1991 reached one of the maximal historical abundances reported globally (Akselman and Negri 2012; Tillmann 2018); and the widespread *Dinophysis* spp., which commonly co-occur with their obligate prey, the mixotrophic ciliate *Mesodinium rubrum* (e.g. Fabro et al. 2016; Guinder et al. 2018, 2020).

3.4.3 Trans-Oceanic Plankton Connectivity and Dispersion of Toxic Species

The transport of HAB species by edge currents along shelf-break fronts affects their dispersion between ocean basins and may affect the regional strain-specific toxic potential. In particular, a notable difference in *Amphidomataceae* assemblages and bloom occurrence is observed in the shelf-break fronts of Chile and Argentina. Both shelves expand along similar latitudinal gradients in the Southeastern Pacific and the Southwestern Atlantic, respectively, and their shelf-breaks are influenced by currents originating in the Southern Ocean (i.e. the Humboldt Current and the Malvinas Current, respectively, running northward as branches of the Circumpolar Antarctic Current). For example, AZAs in Chile have been detected in bivalves, but so far, *Azadinium poporum* is the only AZA producer reported in Chilean waters, where it occurs in low background abundance (Tillmann et al. 2017b) compared to populations in the PSBF.

Compared to blooms of Amphidomataceans in the North Atlantic, where they commonly produce a wide composition of AZA variants, with AZA-1 being one of the most frequent (Tillmann 2018), only AZA-2 is produced by strains found in Argentina (Fabro et al. 2019). Notably, *Azadinium spinosum* was not related to toxin production when found in spring 2015 at the PSBF (Tillmann et al. 2019), but this species is the most important causative agent of AZA poisoning in Europe (Salas et al. 2011). The fact that some toxic species in the PSBF have been detected in low background concentrations (e.g. *Alexandrium ostenfeldii*, producer of spirolides, Guinder et al. 2018) or non-toxic events have been registered associated with these species compared to other temperate seas may indicate differences in population biogeography and toxicity depending on regional conditions.

These findings suggest that more monitoring efforts are required in the PSBF to capture short-lived extraordinary blooms and their potential to cause toxic outbreaks. These natural hazards gain attention under climate change-driven ocean warming, intensified marine heatwaves and changes in the intensity of upwelling and ocean currents, which may lead to more frequent and severe HABs along with shifts in their geographical distribution (Bindoff et al. 2019). The risk of the potential poleward expansion of HAB species in response to ocean warming is of particular concern in the Southwestern Atlantic region, which has been identified as a warming hotspot (Risaro et al. 2022), likely displacing the BMC southward. Still, a large area of the PSBF and adjacent ocean waters remains unexplored, leading to open questions regarding trans-oceanic plankton connectivity and the transport of potentially toxic species by edge currents.

3.5 Concluding Remarks and Future Considerations

The information gathered here about the phytoplankton abundance, structure and biodiversity in different regions of the PSBF in relation to the environmental conditions contributes to a better understanding of the regional behaviour of this community. This information, which certainly needs to be reinforced with new observations and studies, is highly relevant to understanding changes in the carbon budget occurring on a global scale, given the high productivity of the PSBF. Some points to consider for future research are: (i) so far, most field studies of phytoplankton structure and composition carried out in the PSBF have been based only on protistan plankton >10 microns in cell size, mostly determined by light microscopy techniques. To a lesser extent, pigment analyses (HPLC) and bio-optical measurements were performed. In order to be able to address what is known for other areas of the world ocean, i.e. the well-recognized functional and taxonomical diversity of marine microorganisms (e.g. Litchman et al. 2007; Vaulot et al. 2008; Flombaum et al. 2013; De Vargas et al. 2015), more in situ data collection and multiproxy analyses, including flow cytometry, strain cultures and phylogenetic studies, metagenomics, scanning for phycotoxins and eco-physiological experiments (e.g. grazing, primary productivity), are needed for the PSBF. This will bring light to the environmental conditions underpinning their size structure, feeding modes and ecological traits, encompassing prokaryotes (cyanobacteria) and eukaryotes. This complete analysis will also contribute to a more precise remote sensing estimation of chl-*a* in the PSBF. In addition, (ii) more interdisciplinary oceanographic cruises are needed to integrate physical and biogeochemical disciplines, essential to tackle the complex interplay of multiple environmental drivers of phytoplankton bloom phenology, composition and associated carbon fluxes. One example would be the oceanographic cruises with a strong phytoplankton ecology component carried out in the Patagonian Shelf and along the PSBF during spring and summer 2021 and 2022 (e.g. Guinder et al. 2023), from which data are being processed at the time of publication of this book. Finally, (iii) more observational data are needed during austral autumn and winter (March–August) to have an integrated picture of the annual development and composition of phytoplankton in the permanent front. More frequent sampling will bring light to the natural species succession and bloom structure, as well as potential long-term shifts in their seasonal and inter-annual variability. In particular, the growing detection of harmful phytoplankton warns of the importance of sustained monitoring systems in the PSBF to evaluate the risk to ecosystem functioning and seafood security. Historical data and future multidisciplinary studies will contribute to the understanding of the multiscale variability of phytoplankton along the complex PSBF, in the frame of the pervasive climate-related changes in the ocean.

References

Akselman R, Negri RM (2012) Blooms of *Azadinium* cf. *spinosum* Elbrächter et Tillmann (Dinophyceae) in northern shelf waters of Argentina, southwestern Atlantic. Harmful Algae 19:30–38

Akselman R, Krock B, Alpermann TJ, Tillmann U, Borel M, Almandoz GO, Ferrario ME (2015) *Protoceratium reticulatum* (Dinophyceae) in the austral Southwestern Atlantic and the first report on YTX-production in shelf waters of Argentina. Harmful Algae 45:40–52

Almandoz GO, Ferrario ME, Ferreyra GA, Schloss IR, Esteves JL, Paparazzo FE (2007) The genus *Pseudo-nitzschia* (Bacillariophyceae) in continental shelf waters of Argentina (southwestern Atlantic Ocean, 38–55°S). Harmful Algae 6(1):93–103

Almandoz GO, Ferreyra GA, Schloss IR, Dogliotti AI, Rupolo V, Paparazzo FE et al (2008) Distribution and ecology of *Pseudo-nitzschia* species (Bacillariophyceae) in surface waters of the Weddell Sea (Antarctica). Polar Biol 31(4):429–442

Almandoz GO, Fabro E, Ferrario M, Tillmann U, Cembella A, Krock B (2017) Species occurrence of the potentially toxigenic diatom genus *Pseudo-nitzschia* and the associated neurotoxin domoic acid in the Argentine sea. Harmful Algae 63:45–55

Almandoz GO, Cefarelli AO, Diodato S, Montoya NG, Benavides HR, Carignan M et al (2019) Harmful phytoplankton in the Beagle Channel (South America) as a potential threat to aquaculture activities. Mar Pollut Bull 145:105–117

Antacli JC, Silva RI, Jaureguizar AJ, Hernández DR, Mendiolar M, Sabatini ME et al (2018) Phytoplankton and protozooplankton on the southern Patagonian shelf (Argentina, 47°–55° S) in late summer: potentially toxic species and community assemblage structure linked to environmental features. J Sea Res 140:63–80

Balch WM, Drapeau DT, Bowler BC, Lyczkowski ER, Lubelczyk LC, Painter SC, Poulton AJ (2014) Surface biological, chemical, and optical properties of the Patagonian Shelf coccolithophore bloom, the brightest waters of the Great Calcite Belt. Limnol Oceanogr 59(5):1715–1732

Berghoff CF, Pierrot D, Epherra L, Silva RI, Segura V, Negri RM et al (2023) Physical and biological effects on the carbonate system during summer in the Northern Argentine continental Shelf (Southwestern Atlantic). J Mar Syst 237:103828

Bianchi AA, Ruiz Pino D, Perlender HGI, Osiroff AP, Segura V, Lutz V, Clara ML, Balestrini CF, Piola AR (2009) Annual balance and seasonal variability of sea-air CO_2 fluxes in the Patagonia Sea: their relationship with fronts and chlorophyll distribution. J Geophys Res 114:C03018

Bindoff NL, Cheung WWL, Kairo JG, Arístegui J, Guinder VA, … Tagliabue A (2019) Changing ocean, marine ecosystems, and dependent communities. In: Pörtner HO, Roberts DC, Masson-Delmotte V, Zhai P, Tignor M, Poloczanska E et al (eds), IPCC special report on the ocean and cryosphere in a changing climate. Cambridge University Press, Cambridge pp. 447–587

Birchill AJ, Milne A, Woodward EMS, Harris C, Annett A, Rusiecka D et al (2017) Seasonal iron depletion in temperate shelf seas. Geophys Res Lett 44(17):8987–8996

Bogazzi E, Baldoni AN, Rivas A, Martos P, Reta RA, Orensanz JM et al (2005) Spatial correspondence between areas of concentration of Patagonian scallop (*Zygochlamys patagonica*) and frontal systems in the southwestern Atlantic. Fish Oceanogr 14:359–376

Brown CW, Podesta GP (1997) Remote sensing of coccolithophore blooms in the western South Atlantic Ocean. Remote Sens Environ 60(1):83–91

Brunetti NE, Ivanovic ML, Rossi GR, Elena B, Pineda SE (1998) Fishery biology and life history of *Illex argentinus*. In: Okutani T (ed) Large Pelagic Squid. Japan Marine Fishery Resources Center (JAMARC) Special Publication, Tokio, pp 216–231

Campodónico S, Escolar M, García J, Aubone A (2019) Síntesis histórica y estado actual de la pesquería de vieira patagónica Zygochlamys patagonica (King 1832) en la Argentina. Biología, evaluación de biomasa y manejo. MAFIS 32(2):125–148

Carranza MM, Gille ST, Piola AR, Charo M, Romero SI (2017) Wind modulation of upwelling at the shelf-break front off Patagonia: observational evidence. J Geophys Res Oceans 122(3):2401–2421

Carreto JI, Montoya NG, Benavides HR et al (2003) Characterization of spring phytoplankton communities in the Río de La Plata maritime front using pigment signatures and cell microscopy. Mar Biol 143:1013–1027

Carreto JI, Montoya N, Akselman R, Carignan MO, Silva RI, DAC C (2008) Algal pigment patterns and phytoplankton assemblages in different water masses of the Río de la Plata maritime front. Cont Shelf Res 28(13):1589–1606

Carreto JI, Montoya NG, Carignan MO, Akselman R, Acha EM, Derisio C (2016) Environmental and biological factors controlling the spring phytoplankton bloom at the Patagonian shelf-break front – degraded fucoxanthin pigments and the importance of microzooplankton grazing. Prog Oceanogr 146:1–21. https://doi.org/10.1016/j.pocean.2016.05.002

Cefarelli AO, Ferrario ME, Almandoz GO, Atencio AG, Akselman R, Vernet M (2010) Diversity of the diatom genus *Fragilariopsis* in the Argentine Sea and Antarctic waters: morphology, distribution and abundance. Polar Biol 33(11):1463–1484

Charalampopoulou A, Poulton AJ, Bakker DC, Lucas MI, Stinchcombe MC, Tyrrell T (2016) Environmental drivers of coccolithophore abundance and calcification across Drake Passage (Southern Ocean). Biogeosciences 13(21):5917–5935

Combes V, Matano RP (2018) The Patagonian shelf circulation: drivers and variability. Prog Oceanogr 167:24–43

Cosentino NJ, Ruiz-Etcheverry LA, Bia GL, Simonella LE, Coppo R, Torre G et al (2020) Does satellite chlorophyll-a respond to southernmost Patagonian dust? A multi-year, event-based approach. J Geophys Res Biogeosci 125:e2020JG006073

de Oliveira Carvalho A, Kerr R, Tavano VM, Mendes CRB (2022) The southwestern South Atlantic continental shelf biogeochemical divide. Biogeochemistry 159(2):139–158

De Souza MS, Mendes CRB, Garcia VMT, Poller R, Brotas V (2012) Phytoplankton community during a coccolithophorid bloom in the Patagonian Shelf: microscopic and high-performance liquid chromatography pigment analyses. Mar Biolog Assoc UK 92(1):13

De Vargas C, Audic S, Henry N, Decelle J, Mahé F, Logares R et al (2015) Eukaryotic plankton diversity in the sunlit ocean. Science 348(6237):1261605

Delgado AL, Guinder VA, Dogliotti AI, Zapperi G, Pratolongo PD (2019) Validation of MODIS-Aqua bio-optical algorithms for phytoplankton absorption coefficient measurement in optically complex waters of El Rincón (Argentina). Cont Shelf Res 173:73–86

Delgado AL, Hernández-Carrasco I, Combes V, Font-Muñoz J, Pratolongo PD, Basterretxea G (2023) Patterns and trends in chlorophyll-a concentration and phytoplankton phenology in the biogeographical regions of Southwestern Atlantic. J Geophys Res: Oceans 128(9):e2023JC019865

Díaz MV, Marrari M, Casa V, Gattas F, Pájaro M, Macchi GJ (2018) Evaluating environmental forcing on nutritional condition of Engraulis anchoita larvae in a productive area of the south-western Atlantic Ocean. Prog Oceanogr 168:13–22

Dogliotti AI, Lutz VA, Segura V (2014) Estimation of primary production in the southern Argentine continental shelf and shelf-break regions using field and remote sensing data. Remote Sens Environ 140:497–508

Fabro E, Almandoz GO, Ferrario M, Tillmann U, Cembella A, Krock B (2016) Distribution of *Dinophysis* species and their association with lipophilic phycotoxins in plankton from the Argentine Sea. Harmful Algae 59:31–41

Fabro E, Almandoz GO, Ferrario M, John U, Tillmann U, Toebe K, Krock B, Cembella A (2017) Morphological, molecular, and toxin analysis of field populations of *Alexandrium* genus from the Argentine Sea. J Phycol 53(6):1206–1222

Fabro E, Almandoz GO, Krock B, Tillmann U (2019) Field observations of the dinoflagellate genus *Azadinium* and azaspiracid toxins in the South-West Atlantic Ocean. Mar Freshw Res 71(7):832–843

Ferrario ME, Almandoz GO, Cefarelli AO, Fabro E, Vernet M (2012) *Stephanopyxis* species (Bacillariophyceae) from shelf and slope waters of the Argentinean Sea: ultrastructure and distribution. Nova Hedwigia 96(1–2):249–263

Ferreira A, Stramski D, García CA, García VM, Ciotti AM, Mendes CR (2013) Variability in light absorption and scattering of phytoplankton in Patagonian waters: role of community size structure and pigment composition. J Geophys Res Oceans 118(2):698–714

Ferronato C, Guinder VA, Chidichimo MP, López-Abbate C, Amodeo M (2021) Zonation of protistan plankton in a productive area of the Patagonian shelf: potential implications for the anchovy distribution. Food Webs 29:e00211

Ferronato C, Brenden G, Rivarossa M, Guinder VA (2023) Wind-driven currents and water masses shape spring phytoplankton distribution and composition in Northern Patagonian shelf. Limnol Oceanogr 68:2195. https://doi.org/10.1002/LNO.12413

Flombau P, Gallegos JL, Gordillo RA, Rincón J, Zabala LL, Jiao N et al (2013) Present and future global distributions of the marine Cyanobacteria Prochlorococcus and Synechococcus. PNAS 110(24):9824–9829

Franco BC, Ruiz-Etcheverry LA, Marrari M, Piola AR, Matano RP (2022) Climate change impacts on the Patagonian shelf break front. Geophys Res Lett 49(4):e2021GL096513

Frey DI, Piola AR, Krechik VA, Fofanov DV, Morozov EG, Silvestrova KP et al (2021) Direct measurements of the Malvinas current velocity structure. J Geophys Res Oceans 126(4):e2020JC016727

Frey DI, Piola AR, Morozov EG (2023) Convergence of the Malvinas current branches near 44° S. Deep Sea Res Part I: Oceanogr Res Pap 196:104023

García VMT, Garcia CAE, Mata MM, Pollery RC, Piola AR, Signorini SR, McClain CR, Iglesias-Rodriguez MD (2008) Environmental factors controlling the phytoplankton blooms at the Patagonia shelf-break in spring. Deep-Sea Res I Oceanogr Res Pap 55(9):1150–1166

Garcia CAE, Garcia VMT, Dogliotti AI, Ferreira A, Romero SI, Mannino A et al (2011) Environmental conditions and bio-optical signature of a coccolithophorid bloom in the Patagonian shelf. J Geophys Res Oceans 116(C3):C03025
Gayoso AM (1995) Bloom of *Emiliania huxleyi* (Prymnesiophyceae) in the western South Atlantic Ocean. J Plankton Res 17(8):1717–1722
Gayoso AM, Fulco VK (2006) Occurrence patterns of Alexandrium tamarense (Lebour) Balech populations in the Golfo Nuevo (Patagonia, Argentina), with observations on ventral pore occurrence in natural and cultured cells. Harmful Algae 5(3):233–241
Gayoso AM, Podestá GP (1996) Surface hydrography and phytoplankton of the Brazil-Malvinas currents confluence. J Plankton Res 18(6):941–951
Gómez MI (2011) Estructura, distribución y fluctuación temporal del fitoplancton en aguas del ecosistema sud Patagónico (47–55° S, Mar Argentino). Ph.D. Thesis, Universidad de Buenos Aires, p 175
Gonçalves-Araujo R, De Souza MS, Mendes CRB, Tavano VM, Pollery RC, Garcia CAE (2012) Brazil-Malvinas confluence: effects of environmental variability on phytoplankton community structure. J Plankton Res 34(5):399–415
Goncalves-Araujo R, De Souza MS, Mendes CRB, Tavano VM, Garcia CAE (2016) Seasonal change of phytoplankton (spring vs. summer) in the southern Patagonian shelf. Cont Shelf Res. https://doi.org/10.1016/j.csr.2016.03.023
Guinder VA, Tillmann U, Krock B, Delgado AL, Krohn T, Garzón Cardona JE, Katja Metfies K, López-Abbate C, Silva R, Lara R (2018) Plankton multiproxy analyses in the Northern Patagonian Shelf, Argentina: community structure, phycotoxins, and characterization of toxic *Alexandrium* strains. Front Mar Sci 5:394
Guinder VA, Malits A, Ferronato C, Martin J, Krock B, Garzón Cardona JE, Martínez A (2020) Microbial plankton configuration in the epipelagic realm from the Beagle Channel to the Burdwood Bank, a Marine Protected Area in Sub-Antarctic waters. PLoS One 15(5):e0233156
Guinder VA, Tillmann U, Ramírez F, Ferronato C, Rivarossa M, Krock B (2023) Long-lasting extraordinary spring bloom of Amphidomataceae (Dinophyceae) and AZA-2 in the Argentine Sea. In: 20th international conference on Harmful Algae (ICHA). Hiroshima, Japan
Jickells TD, An ZS, Andersen KK, Baker AR, Bergametti G, Brooks N et al (2005) Global iron connections between desert dust, ocean biogeochemistry, and climate. Science 308(5718):67–71
Kahl LC, Bianchi AB, Osiroff AP, Ruiz Pino D, Piola AR (2017) Distribution of sea-air CO2 fluxes in the Patagonian Sea: seasonal, biological and thermal effects. Cont Shelf Res 143:18–28
Laruelle GG, Cai WJ, Hu X, Gruber N, Mackenzie FT, Regnier P (2018) Continental shelves as a variable but increasing global sink for atmospheric carbon dioxide. Nat Commun 9(1):454
Litchman E, Klausmeier CA, Schofield OM, Falkowski PG (2007) The role of functional traits and trade-offs in structuring phytoplankton communities: scaling from cellular to ecosystem level. Ecol 10(12):1170–1181
Lucas AJ, Guerrero RA, Mianzan HW, Acha EM, Lasta CA (2005) Coastal oceanographic regimes of the northern Argentine continental shelf (34–43°S). Estuar Coast Shelf Sci. https://doi.org/10.1016/j.ecss.2005.06.015
Lutz VA, Carreto JI (1991) A new spectrofluorometric method for the determination of chlorophylls and degradation products and its application in two frontal areas of the Argentine Sea. Cont Shelf Res 11(5):433–451
Lutz VA, Segura V, Dogliotti AI, Gagliardini DA, Bianchi AA, Balestrini CF (2010) Primary production in the Argentine Sea during spring estimated by field and satellite models. J Plankton Res. https://doi.org/10.1093/plankt/fbp117
Lutz V, Segura V, Dogliotti A, Tavano V, Brandini FP, Calliari DL et al (2018) Overview on primary production in the Southwestern Atlantic. In: Hoffmeyer MS, Sabatini ME, Brandini FP, Calliari DL, Santinelli NH (eds) Plankton ecology of the Southwestern Atlantic. Springer, Cham, pp 101–126
Marrari M, Piola AR, Valla D (2017) Variability and 20-year trends in satellite-derived surface chlorophyll concentrations in large marine ecosystems around South and Western Central America. Front Mar Sci 4:372

Martinetto P, Alemany D, Botto F, Mastrángelo M, Falabella V, Acha EM et al (2020) Linking the scientific knowledge on marine frontal systems with ecosystem services. Ambio 49(2):541–556
Matano RP, Palma ED (2008) On the upwelling of downwelling currents. J Phys Oceanogr 38(11):2482–2500
Matano RP, Palma ED, Piola AR (2010) The influence of the Brazil and Malvinas currents on the Southwestern Atlantic Shelf circulation. Ocean Sci 6(4):983–995
Matano RP, Palma ED, Combes V (2019) The Burdwood bank circulation. J Geophys Res Oceans 124(10):6904–6926
Mitra A, Flynn KJ (2021) HABs and the mixoplankton paradigm. Harmful Algae News 67:4–6
Moreno DV, Marrero JP, Morales J, García CL, Úbeda MGV, Rueda MJ, Llinás O (2012) Phytoplankton functional community structure in Argentinian continental shelf determined by HPLC pigment signatures. Estuar Coast Shelf Sci. https://doi.org/10.1016/j.ecss.2012.01.007
Negri RM, Carreto JI, Benavides HR, Akselman R, Lutz VA (1992) An unusual bloom of *Gyrodinium cf. aureolum* in the Argentine sea: community structure and conditioning factors. J Plankton Res 14(2):261–269
Negri RM, Silva RI, Valinas M (2003) Gephyrocapsa oceanica (Haptophyta) distribution in a sector of the Argentine continental shelf (SW Atlantic Ocean, 37°-40°S). Bol Soc Argent Bot 38(1/2):131–137
Olguín Salinas HF, Brandini F, Boltovskoy D (2015) Latitudinal patterns and interannual variations of spring phytoplankton in relation to hydrographic conditions of the southwestern Atlantic Ocean (34-62 S). Helgoland Mar Res 69(2):177
Palma ED, Matano RP, Piola AR (2008) A numerical study of the southwestern Atlantic shelf circulation: Stratified Ocean response to local and offshore forcing. J Geophys Res. https://doi.org/10.1029/2007jc004720
Palma ED, Matano RP, Combes V (2021) Circulation and cross-shelf exchanges in the Malvinas Islands Shelf region. Prog Oceanogr 198:102666
Paparazzo FE, Esteves JL (2018) Surface macronutrient dynamics of the drake passage and the Argentine Sea. In: Plankton ecology of the Southwestern Atlantic. Springer, Cham, pp 71–86
Piola AR, Franco BC, Palma ED, Saraceno M (2013) Multiple jets in the Malvinas current. J Geophys Res Oceans 118(4):2107–2117
Poulton AJ, Painter SC, Young JR, Bates NR, Bowler B, Drapeau D et al (2013) The 2008 *Emiliania huxleyi* bloom along the Patagonian Shelf: ecology, biogeochemistry, and cellular calcification. Glob Biogeochem Cycles 27(4):1023–1033
Ramírez FJ, Guinder VA, Ferronato C, Krock B (2022) Increase in records of toxic phytoplankton and associated toxins in water samples in the Patagonian Shelf (Argentina) over 40 years of field surveys. Harmful Algae 118:102317
Risaro DB, Chidichimo MP, Piola AR (2022) Interannual variability and trends of sea surface temperature around southern South America. Front Mar Sci 9:829144
Rivas AL, Dogliotti AI, Gagliardini DA (2006) Seasonal variability in satellite measured surface chlorophyll in the Patagonian shelf. Cont Shelf Res 26:703–720
Rivero-Calle S, Gnanadesikan A, Del Castillo CE, Balch WM, Guikema SD (2015) Multidecadal increase in North Atlantic coccolithophores and the potential role of rising CO2. Science 350(6267):1533–1537
Romero SI, Piola AR, Charo M, Garcia CAE (2006) Chlorophyll-a variability off Patagonia based on SeaWiFS data. J Geophys Res. https://doi.org/10.1029/2005jc003244
Sabatini M, Akselman R, Reta R, Negri R, Lutz V, Silva R, Segura V, Gil MN, Santinelli NH, Sastre AV, Daponte MC, Antacli JC (2012) Spring plankton communities in the southern Patagonian shelf: hydrography, mesozooplankton patterns and trophic relationships. J Mar Syst 94:33–51
Salas R, Tillmann U, John U, Kilcoyne J, Burson A, Cantwell C, Hess P, Jauffrais T, Silke J (2011) The role of Azadinium spinosum (Dinophyceae) in the production of Azaspiracid shellfish poisoning in mussels. Harmful Algae 10:774–783

Saraceno M, Provost C, Piola AR (2005) On the relationship between satellite-retrieved surface temperature fronts and chlorophyll a in the western South Atlantic. J Geophys Res Oceans 110(C11)

Sastre AV, Santinelli NH, Solís ME, Pérez LB, Ovejero SD, Gracia-Villalobos L, Cadaillón A, D'Agostino VC (2018) Harmful marine microalgae in coastal waters of Chubut (Patagonia, Argentina). In: Hoffmeyer MS, Sabatini ME, Brandini FP, Calliari DL, Santinelli NH (eds) Plankton ecology of the Southwestern Atlantic. Springer, Cham, pp 495–518

Schloss IR, Ferreyra GA, Ferrario ME, Almandoz GO, Codina R, Bianchi AA et al (2007) Role of phytoplankton communities in the sea–air variation of pCO2 in the SW Atlantic Ocean. Mar Ecol Prog Ser 332:93–106

Segura V, Lutz VA, Dogliotti A, Silva RI, Negri RM, Akselman R, Benavides H (2013) Phytoplankton types and primary production in the Argentine Sea. Mar Ecol Prog Ser. https://doi.org/10.3354/meps10461

Signorini SR, Garcia VMT, Piola AR, Garcia CAE, Mata MM, McClain CR (2006) Seasonal and interannual variability of calcite in the vicinity of the Patagonian shelf break (38°S–52°S). Geophys Res Lett. https://doi.org/10.1029/2006gl026592

Smayda TJ (2002) Turbulence, watermass stratification and harmful algal blooms: an alternative view and frontal zones as "pelagic seed banks". Harmful Algae 1(1):95–112

Smith HE, Poulton AJ, Garley R, Hopkins J, Lubelczyk LC, Drapeau DT et al (2017) The influence of environmental variability on the biogeography of coccolithophores and diatoms in the Great Calcite Belt. Biogeosciences 14(21):4905–4925

Thuróczy CE, Gerringa LJA, Klunder MB, Laan P, De Baar HJW (2011) Observation of consistent trends in the organic complexation of dissolved iron in the Atlantic sector of the Southern Ocean. Deep-Sea Res II Top Stud Oceanogr 58(25–26):2695–2706

Tillmann U (2018) Amphidomataceae. In: Shumway SE, Burkholder JM, Morton SL (eds) Harmful algal blooms: a compendium desk reference. Wiley, Singapore, pp 575–582

Tillmann U, Akselman R (2016) Revisiting the 1991 algal bloom in shelf waters off Argentina: *Azadinium luciferelloides* sp. nov. (Amphidomataceae, Dinophyceae) as the causative species in a diverse community of other amphidomataceans. Phycol Res 64(3):160–175

Tillmann U, Gottschling M, Guinder V, Krock B (2017a) *Amphidoma parvula* (Amphidomataceae), a new planktonic dinophyte from the Argentine Sea. Eur J Phycol 53(1):14–28

Tillmann U, Trefault N, Krock B, Parada-Pozo G, De la Iglesia R, Vásquez M (2017b) Identification of *Azadinium poporum* (Dinophyceae) in the Southeast Pacific: morphology, molecular phylogeny, and azaspiracid profile characterization. J Plankton Res 39(2):350–367

Tillmann U, Gottschling M, Krock B, Smith KF, Guinder V (2019) High abundance of Amphidomataceae (Dinophyceae) during the 2015 spring bloom of the Argentinean Shelf and a new, non-toxigenic ribotype of *Azadinium spinosum*. Harmful Algae 84:244–260

Tréguer P, Bowler C, Moriceau B, Dutkiewicz S, Gehlen M, Aumont O et al (2018) Influence of diatom diversity on the ocean biological carbon pump. Nat Geosci 11(1):27–37

Tyrrell T, Merico A (2004) *Emiliania huxleyi*: bloom observations and the conditions that induce them. In *Coccolithophores*. In: Thierstein HR, Young JR (eds) Coccolithophores. Springer, Berlin, Heidelberg, pp 75–97

Valiadi M, Painter SC, Allen JT, Balch WM, Iglesias-Rodriguez MD (2014) Molecular detection of bioluminescent dinoflagellates in surface waters of the Patagonian shelf during early austral summer 2008. PLoS One 9(6):e98849

Vaulot D, Eikrem W, Viprey M, Moreau H (2008) The diversity of small eukaryotic phytoplankton (≤ 3 μm) in marine ecosystems. FEMS Microbiol Rev 32(5):795–820

Viljoen JJ, Weir I, Fietz S, Cloete R, Loock J, Philibert R, Roychoudhury AN (2019) Links between the phytoplankton community composition and trace metal distribution in summer surface waters of the Atlantic southern ocean. Front Mar Sci 6:295

Chapter 4
Zooplanktonic Crustacea and Ichthyoplankton of the Patagonian Shelf-Break Front

Georgina D. Cepeda, Martín D. Ehrlich, Carla M. Derisio, Ayelén Severo, Laura Machinandiarena, Mariana Cadaveira, Paola Betti, Marina Do Souto, Carolina Pantano, and E. Marcelo Acha

Abstract This chapter reviews information from the 1970s to the present day about zooplankton and ichthyoplankton occurring at the Patagonian shelf-break front (PSBF), aiming to establish a baseline of knowledge for both groups. Zooplankton information focuses on copepods, hyperiid amphipods, and euphausiids because of their abundance and relevance to the food web. General patterns of species diversity and biogeographic zonation of these zooplanktonic groups are described. Copepod diversity roughly increases northward, and the shelf-break front does not represent a sharp boundary between shelf and oceanic zooplankton species. Key species are defined, and data about their seasonal distribution, abundance, and life-history traits indicate that the same few species dominate the composition of these taxa all year round. The presence of larvae of some demersal and mesopelagic fish typical of cold sub-Antarctic waters denotes reproductive activity in the PSBF surroundings. Ichthyoplankton can be classified, according to bottom depths, as pertaining to the shelf-edge (80–200 m); slope (200–800 m); or oceanic regions (800–3100 m). Larvae show higher abundances immediately offshore of the front, especially above

Supplementary Information The online version contains supplementary material available at https://doi.org/10.1007/978-3-031-71190-9_4.

G. D. Cepeda (✉) · A. Severo · M. Do Souto · E. M. Acha
Instituto de Investigaciones Marinas y Costeras (IIMyC), Universidad Nacional de Mar del Plata, (UNMdP), Consejo Nacional de Investigaciones Científicas y Técnicas (CONICET), Mar del Plata, Argentina

Instituto Nacional de Investigación y Desarrollo Pesquero (INIDEP), Mar del Plata, Argentina
e-mail: gcepeda@inidep.edu.ar

M. D. Ehrlich · C. M. Derisio · L. Machinandiarena · M. Cadaveira · P. Betti
Instituto Nacional de Investigación y Desarrollo Pesquero (INIDEP), Mar del Plata, Argentina

C. Pantano
Facultad de Ciencias Exactas y Naturales (FCEN), Departamento de Ecología, Genética y Evolución (EGE), Universidad de Buenos Aires (UBA), Buenos Aires, Argentina

E. M. Acha et al. (eds.), *The Patagonian Shelfbreak Front*, Aquatic Ecology Series 13, https://doi.org/10.1007/978-3-031-71190-9_4

the thermocline. Copepods, especially their larval stages, constitute the main feeding source for fish larvae inhabiting along the shelf-break. Zooplankton and ichthyoplankton composition and abundance along the shelf-break are compared with those of the adjacent shelf. We finally go over open questions and prospects for future work on zooplankton and ichthyoplankton along the shelf-break.

Acronyms and Abbreviations

ACS	Argentine continental shelf
DVM	daily vertical migrations
FAO	Food and Agriculture Organization
OF	occurrence frequency
PSBF	Patagonian shelf-break front
INIDEP	National Institute for Fisheries Research and Development
MCW	Malvinas Current Waters
RII	relative importance index
SASW	Subantarctic Shelf Water

4.1 Introduction

The shelf-break off Argentina conforms a remarkable ecosystem due to its physical and dynamic characteristics and the associated fishery resources. Along the region composed of the outer continental shelf, the continental slope, and the adjacent oceanic waters, seasonal, extended, and intense chlorophyll-*a* hotspots occur (e.g., Romero et al. 2006), indicative of high primary production (e.g., Lutz et al. 2010; Segura et al. 2013), creating a favorable feeding habitat for zooplankton and ichthyoplankton.

The transition between the relatively low salinity waters of the Argentine Continental Shelf (ACS) and those colder, saltier, and nutrient-rich waters of the Malvinas Current produces a permanent thermo-haline front: the Patagonian Shelf-Break Front (PSBF hereafter), which appears broader and more intense during spring and summer than in the colder season (Rivas and Pisoni 2010). Along this front, the combination of the processes of Bakun's triad—i.e., nutrient enrichment, water column stability, and a circulation pattern that secures the retention of early life history stages—favors larval survival of fishes and invertebrates (Bakun and Parrish 1991). In this regard, the need to find an adequate food supply for maintenance, growth, and reproduction has contributed to delineating the timing of the migratory patterns of the commercial species and others, which make use of the PSBF as feeding and/or as spawning grounds (e.g., Angelescu 1982). Zooplankton inhabiting the PSBF constitute a key link through which energy and matter transfer from the basal trophic levels to those upper ones, supporting economically relevant invertebrates such as *Illex argentinus*; and fish stocks such as *Salilota australis*, *Micromesistius australis*, *Engraulis anchoita*, *Merluccius hubbsi*, or *Dissostichus eleginoides* (Bertolotti et al. 1996 and references), and also top-predator populations (i.e., marine birds and mammals, Falabella et al. 2009).

Zooplanktonic Crustacea and ichthyoplankton of the PSBF have been scarcely studied when compared with the information produced in the last four decades on the ecology of the neighboring ACS (see Acha et al. 2018; Cepeda et al. 2018 for reviews). Reasons for this scarce exploration are varied and include its relatively harsh hydrometeorological conditions, besides its remoteness and depth (~4000–5000 m in some latitudes), which jointly represent great financial and logistical challenges for its research. The first investigations dealing with zooplankton and ichthyoplankton of the PSBF were carried out in the mid-1960s, in cooperation between Argentina and other countries in the context of a fishing prospection supported by the Food and Agriculture Organization (FAO). These efforts contributed to generating faunal inventories, the first knowledge about the spatial distribution patterns along the great extension of the shelf-break, and to a lesser extent, about the abundance of zooplankton and ichthyoplankton (Ramírez 1971a, b, 1973; Montú 1977; Ramírez 1981; Ciechomski and Sánchez 1983; Ramírez and Viñas 1985). In the 1980s, the acquisition of research vessels by the National Institute for Fisheries Research and Development (INIDEP, Argentina) increased the possibility to investigate the target species of Argentina's fisheries, mainly on the shelf and less frequently at the shelf-break. Fisheries research cruises enlarged plankton knowledge, generating a valuable although heterogeneous dataset (i.e., presence–absence, semiquantitative, and quantitative data), biased mainly toward the mesozooplankton and macrozooplanktonic size fractions, including ichthyoplankton.

Current knowledge about the great diversity of zooplankton and ichthyoplankton along the PSBF is spatiotemporally patchy, and information is fragmented, dispersed, and even unpublished. Hence, our goal here is to compile all the available information on both groups, focusing mainly on diversity and spatial distribution. Ecological aspects (i.e., life-history traits) will also be dealt with, but to a lesser extent. Albeit reports show a highly diverse zooplankton community inhabiting the PSBF (e.g., Copepoda, Amphipoda, Euphausiacea, Ostracoda, Cladocera, Chaetognatha, Cnidaria, Tunicata, and Ctenophora, Boltovskoy 1999), we will focus on the most abundant and best-studied crustaceans, in particular copepods, hyperiid amphipods, and euphausiids, which are known to play a key role in the pelagic and mesopelagic food webs along the shelf-break.

This review compiles and reanalyzes most of the data collected from the 1970s to the present to establish baseline knowledge on crustacean zooplankton and ichthyoplankton of the PSBF. Aspects concerning both groups will be addressed separately. A common section linking crustacean zooplankton and ichthyoplankton with higher trophic levels is included. Finally, information gaps and prospects for future work on both groups in this region will be highlighted. The diversity of crustacean zooplankton and its abundance patterns are addressed; key species are recognized based on an up-to-date species inventory, and seasonal distribution patterns and life-history traits along two cross-shelf-break sections will be briefly described.

Available information on ichthyoplankton is much less than that referred to zooplankton, but the general abundance and distribution patterns of eggs and fish larvae south of 46 °S are established, as well as their relationship with bottom depth and temperature. Three ichthyoplankton groups are defined according to bottom depths and then characterized: (i) shelf-edge, (ii) slope, and (iii) oceanic. Existing information about spawning activity for some species inhabiting the PSBF is also provided.

4.2 Data Source

An historical data set, including published and unpublished information on copepods, amphipods, and euphausiids obtained from the 1970s to the present nearest to the PSBF, was gathered, providing a background to determine spatial distribution patterns for zooplanktonic Crustacea. The full set of plankton samples—largely biased toward the spring–summer period except for the northern region, 37–41 °S (Table 4.1)—was examined to determine zooplankton species richness (i.e., the number of species) and their mean abundance. The study area was divided into a grid of 0.5 × 0.5 degrees to analyze spatial patterns. Quantitative (ind.m^{-3}) and semiquantitative (scarce, frequent, abundant, very abundant) copepod data were included; therefore, four categories—commonly used in the literature—were used to homogenize both types of data. Amphipods and euphausiids abundance values were divided into quartiles. Extreme values were estimated as those larger than the third quartile by at least 1.5 times the interquartile range (i.e., 1.5(Q3-Q1) > Q3). Species richness (number of species present in each 0.5 × 0.5 degrees box) and mean abundance of each group were mapped for the study area. To determine the key species, an importance index *I* was calculated as the average between the relative abundance (i.e., the average contribution of each species to the total abundance) and the frequency of occurrence (i.e., the number of stations in which a given species occurs related to the total number of stations). Those copepod species with relative abundance ranging between abundant and very abundant in more than 50% of the stations where they occurred, and euphausiid and amphipod species with *I* values larger than 50%, were recognized as key species.

General patterns of ichthyoplankton are based on historical data obtained during seasonal research cruises carried out onboard RV Walther Herwig (Ciechomski et al. 1981). These cruises covered all the Argentine shelf and slope, performing

Table 4.1 Crustacean zooplankton sampling effort expressed as the number of sampled stations during the spring–summer and autumn–winter periods in three latitudinal ranges along the shelf-break

	Latitudinal range		
Sampling period	36–41 °S	41–47 °S	47–55 °S
Spring-summer	107	88	235
Autumn-winter	104	44	104

plankton tows employing a Bongo net with a 330 µm mesh size. Moreover, novel data from a cruise performed in the context of the Pampa Azul initiative (https://www.pampazul.gob.ar/) were included. An exhaustive plankton sampling, focused between 42 °S and 47 °S, was conducted in 47 oceanographic stations (25 during the day and 22 at night) employing a multinet sampler equipped with 5 nets of 300 µm mesh aperture, covering the outer portion of the shelf, the slope, and the adjacent oceanic area.

4.3 Zooplankton

4.3.1 General Diversity and Abundance Patterns

Copepods diversity roughly increases northwards along the shelf-break (Fig. 4.1), which agrees with previous reports for the ACS (Cepeda et al. 2018 and references therein). This pattern possibly results from the mixing of subantarctic species with those of subtropical origin near the Brazil-Malvinas Confluence, together with the occurrence, near the shelf-break, of species that typically occur in the inner and mid-shelf waters in southern Patagonia, which could be advected by the northward displacement of subantarctic waters (Ramírez et al. 1990). Amphipods and euphausiids diversity does not present any evident pattern. In fact, amphipod diversity seems to slightly increase around 48 °S and 52 °S (Fig. 4.1b). Patches of high euphausiid diversity occur along the entire shelf-break (Fig. 4.1c), which strongly agrees with that reported on a global scale, establishing that the shelf-break and oceanic euphausiids assemblages are relatively species-rich as compared to those from the shelf, emphasizing thus their oceanic character (e.g., Gibbons et al. 1999).

The abundance of these three taxa does not present any clear latitudinal pattern; though any tendency could be masked due to the employment of semiquantitative data (Fig. 4.1).

4.3.2 Key Species and Distribution Patterns

Considering the up-to-date species inventory for the shelf-break, some of them can be recognized as key species (see Supplementary material). The copepods *Oithona atlantica*, *Clausocalanus brevipes*, *Clausocalanus laticeps*, *Calanus simillimus*, *Rhincalanus nasutus*, *Subeucalanus longiceps*, and *Metridia lucens*, which become best represented toward the shelf-break and oceanic waters, show the highest abundances. In addition, some conspicuous species over the ACS, such as *Oithona* aff. *helgolandica*, *Ctenocalanus vanus*, *Drepanopus forcipatus*, and *Centropages brachiatus*, are also important near the slope but scarcely occur in oceanic waters (Ramírez 1970; Ramírez 1971a; Cepeda et al. 2018; Acha et al. 2020). Among these

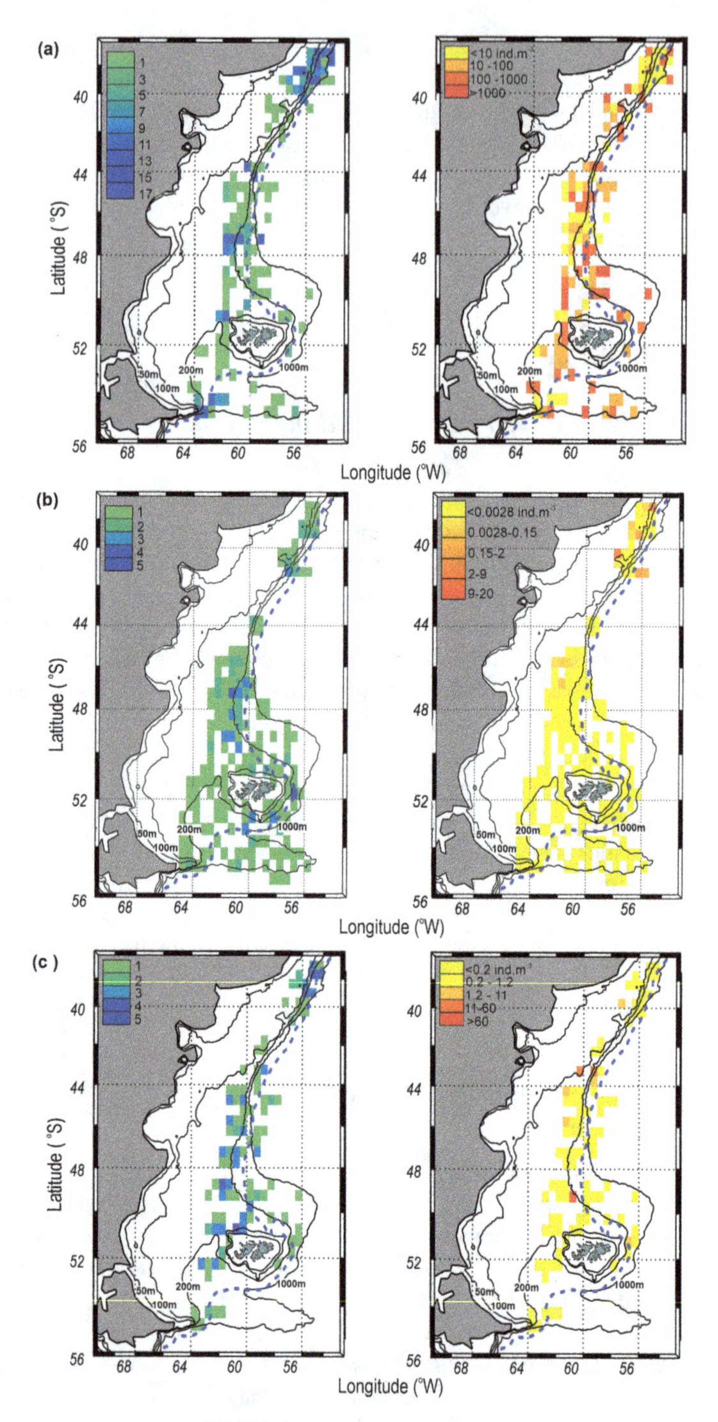

Fig. 4.1 Species richness of copepods (**a**), hyperiid amphipods (**b**), and euphausiids (**c**); and mean abundance (right panels) over the Patagonian Shelf-break Front. Maps are based on 0.5° × 0.5° grid box data. Note that scales are different for each map. Blue dotted line: mean position of the Patagonian Shelf-break front

key species, *O. atlantica*, *O.* aff. *helgolandica*, and *C. vanus* are worldwide distributed, while the others are restricted to the Antarctic-Subantarctic regions (Bradford-Grieve et al. 1999 and references).

The hyperiid *Themisto gaudichaudii* and *Primno macropa*, and the euphausiid *Euphausia lucens*, are key macrozooplanktonic species. *Themisto gaudichaudii* and *E. lucens* are widely distributed over the ACS, frequently occurring along the shelf-break, but with abundances usually decreasing toward the oceanic region (Ramírez 1971b, 1973; Ramírez and Viñas 1985; Padovani 2013). Their occurrence far off the shelf-break is possibly the result of being advected by the Malvinas Current Waters (MCW) and then transported eastwards through the Brazil-Malvinas Confluence (Tarling et al. 1995; Burridge et al. 2017). The subantarctic-antarctic *P. macropa* is distributed in the Atlantic and Pacific Oceans of both hemispheres and is characteristic of polar regions (Tarling et al. 1995; Zeidler and De Broyer 2009); it is less abundant than *T. gaudichaudii* and occurs almost exclusively in the shelf-break and oceanic waters off the ACS (Ramírez and Viñas 1985; Padovani 2013 and references therein).

Distribution patterns of most zooplankters are mainly driven by the water masses and the ocean currents that transport them (e.g., Björnberg 1981). Some species are closely related to the core of the MCW, which rarely occurs in shelf waters, possibly because of the thermal increase in the neritic region. Overlapping along the shelf-break between the shelf and the oceanic species has been reported (Ramírez 1970, 1971a, b, 1973; Padovani 2013; Acha et al. 2020). This blurred transition can be due to sporadic intrusions of the MCW onto the continental shelf (e.g., Piola et al. 2010; Carreto et al. 2016 and references; Acha et al. 2020 and references; Piola et al. current issue). The degree of this overlapping becomes evident from the reanalysis of the cross-shelf-break synoptic seasonal data further presented in Sect. 4.2, which does not detect species assemblages across the shelf-break (MDS stress along each section =0.00). Our analysis, based on semiquantitative and historical data (i.e., data from several years and seasons pooled), roughly indicates that the PSBF does not represent a sharp boundary for zooplankton between shelf and oceanic species. However, a recent analysis based on synoptic data shows the PSBF as a clear-cut boundary for copepods (Severo et al. 2024).

4.3.3 Seasonal Distribution Patterns and Life-History Traits Along the Shelf-Break

The seasonal pattern indicates that the compositions of copepods, euphausiids, and amphipods, in general, remain similar and are dominated by the same few species all year round (e.g., Ramírez and Santos 1994; Carreto et al. 2016; Sabatini et al. 2016). On the other hand, appreciable variations in seasonal abundances exist and are presumably driven by life-history traits and niche adaptations primarily governed by prey size (Kiørboe 2008). The timing of the seasonal progression of the primary production bloom in a north-south direction, which also encompasses the succession from large diatoms to small dinoflagellates, could be key factors for the spatiotemporal heterogeneities observed in the different copepods sizes (treated in the following sections as small <1 mm total length, medium 1–2 mm, and large-sized >2 mm for better description), as well as in hyperiid amphipods and euphausiids (Ramírez and Santos 1994; Sabatini et al. 2012, 2016; Carreto et al. 2016).

4.3.4 Copepods

4.3.4.1 Small-Sized Species

Oithona aff. *helgolandica* and *O. atlantica* are the most abundant small-sized species along the shelf-break, building up the copepod abundance mainly during winter and late spring at southern and northern latitudes, respectively (Ramírez and Santos 1994; Sabatini et al. 2016, Fig. 4.2). Their noteworthy—though surely undersampled (300 μm mesh size employed)—abundances probably happen because of the inadequate food conditions for any other co-occurring large-sized calanoids. Such trophic conditions occur mainly during pre- and post-blooming scenarios when nanoplanktonic heterotrophs, microplanktonic dinoflagellates, and ciliates dominate in the plankton, suggesting that energy pathways mainly occur through a microbial food web (e.g., Segura et al. 2013; Carreto et al. 2016; Sabatini et al. 2016). Under these relatively low food concentrations, these omnivore and passive ambush feeders can successfully feed and even reproduce, having an ecological advantage over the large-sized copepods (e.g., *C. simillimus*) that require abundant and large-sized food supply (i.e., mainly diatoms) for the onset of reproduction (Runge 1985; Turner 2004; Benedetti et al. 2018). Because of their life-history trait adaptations, both cyclopoid species can significantly contribute to the epipelagic secondary production. Together with these *Oithona* species, the small-sized oceanic and deep-water detritivore *Scolecithricella minor* occurs along the PSBF, though in lower abundances. This species can also take advantage of the heterotrophic food sources available (e.g., Yamaguchi et al. 1999). Both species would play a key role in particulate organic matter recycling (e.g., Benedetti et al. 2018 and references) (<5 ind.m^{-3}, Fig. 4.2).

4.3.4.2 Medium-Sized Species

Clausocalanidae is the most abundant medium-sized group inhabiting the PSBF (Fig. 4.2). Four species typically occur, amongst which *C. brevipes* (up to 1500 ind m^{-3}, Fig. 4.2) and *C. vanus* (up to 450 ind.m^{-3}, Fig. 4.2) are dominant. *Drepanopus forcipatus* is, in general, less numerous (up to 150 ind.m^{-3}); however, it is abundant in coastal and middle-shelf waters at southern latitudes, where it is the most conspicuous (1000–5000 ind.m^{-3}) mesozooplankton component (Sabatini 2008; Cepeda et al. 2018

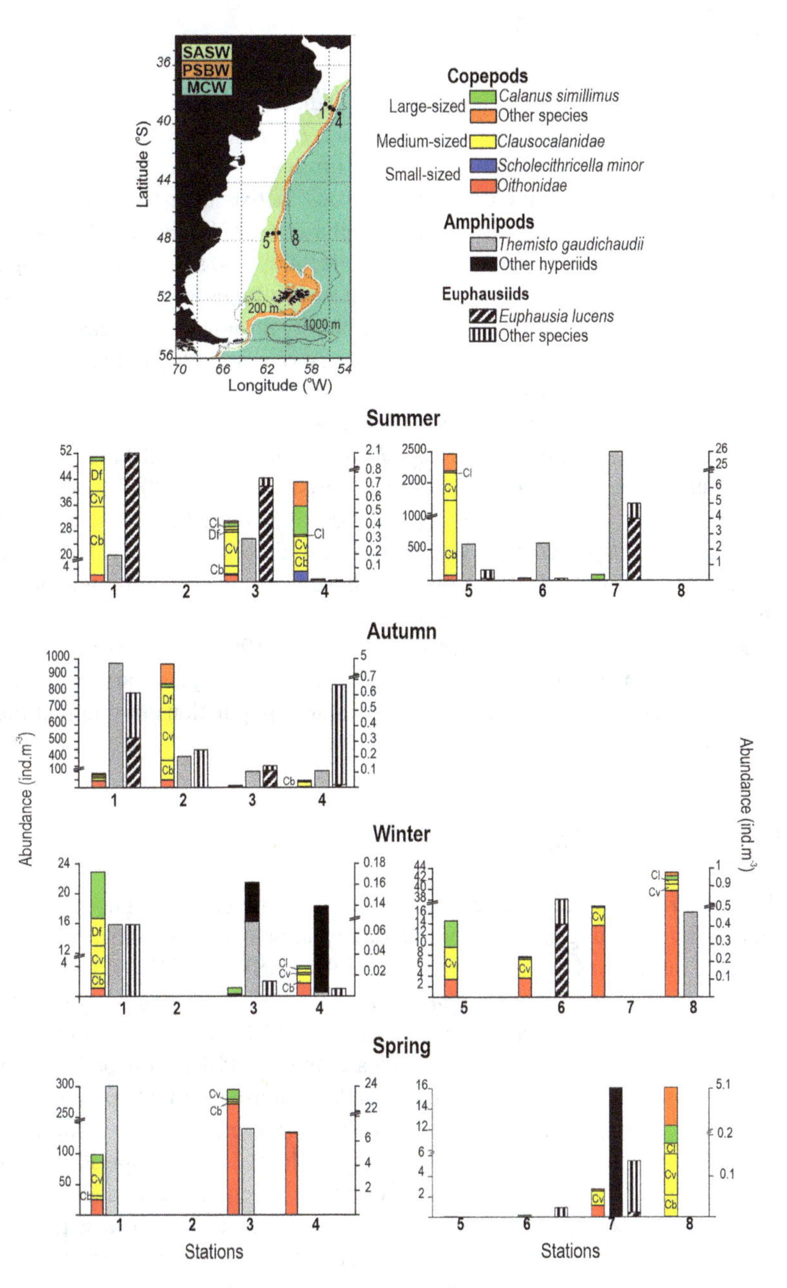

Fig. 4.2 Seasonal abundance of copepods, hyperiid amphipods, and euphausiids species along two across shelf-break transects (38 °S, left panels, and 47 °S, right panels). The map shows the average annual horizontal distribution of the Subantarctic Shelf Water (SASW, 33.6–33.8), the Patagonian Shelf-break Water (PSBW, 33.8–33.9), and the Malvinas Current and oceanic waters (MCW, >33.9) (Modified from Baldoni et al. 2015). The 33.9 isohaline (white solid line) indicates the position of the Patagonian Shelf-break Front (*sensu* Romero et al. 2006). Abundance records correspond to the original values for each cruise published by Ramírez and Santos (1994) and Sabatini et al. (2016), which are comparable through sections among seasons since they were developed during the same year employing the same sampler. Main *y*-axis: copepods abundance, secondary *y*-axis: amphipods and euphausiids abundance, please note differences in scales; Cb: *Clausocalanus brevipes*, Cv: *Ctenocalanus vanus*, Cl: *Clausocalanus laticeps*, Df: *Drepanopus forcipatus*

and references). The oceanic *C. laticeps*, an indicator of Subantarctic and Antarctic waters (Björnberg 1981), is barely represented (*C. laticeps* < 1 ind.m^{-3}, Fig. 4.2), increasing its presence and abundance in the surroundings of the MCW.

Clausocalanids are omnivorous-herbivorous, active filter-feeding species that prey upon a wide prey spectrum (e.g., Peralba et al. 2017; Benedetti et al. 2018). These species reproduce even at low food concentrations, suggesting a strong adaptation to oligotrophic conditions (Mazzocchi and Paffenhöfer 1998, 1999; Bi 2005). These ecological traits could explain their ubiquity throughout the seasonal cycle of primary production (Fig. 4.2), suggesting a key role in both the microbial and the classical trophic food webs.

Clausocalanids occurring along the shelf-break exhibit two main reproductive strategies, with egg-carrier species (e.g., *D. forcipatus*, Antacli 2011) and others that are broadcasting spawners (e.g., *C. vanus*, Derisio and Acha 2015). Because these species could be preyed upon by larger zooplankters and fishes, the broadcasting strategy possibly represents an adaptation to avoid being eaten together with their eggs (Bollens and Frost 1991). However, those eggs carried by the females would eventually have lower mortality than the free-spawned eggs, and consequently, it would pay off to carry them (Kiørboe and Sabatini 1994). Presumably, the coexistence of both spawning behaviors—as alternative strategies—among congeners might also guarantee the success of clausocalanid populations throughout the seasons nearest the PSBF.

4.3.4.3 Large-Sized Species

Calanus simillimus is the unique large-sized species present throughout the seasons, though with fluctuating abundances (Ramírez and Santos 1994; Sabatini et al. 2016, Fig. 4.2). This species is well represented in the SASW, the PSBW, and the MCW, contrary to *C. carinatus* and *Calanus australis*, whose abundances increase toward middle and inner shelf waters (S < 33.6, Ramírez 1981; Ramírez and Sabatini 2000; Cepeda et al. 2018). At northern latitudes along the PSBF, copepodite stages of *C. simillimus* occur in plankton in high abundance during spring (ca. 100 ind.m^{-3} on average), decreasing in autumn (ca. 20 ind.m^{-3} on average) and becoming practically absent in winter (< 1 ind.m^{-3}) at northern latitudes along the PSBF (Verona et al. 1972; Santos unpublished data). Contrarily, significant copepodites densities have been found only at the end of summer over the outer shelf near the slope at southern latitudes (Sabatini et al. 2016). Regional life cycle characteristics and ecological traits of this species are practically unknown, but this seasonal pattern of copepodites matches the general life cycle described in other subantarctic ecosystems (e.g., Atkinson 1998). The reproductive period occurs mainly from early spring to autumn, and possibly two generations (spring and summer) occur per year; the late copepodites resulting from the summer breeding would enter diapause in autumn (Mauchline 1998), explaining their scarce presence during the colder season. Adults of *C. simillimus* would preferably be filter-feeding herbivorous, such as other congeners (Santos unpublished data; Benedetti et al. 2015, 2018 and references), attaining higher representation during spring and late summer at northern and southern latitudes of the PSBF, respectively (Fig. 4.2). Abundances peak after the phytoplankton bloom, when large-sized diatoms are available (Carreto et al. 1995; Segura et al. 2013; Sabatini et al. 2016). In southern latitudes, *C. simillimus* abundances increase during early spring (Fig. 4.2), when the thermocline is not yet well established and may likely respond to a bloom of large cells caused by new production (nitrate) from the nutrient-rich Malvinas Current (Sabatini et al. 2016). Taking all these characteristics into account, *C. simillimus* would be mainly linked to the classical trophic web.

Other large-sized and oceanic species (i.e., *M. lucens*, *S. longiceps*, *R. nasutus*, *Neocalanus tonsus*, *Euchaeta marina*, *R. gigas*, *Euchirella rostrata*, *Heterorhabdus austrinus*, *Haloptilus oxycephalus*, *Pleuromamma robusta*, and *P. gracilis*) are associated mainly with the core of the MCW (offshore from the PSBF). Such species sporadically occur along the PSBF and in the outer shelf, possibly associated with intrusions of the MCW onto the shelf (Ramírez and Björnberg 1981; Piola et al. 2010). In the low-chlorophyll, less productive, and populated oceanic waters, where the core of their populations develop, predation risks would be relatively low, promoting the development of large-sized copepod populations. The large size certainly drives survival and the competitive advantage of these species, improving their active (cruise or filter current) feeding performance, and increasing the chance of mate finding and potential fertility (Kiørboe 2011). Furthermore, these large copepods commonly show greater tolerance to starvation than the smaller ones, which becomes an additional beneficial factor in the less productive oceanic habitats (Kiørboe 2008).

Some of the above-mentioned species present a seasonal (summer to autumn) occurrence along the PSBF (Fig. 4.2), possibly linked to their life cycles, which involve the vertical migration toward epipelagic waters in summer and a deeper distribution in winter, explaining the progressive depopulation of the surface layers of these species during the colder season (Atkinson and Sinclair 2000).

Chlorophyll-*a* abundance remains high at the PSBF still during summer, due to its richness in microplanktonic Bacillariophyceae (Segura et al. 2013) and small pico- and nanoplankton cells (Carreto et al. 2003a, b). All this would be a proper food resource for large-sized herbivorous such as *M. lucens*, *R. nasutus*, *R. gigas*, *P. robusta*, and *P. gracilis*, which are well adapted to a diatom-based diet (e.g., Schnack-Schiel et al. 2008; Benedetti et al. 2015). At southern latitudes, the increased representation of these large herbivores is related to the coming onset of the spring bloom that propagates from the shelf-break to the inner shelf (Carreto et al. 2016; Sabatini et al. 2016). The large-sized copepod abundances are higher in spring at the shelf-break, increasing toward the outer shelf in late summer (Fig. 4.2). These calanoids are reported as strong migrants and may play a key role in carbon cycling as they graze in the euphotic zone and then migrate down where they excrete their lipid reserves (Jónasdóttir et al. 2015). *Euchaeta marina*, *H. austrinus*, and *H. oxycephalus*, on the contrary, are carnivorous species that primarily prey on nauplii and small-sized copepods, thus acting as a structuring force in the top-down control of mesozooplankton along the shelf-break (Benedetti et al. 2015).

4.3.5 Amphipods and Euphausiids

Themisto gaudichaudii is the most abundant hyperiid species both seasonally and spatially (Ramírez and Viñas 1985, Fig. 4.2). Adults of *T. gaudichaudii* show, on average, seasonally variable concentrations but high occurrence frequency throughout all seasons along the PSBF. This species is relatively best represented at northern latitudes of the PSBF during summer and autumn (Ramírez and Santos 1994, Fig. 4.2), while further south, the highest abundance also occurs in late summer, decreasing in winter and spring (Sabatini et al. 2016, Fig. 4.2). This seasonal pattern could be related to the mechanisms of enrichment and accumulation of planktonic organisms along the slope, which may guarantee an adequate copepod food supply to prey upon.

Other species such as *P. macropa*, *Hyperiella antarctica*, *Vibilia antarctica*, and *Cyllopus magellanicus* seasonally occur nearest the slope. They are mainly present during winter and spring in northern latitudes (up to 47 °S, Ramírez and Viñas 1985; Ramírez and Santos 1994, Fig. 4.1) and occasionally occur further south, surrounding the neritic sector (Ramírez and Viñas 1985, Fig. 4.2).

Euphausiids are seasonally and spatially well represented (Fig. 4.2). At northern latitudes, adults are abundant in summer and autumn, decrease in winter, and are absent in spring (Ramírez and Santos 1994, Fig. 4.2). Further south, higher abundances are recorded in summer, with the lowest densities occurring at the end of winter (Montú 1977; Sabatini et al. 2016, Fig. 4.2). Among them, *E. lucens* is the best represented during summer, while other cryophilic species such as *Euphausia vallentini*, *Thysanoessa gregaria*, and *Nematocelis megalops* become relatively abundant during the coldest seasons (Ramírez and Santos 1994; Sabatini et al. 2016, Fig. 4.2). *Nematocelis megalops*, in particular, thrives below 200 m during the daytime (Curtolo et al. 1990); therefore, its overall absence may be a bias because most available data came from diurnal tows, which did not reach that depth.

Adult euphausiids are opportunistic omnivores (Mauchline and Fisher 1969) whose seasonal spawning peak could be coupled with the phytoplankton biomass increase (e.g., Sabatini et al. 2012). In the region, *T. gregaria* and *E. lucens* reach ovarian maturity by the end of winter (Ramírez and Dato 1983), while ripe females of *E. vallentini* increase during early spring at southern latitudes nearest the PSBF (Sabatini et al. 2016). Coupled with this, densities of the herbivorous furcilia larvae noticeably peak in early summer (Montú 1977; Ramírez and Santos 1994; Sabatini et al. 2016). Given that adult euphausiids can also exert a considerable predatory impact on mesozooplankton (Pillar et al. 1992), they could play a key role in the PSBF food web by competing directly with pelagic fish.

4.4 Ichthyoplankton

Of the more than 600 fish species occurring along the ACS (Figueroa 2019), 524 are bony fishes that distribute following temperature, salinity, and depth patterns, as well as bottom types, and various biological production mechanisms such as fronts that constitute appropriate places for larval feeding and growth. According to these ecological constraints, five fish assemblages were defined for all the shelf and shelf-break, including one of them approximately 46 species that inhabit the PSBF (Angelescu and Prenski 1987). This ichthyofauna, belonging to the Magellanic Province (Balech and Ehrlich 2008), is quite different from that occurring on the shelf, encompassing a group of demersal fish typical of cold sub-Antarctic waters occurring at depths greater than 200 m with bottom temperatures between 4.0 and 5.5 °C, which Menni and Gosztonyi (1982) designated as "Magellanic Fauna," and Prenski and Sánchez (1988) as the "Slope fish assemblage." Reproductive favorable conditions along the PSBF, such as nutrient enrichment, prey concentration, and eggs/larvae retention, conform to the Bakun triad (Bakun 1996), allowing many of these species to close their life cycles there. Therefore, it is common to find eggs and larvae of demersal and mesopelagic fish along the PSBF (Table 4.2).

Even though a great disparity exists in knowledge about different adult fishes and their early life stages along the PSBF, information from cruises detailed in Sect. 4.2, together with fishery research cruises later carried out by INIDEP, generated the basic knowledge on ichthyoplankton distribution and abundance at the PSBF and beyond.

Based on the cruises developed in 1978, five ichthyoplankton ecological groups were established according to latitude, bottom depth, and water surface temperature (Ciechomski et al. 1981). The distribution and abundance of eggs and larvae of the different species obtained on these cruises are shown in Fig. 4.3, while their relative abundances related to bottom depth and surface temperatures are presented in Table 4.2. Results from the analyses of the samples obtained in the 2017 cruise indicated that the highest larval densities were found in the oceanic area (Fig. 4.4) and that they were distributed differently according to depth. The highest records occurred in the shallower layers of both the shelf/slope waters and the oceanic ones. These results are expected since temperatures in the shallower levels were relatively higher (7.2–5.2 °C) than in the deeper ones (5.2–4.4 °C), and also because the presence of a thermocline separated the shallower levels from the deepest ones. These physical characteristics occurring in the surface layers usually represent a favorable environment for larval development (Olivar et al. 2018).

The ichthyoplankton obtained from the available information could be classified according to bottom depths as shelf-edge, slope, and oceanic.

Table 4.2 Abundance of fish eggs and larvae related to bottom depth and sea surface temperature. Data from Ciechomski et al. (1981). Scarce (S), present (P), frequent (F), and very abundant (VA)

STAGE-Order/	Depth range						Surface temperature		
Family-*Species*	201–300	301–400	401–500	501–600	601–700	701–800	4.0–5.9	6.0–7.9	8.0–9.9
EGGS									
Gadiformes									
Macrourus whitsoni					VA	F	VA		
Coelorhinchus fasciatus		F	P	A		F	VA		
Micromesistius australis	S	P	VA				VA		
Salilota australis	VA		S					VA	
Clupeiformes									
Sprattus fuegensis			F				A	F	
LARVAE									
Stomiiformes									
Idiacanthus atlanticus					VA			VA	
Gadiformes									
Micromesistius australis			VA				VA	P	
Perciformes									
Dissostichus eleginoides	S		S	P			P	VA	
Nototheniidae		P		S			F	S	
Zoarcidae	A	S	P				P	S	F
Myctophiformes									
Myctophidae	S	P	S	S	F	F	F	S	
Clupeiformes									
Sprattus fuegensis	P							S	F

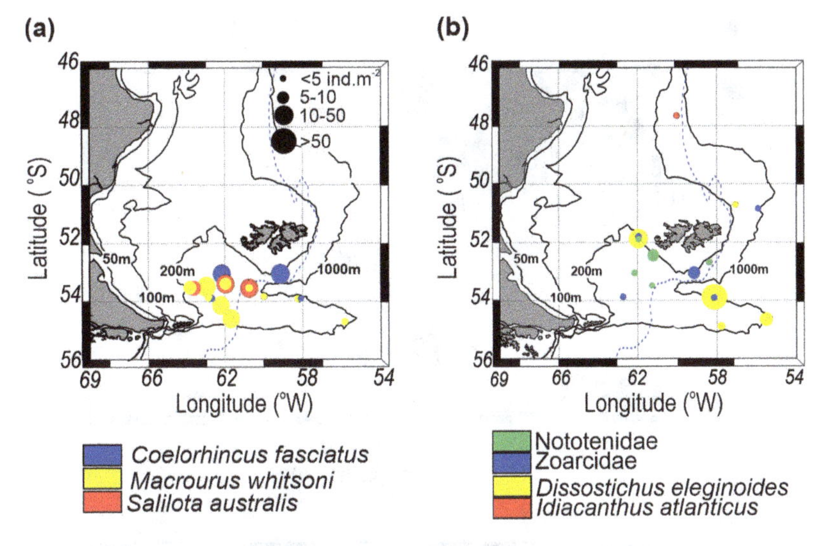

Fig. 4.3 Distribution and abundance of eggs (**a**) and fish larvae (**b**) sampled south of 46 °S, expressed as individuals per 10 m^{-2}. The scale inside map (**a**) also corresponds to (**b**). (Data from Ciechomski et al. 1981)

4.4.1 Shelf-Edge Ichthyoplankton

This group corresponds to that occurring at intermediate depths, following Ciechomski et al. (1981), and is composed of larvae captured between 48 °S and 54 °S, at bottom depths between 80 and 200 m, and over a thermal range of 5–8 °C. Species included in this group are the southern blue whiting *M. australis*, the Fuegian sprat *S. fuegensis*, and the red cod *S. australis*. Other larvae, such as those of the grenadiers *Macrourus whitsoni* and *Coelorhinchus fasciatus*, the long-tail southern cod *Patagonothoten ramsayi*, and the Patagonian redfish *Sebastes oculatus*, also occurred in the shelf-edge.

The anchovy *E. anchoita* and the hake *M. hubbsi*, typically associated with shelf waters, are also able to spawn along the shelf-break after the peak of intense coastal spawning is completed and conditions in those neritic areas become unfavorable (Bertolotti et al. 1996).

4.4.2 Slope Ichthyoplankton

The slope ichthyoplankton includes those shelf and oceanic species obtained between 200 and 800 m bottom depth, with surface temperatures ranging from 4 to 6 °C. It is characterized by the presence of grenadiers *C. fasciatus*, *M. whitsoni*, and *Hymenocephalus* sp., a deep-sea sole *Mancopsetta maculata*, a snake mackerel *Paradiplospinus antarcticus*, and a deep-water fish *Echiodon cryomargarites*.

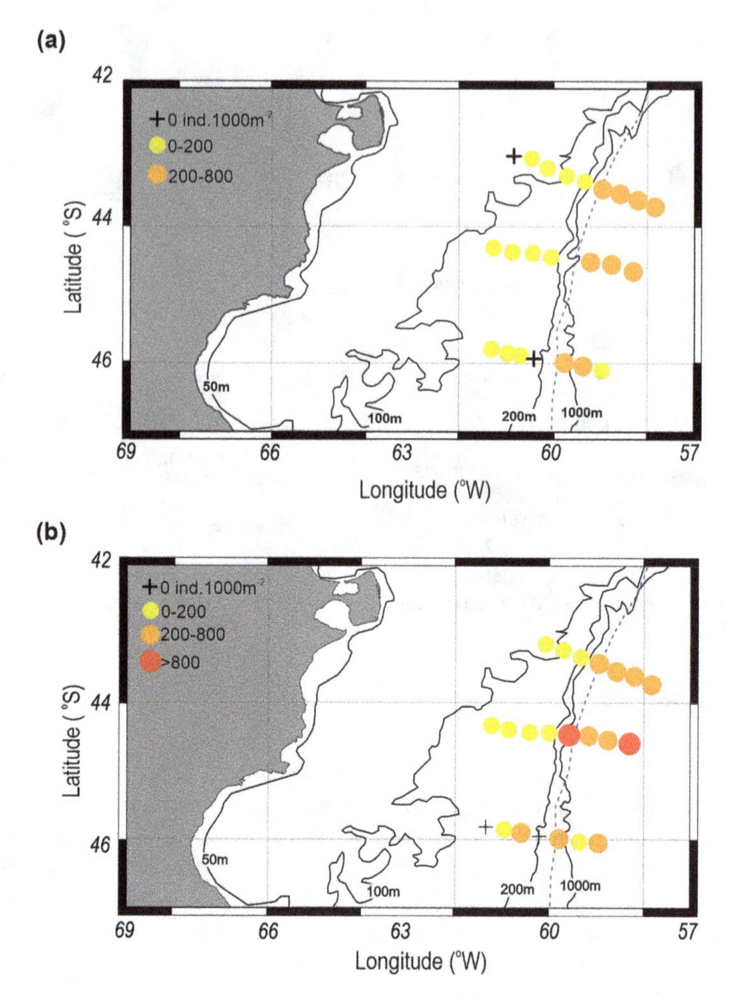

Fig. 4.4 Distribution and abundance of total fish larvae expressed as individuals per 1000 m^{-2} during daytime (**a**) and nighttime (**b**), obtained during the 2017 cruise at "Agujero Azul" and surroundings. Blue dotted line: mean position of the Patagonian Shelf Break

Among the oceanic species, larvae of mesopelagic fish, mainly myctophids, such as *Gymnoscopelus nicholsi*, *Gymnoscopelus braueri*, *Gymnoscopelus opisthopterus*, *Krefftichthys anderssoni*, and species of the genus *Protomyctophum*, occurred.

4.4.3 *Oceanic Ichthyoplankton*

It is mainly composed of cold-water mesopelagic fish larvae of Myctophidae occurring at bottom depths between 800 and 3160 m and surface temperatures of about 6 °C. Sánchez and Ciechomski (1995), through two cruises performed in 1988 by

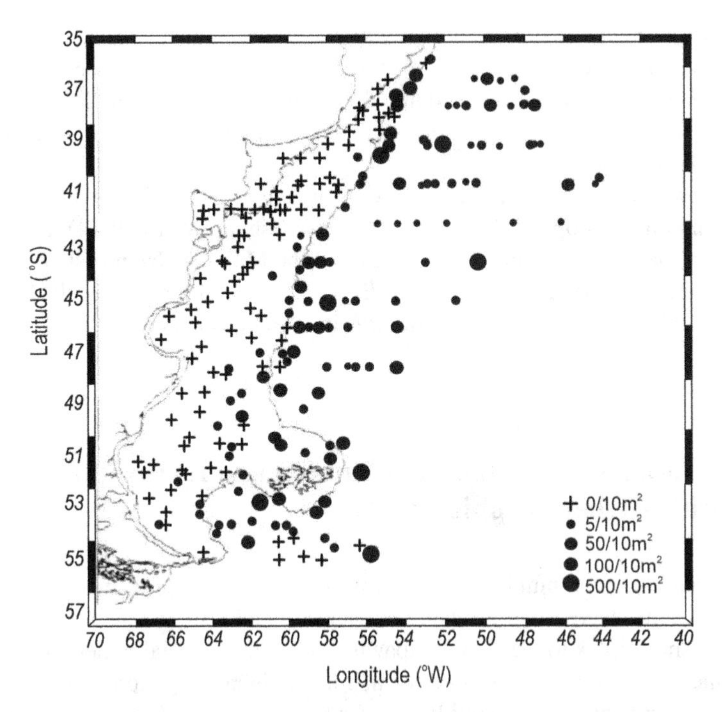

Fig. 4.5 Occurrence of myctophid larvae on the Argentine Continental shelf, shelf-break, and beyond during winter. (Redrawn from Sánchez and Ciechomski 1995)

the URSS Ministry of the Fishing Industry, indicate that myctophids are a predominant group of epipelagic and mesopelagic fish. Figure 4.5 shows the distribution and abundance of larvae of this family in a large offshore area during winter. In general, larvae are located offshore of the 200 m isobath throughout the year, except for spring, when they occupy part of the shelf south of 47 °S (Sánchez and Ciechomski 1995).

Among Myctophidae, *G. nicholsi*, *G. braueri*, *G. opisthopterus*, *K. anderssoni*, *Electrona antarctica*, and *Protomyctophum* spp. have been recorded. Other species, such as the Antarctic deep-sea smelt *Bathylagus antarcticus*, a macrurid *Coryphaenoides* sp., and a species of Paralepipidae, *Notolepis coatsi*, also occurred.

4.4.4 Early Fish Stages and Spawning Activity

The presence of eggs and larvae at different ontogenic stages in plankton along the PSBF allows for inferring the spawning season of those species inhabiting there. The southern blue whiting *M. australis* has been observed to spawn in late winter and spring (Weiss 1974), as well as the red cod *S. australis* (Brickle et al. 2011) and the grenadier *C. fasciatus* (Cousseau and Perrotta, 1998). The Fuegian sprat

S. fuegensis spawns in spring and early summer, taking advantage of the favorable conditions that develop during both seasons along the PSBF. Larvae with a longer planktonic life span were observed mainly in spring, except for the Fuegian sardine larvae, which are also present in summer, and those of the Myctophidae, which are present all year round.

All these species have different spatial spawning patterns. Thus, while *M. australis* spawns on the slope and in deeper waters offshore (Weiss 1974), *S. fuegensis* done it on the shelf near the coast and around the Malvinas Islands shelf and slope (Cousseau and Perrotta 1998), *S. australis* on the slope around the Malvinas (Weiss 1975; Brickle et al. 2011), and *C. fasciatus* in oceanic waters between 300 and over 1000 m (Ciechomski et al. 1981).

4.5 Linking Crustacean Zooplankton and Ichthyoplankton with Higher Trophic Levels

The zooplankton community described in this chapter directly fuels the different developmental stages of important invertebrate and fish stocks that concentrate along the shelf-break to feed and/or spawn. Copepods, especially their larval stages, constitute the main feeding source for fish larvae inhabiting along the shelf-break. Large-sized copepods, euphausiids, and the hyperiid *T. gaudichaudii* constitute the main prey items for juveniles and adult stages of small planktivorous fish such as *E. anchoita* (e.g., Angelescu 1982), while high mean percentages of the relative importance index (RII) estimated for copepods (up to 74.3%), amphipods (14.51–47.61%), and euphausiids (8.29–20.5%) indicate their relevance also for several myctophid species (Saunders et al. 2018).

Euphausia spp. and *T. gaudichaudii* are the most frequent prey in the stomach contents of some demersal fish, such as the Patagonian toothfish *D. eleginoides* (occurrence frequency, % FO, up to 96.2%), the hoki *M. magellanicus* (% FO up to 96.6%), and the Argentine hake *M. hubbsi* (% FO up to 83.3%) (Marí and Sánchez 2002). Both prey items also constitute between 71% and 99% of the prey ingested by squids (Ivanovic 2010). Therefore, the composition and abundance of zooplankton along the shelf-break play a key role in ensuring optimal prey availability for higher trophic levels, and their spatiotemporal heterogeneities probably influence the seasonal migration patterns of fish and cephalopods (e.g., Prenski and Sánchez 1988).

Fish larvae, as well, are also preyed upon by other organisms. For example, Macrouridae larvae are preyed upon by the squid *I. argentinus* in the Brazilian continental slope (Gomes Fisher 2012), and that of the Fuegian sprats are preyed upon by juvenile hoki *M. magellanicus* (Brickle et al. 2009).

4.6 Gaps and Prospects for Future Work on Zooplankton and Ichthyoplankton

Compiled information indicates that zooplankton and ichthyoplankton of the shelf-break are, in general, reasonably well known. There are, however, several important issues altogether unexplored.

Information on crustacean zooplankton composition is almost totally based on coarse mesh sizes. When finer mesh sizes (less than 150 μm) have been employed in other regions of the ACS (and also worldwide), different community patterns have arisen, highlighting the importance of smaller species. Combining both fine and coarse mesh size nets for future sampling will be useful to complete (and/or confirm) the current species inventories, especially for copepods.

Much less work has been conducted at the population level; therefore, almost nothing is known about the crustacean zooplankton population dynamics at the shelf-break. Future directions should focus on the details of their life cycles and ecological traits. To attain these goals, it will be necessary to increase sampling depths to confirm, for example, the assumptions described in Sect. 4.3.4.3 about the vertical distribution of earlier stages of the largest copepods. Coupled with this, seasonal cruises should be developed to depict their seasonal vertical migration.

Moreover, information on daily vertical migrations (DVM) of zooplankton is lacking for the shelf-break. In general, it is proposed that DVMs, coupled with physical mechanisms, could retain organisms at the frontal zones (Derisio et al. 2014). However, for many herbivorous and omnivorous zooplankters that feed in surface layers, the primary factor driving the DVMs seems to be reducing the predation risk ("dark shelter" hypothesis). Besides, several congeneric zooplankton species (e.g., *Clausocalanus* spp., *Euphausia* spp.) by vertically segregating during the night avoid competing for food with each other. Future cruises aimed at investigating the DVM process and the degree of niche diversification would be useful for gaining insight into ecological mechanisms and species interactions occurring along the PSBF. Moreover, they would allow to establish relationships between migrant species and the different developmental stages of fish that prey upon them.

Although the importance of copepods, hyperiid amphipods, and euphausiids as main prey for higher trophic levels is recognized, nutritional studies focused on their quality as prey are still scarce along the PSBF. These kinds of studies are welcome to establish the energetic density available in the lipid content of each one of these major zooplankton prey, as well as their contents of long-chain polyunsaturated fatty acids.

Still further is the development of studies on the effects of environmental conditions on secondary production and biological diversity, as well as the application of ecotrophic models to understand energy flow through the trophic web, its variability, and its impact on secondary productivity.

Present knowledge on ichthyoplankton demonstrates the potential of this area to offer an appropriate environment for fish larvae, favoring the closure of life cycles of demersal and pelagic fishes. Anyway, it is mandatory to improve this basic knowledge by increasing the sampling effort seasonally to know the spawning and rearing periods. As well, the PSBF ichthyoplankton inventory should be completed, in particular, with the identification of myctophids and other poorly known larval groups.

The connectivity between the shelf larval assemblages and those inhabiting the oceanic waters remains to be determined. Moreover, studies on the vertical migrations of the different larval species and their interactions with both prey and predators must be intensified. The role of ichthyoplankton as consumers of zooplankton and as prey for larger fish and other organisms remains an open question, as well as the environmental processes that regulate their occurrence, abundance, and retention in this hydrographically complex area.

Implementing broadly used methodologies to determine biochemical indexes to study the quality and condition of the larvae, as well as using stable isotopes to complement and reinforce gut content analysis, is necessary to study the role of fish larvae in the trophic web. Moreover, there is a need to increase knowledge on larval growth and mortality, essential parameters for understanding the recruitment process of species inhabiting the shelf-break and oceanic areas.

Acknowledgments The authors thank infinitely to Drs. F.C. Ramírez, M.E. Sabatini, M.D. Viñas, and J.D. de Ciechomski for all the knowledge transmitted over so many years along diverse academic instances, which was essential for tackling this chapter. Also, we wish to thank the scientists, technicians, and research vessel crews of INIDEP, CONICET, and other international agencies who got the plankton samples, which were indispensable and valuable material for building our present understanding of the crustacean zooplankton and ichthyoplankton along the Patagonian shelf-break. Special thanks must be given to G.L. Álvarez-Colombo and L.N. Padovani for amphipods data, to the heads of the several assessment groups at INIDEP for receiving and helping plankton researchers on board their stock assessment cruises, to INIDEP facilities, and to the financial support from all the institutions involved and the grants held by all the authors. This is INIDEP contribution No. 2308.

References

Acha EM, Ehrlich MD, Muelbert JH et al (2018) Ichthyoplankton associated to the frontal regions of the Southwestern Atlantic. In: Plankton ecology of the Southwestern Atlantic. Springer, Cham, pp 219–246. https://doi.org/10.1007/978-3-319-77869-3_11

Acha EM, Viñas MD, Derisio C et al (2020) Large-scale geographic patterns of pelagic copepods in the southwestern South Atlantic. J Mar Syst 204:103–281. https://doi.org/10.1016/j.jmarsys.2019.103281

Angelescu V (1982) Ecología trófica de la anchoíta del Mar Argentino (Engraulidae, *Engraulis anchoita*): parte II, Alimentación, comportamiento y relaciones tróficas en el ecosistema. Ser Contrib Inst Nac Invest Desarr Pesq 409:83

Angelescu V Prenski LB (1987) Ecología trófica de la merluza común del Mar Argentino (Merlucciidae, Merluccius hubbsi). Parte 2. Dinámica de la alimentación analizada sobre la base de las condiciones ambientales, la estructura y las evaluaciones de los efectivos en su área de distribución. Contrib. INIDEP, Mar del Plata, 561, p 205

Antacli JC (2011) Estrategias de vida de los copépodos *Drepanopus forcipatus* y *Calanus australis* en relación con los recursos tróficos en la plataforma patagónica austral (Argentina, 47°-55°S). Dissertation, Universidad Nacional de Mar del Plata

Atkinson A (1998) Life cycle strategies of epipelagic copepods in the Southern Ocean. J Mar Syst 15(1–4):289–311. https://doi.org/10.1016/S0924-7963(97)00081-X

Atkinson A, Sinclair JD (2000) Zonal distribution and seasonal vertical migration of copepod assemblages in the Scotia Sea. Polar Biol 23(1):46–58

Bakun A (1996) Patterns in the ocean. Ocean processes and marine population dynamics. University of California Sea Grant, California

Bakun A, Parrish RH (1991) Comparative studies of coastal pelagic fish reproductive habitats: the anchovy (*Engraulis anchoita*) of the southwestern Atlantic. ICES J Mar Sci 48(3):343–361. https://doi.org/10.1093/icesjms/48.3.343

Baldoni AG, Molinari G, Reta R, Guerrero RA (2015) Atlas de temperatura y salinidad de la plataforma continental del Atlántico Sudoccidental: períodos cálido y frío. INIDEP, Mar del Plata

Balech E, Ehrlich MD (2008) Esquema Biogeográfico del Mar Argentino. Rev Invest Des Pesq 19:45–75

Benedetti F, Gasparini S, Ayata SD (2015) Identifying copepod functional groups from species functional traits. J Plankton Res 38(1):159–166. https://doi.org/10.1093/plankt/fbv096

Benedetti F, Vogt M, Righetti D et al (2018) Do functional groups of planktonic copepods differ in their ecological niches? J Biogeogr 45(3):604–616. https://doi.org/10.1111/jbi.13166

Bertolotti MI, Brunetti NE, Carreto JI et al (1996) Influence of shelf-break fronts on shellfish and fish stocks off Argentina. Contribution to ICES annual science conference. Reykjavik, Iceland

Bi H (2005) Population dynamics of *Clausocalanus furcatus* (Copepoda, Calanoida) in the northern Gulf of Mexico. Dissertation, Louisiana State University and Agricultural & Mechanical College

Björnberg T (1981) Copepoda. In Boltovskoy D (ed) Atlas del Zooplancton del Atlántico Sudoccidental y Métodos de Trabajo con el Zooplancton Marino. Publicaciones Especiales INIDEP, Mar del Plata, Argentina, p 587–679

Bollens SM, Frost BW (1991) Diel vertical migration in zooplankton: rapid individual response to predators. J Plankton Res 13(6):1359–1365

Boltovskoy D (ed) (1999) South Atlantic Zooplankton. Backhuys Publishers, Leiden

Bradford-Grieve JM, Markhaseva EL, Rocha CEF et al (1999) Copepoda. In: Boltovskoy D (ed) South Atlantic Zooplankton. Backhuys Publishers, Leiden, pp 869–1098

Brickle P, Arkhipkin AI, Laptikhovsky V et al (2009) Resource partitioning by two large planktivorous fishes *Micromesistius australis* and *Macruronus magellanicus* in the Southwest Atlantic. Estuar Coast Shelf Sci 84(1):91–98. https://doi.org/10.1016/J.ECSS.2009.06.007

Brickle P, Laptikhovsky V, Arkhipkin A (2011) The reproductive biology of a shallow water morid (*Salilota australis* Günther, 1878), around The Falkland Islands. Est Coast Shelf Sci 94:102–110

Burridge AK, Tump M, Vonk R et al (2017) Diversity and distribution of hyperiid amphipods along a latitudinal transect in the Atlantic Ocean. Progr Oceanogr 158:224–235. https://doi.org/10.1016/J.POCEAN.2016.08.003

Carreto JI, Lutz V, Carignan MO et al (1995) Hydrography and chlorophyll a in a transect from the coast to the shelf-break in the Argentinian Sea. Cont Shelf Res 15(2–3):315–336. https://doi.org/10.1016/0278-4343(94)E0001-3

Carreto JI, Montoya NG, Benavides HR et al (2003a) Characterization of spring phytoplankton communities in the Río de La Plata maritime front using pigment signatures and cell microscopy. Mar Biol 143(5):1013–1027. https://doi.org/10.1007/s00227-003-1147-z

Carreto JI, Montoya NG, Carignan MO et al (2003b) La campaña "Prospección ambiental del Río de la Plata y su frente marítimo (EH-09/01, 2da etapa)". I. Caracterización de las comunidades fitoplanctónicas utilizando marcadores pigmentarios (HPLC-CHEMTAX). Informe técnico del Proyecto "Protección ambiental del Río de la Plata y su Frente Marítimo. Prevención y control de la contaminación y restauración de hábitats", p 56

Carreto JI, Montoya NG, Carignan MO et al (2016) Environmental and biological factors controlling the spring phytoplankton bloom at the Patagonian shelf-break front – degraded fucoxan-

thin pigments and the importance of microzooplankton grazing. Progr Oceanogr 146:1–21. https://doi.org/10.1016/j.pocean.2016.05.002
Cepeda GD, Temperoni B, Sabatini ME et al (2018) Zooplankton communities of the Argentine continental shelf (SW Atlantic, ca. 34°-55°S), an overview. In: Plankton ecology of the Southwestern Atlantic. Springer, Cham, pp 171–199. https://doi.org/10.1007/978-3-319-77869-3_9
Ciechomski JD, Sánchez RP (1983) Relationship between ichthyoplankton abundance and associated zooplankton biomass in the shelf waters off Argentina. Biol Oceanogr 3(1):77–101. https://doi.org/10.1080/01965581.1983.10749472
Ciechomski JD, Ehrlich MD, Lasta CA et al (1981) Distribución de huevos y larvas de peces en el Mar Argentino y evaluación de los efectivos desovantes de anchoíta y de merluza. Contrib Inst Nac Invest Des Pesq 383:59–79
Cousseau MB, Perrotta RG (1998) Peces marinos de Argentina. INIDEP, Mar del Plata
Curtolo LM, Dadon JR, Mazzoni HE (1990) Distribution and abundance of euphasids off Argentina in spring 1978. Revista Nerítica 5(1):1–14
Derisio C, Acha M (2015) Efecto del sistema frontal de Península Valdés sobre la actividad reproductiva del copépodo Calaniodeo *Ctenocalanus vanus*. Inf Invest INIDEP N° 87/2015, p 8
Derisio C, Alemany D, Acha M et al (2014) Influence of a tidal front on zooplankton abundance, assemblages and life histories in Península Valdés, Argentina. J Mar Syst 139:475–482. https://doi.org/10.1016/j.jmarsys.2014.08.019
Falabella V, Campagna C, Croxall J (2009) Atlas of the Patagonian Sea: Species and spaces. Buenos Aires: *Wildlife Conservation Society and BirdLife International*, Buenos Aires, Argentina, 15, 303
Figueroa DE (2019) Clave de peces marinos del Atlántico Sudoccidental, entre los 33°S y 56°S. INIDEP, Mar del Plata
Gibbons MJ, Gugushe N, Boyd AJ et al (1999) Changes in the composition of the non-copepod zooplankton assemblage in St Helena Bay (southern Benguela ecosystem) during a six day drogue study. Mar Ecol 180:111–120
Gomes Fisher L (2012) Distribuição, biomassas e ecologia de Macrouridae (Teleostei, Gadiformes) no Talude Continental do Sul do Brasil, com ênfase em *Coelorinchus marinii* Hubbs 1934 e *Malacocephalus occidentalis* Goode & Bean 1885. Dissertation, Universidade Federal do Rio Grande
Ivanovic ML (2010) Alimentación del calamar *Illex argentinus* en la región patagónica durante el verano de los años 2006, 2007 y 2008. Rev Invest Desarr Pesq 20:51–63
Jónasdóttir S, Visser AW, Richardson K et al (2015) Seasonal copepod lipid pump promotes carbon sequestration in the deep North Atlantic. Proc Natl Acad Sci 112(39):12122–12126. https://doi.org/10.1073/pnas.1512110112
Kiørboe T (2008) Optimal swimming strategies in mate-searching pelagic copepods. Oecologia 155(1):179–192. https://doi.org/10.1007/s00442-007-0893-x
Kiørboe T (2011) How zooplankton feed: mechanisms, traits and trade-offs. Biol Rev 86(2):311–339. https://doi.org/10.1111/j.1469-185X.2010.00148.x
Kiørboe T, Sabatini M (1994) Reproductive and life cycle strategies in egg-carrying cyclopoid and free-spawning calanoid copepods. J Plankton Res 16(10):1353–1366. https://doi.org/10.1093/plankt/16.10.1353
Lutz VA, Segura V, Dogliotti AI et al (2010) Primary production in the Argentine sea during spring estimated by field and satellite models. J Plankton Res 32(2):181–195. https://doi.org/10.1093/plankt/fbp117
Marí N, Sánchez F (2002) Espectros tróficos específicos de varias especies de peces demersales de la región austral y sus variaciones anuales entre 1994 y 2000. Inf Téc Int INIDEP 88:19
Mauchline J (1998) The biology of calanoid copepods. Academic, London
Mauchline J, Fisher LR (1969) The biology of euphausiids. Adv Mar Biol 7:1–421
Mazzocchi MG, Paffenhöfer GA (1998) First observations on the biology of *Clausocalanus furcatus* (Copepoda, Calanoida). J Plankton Res 20(2):331–342
Mazzocchi MG, Paffenhöfer GA (1999) Swimming and feeding behaviour of the planktonic copepod *Clausocalanus furcatus*. J Plankton Res 21(8):1501–1518

Menni RC, Gosztonyi AE (1982) Benthic and semidemersal fish associations in the Argentine Sea. Studies Neotrop Fauna Environ 17:1–29. https://doi.org/10.1080/01650528209360599

Montú M (1977) Eufáusidos de la plataforma Argentina y adyacencias. I. Distribución estacional en el sector patagónico. Ecosur 4(8):187–225

Olivar MP, Contreras T, Hulley PA et al (2018) Variation in the diel vertical distributions of larvae and transforming stages of oceanic fishes across the tropical and equatorial Atlantic. Prog Oceanogr 160:83–100

Padovani LN (2013) Biodiversidad y ecología de los anfípodos hiperideos del Mar Argentino y aguas adyacentes: *Themisto gaudichaudii*, una especie clave. Dissertation, Universidad Nacional de Mar del Plata

Peralba À, Mazzocchi MG, Harris RP (2017) Niche separation and reproduction of *Clausocalanus* species (Copepoda, Calanoida) in the Atlantic Ocean. Prog Oceanogr 158:185–202. https://doi.org/10.1016/J.POCEAN.2016.08.002

Pillar SC, Stuart V, Barange M et al (1992) Community structure and trophic ecology of euphausiids in the Benguela ecosystem. Afr J Mar Sci 12(1):3935–3409. https://doi.org/10.2989/02577619209504714

Piola AR, Martínez Avellaneda N, Guerrero RA et al (2010) Malvinas-slope water intrusions on the northern Patagonia continental shelf. Ocean Sci 6(1):345–359. https://doi.org/10.5194/os-6-345-2010

Prenski LB, Sánchez F (1988) Estudio preliminar sobre asociaciones ícticas en la Zona Común de Pesca Argentino-Uruguaya. Publicaciones de la Comisión Técnica Mixta del Frente Marit 4:75–87

Ramírez FC (1970) Copépodos planctónicos del sector patagónico. Resultados de la Campaña Pesquería XI. Physis 79:473–476

Ramírez FC (1971a) Copépodos planctónicos de los sectores bonaerense y norpatagónico. Resultados de la Campaña Pesquería III. Rev Mus La Plata Sec Zool 11:73–94

Ramírez FC (1971b) Eufáusidos de algunos sectores del Atlántico Sudoccidental. Physis 30(81):385–405

Ramírez FC (1973) Eufáusidos de la expedición oceanográfica "Walther Herwig" 1966. Buenos Aires: Asociación Argentina de Ciencias Naturales. Physis 32(84):105–114

Ramírez FC (1981) Zooplancton y producción secundaria. Parte I. Distribución y variación estacional de los copépodos. In: Angelescu V (ed) Campañas de investigación pesquera realizadas en el Mar Argentino por los B/I "Shinkai Maru" y "Walter Herwig" y el B/P "Marburg", años 1978 y 1979. Resultados de la parte Argentina. Ser Contrib Inst Nac Invest Desarr Pesq, vol 383, p 202–212

Ramírez FC, Björnberg T (1981) Distribución horizontal por masas de agua de los copépodos más frecuentes y abundantes (indicadores hidrológicos) en el Atlántico Sudoccidental. In: Boltovskoy D (ed) Atlas del Zooplancton del Atlántico Sudoccidental y Métodos de Trabajo con el Zooplancton Marino. Publicaciones Especiales INIDEP, Mar del Plata, Argentina, p 595

Ramírez FC, Dato C (1983) Seasonal changes in population structure and gonadal development of three euphausiid species. Oceanol Acta 6(4):427–433

Ramírez FC, Sabatini ME (2000) The occurrence of Calanidae species in waters off Argentina. Hydrobiologia 439(1):21–42. https://doi.org/10.1023/A:1004193401931

Ramírez FC, Santos BA (1994) Análisis del zooplancton de la plataforma bonaerense en relación con algunas variables ambientales: campañas "transección" de 1987. Frente Marit 15:141–156

Ramírez FC, Viñas MD (1985) Hyperiid amphipods found in Argentine shelf waters. Physis 43(104):25–37

Ramírez, FC, Mianzan H, Santos B (1990) Synopsis on the reproductive biology and early life of *Engraulis anchoita*, and related environmental conditions in vertical distribution and mortality of *Engraulis anchoita* eggs and larvae. 10C Worksh. Rep., 65, Annex V:5–49

Rivas AL, Pisoni JP (2010) Identification, characteristics and seasonal evolution of surface thermal fronts in the Argentinean Continental Shelf. J Mar Syst 79(1–2):134–143. https://doi.org/10.1016/j.jmarsys.2009.07.008

Romero SI, Piola AR, Charo M et al (2006) Chlorophyll-*a* variability off Patagonia based on SeaWiFS data. J Geophys Res 111:CO5021. https://doi.org/10.1029/2005JC003244

Runge JA (1985) Relationship of egg production of *Calanus pacificus* to seasonal changes in phytoplankton availability in Puget Sound, Washington. Limnol Oceanogr 30(2):382–396

Sabatini ME (2008) Life history trends of copepods *Drepanopus forcipatus* (Clausocalanidae) and *Calanus australis* (Calanidae) in the southern Patagonian shelf (SW Atlantic). J Plankton Res 30:981–996. https://doi.org/10.1093/plankt/fbn062

Sabatini ME, Akselman R, Reta R et al (2012) Spring plankton communities in the southern Patagonian shelf: hydrography, mesozooplankton patterns and trophic relationships. J Mar Syst 94:33–51. https://doi.org/10.1016/j.jmarsys.2011.10.007

Sabatini ME, Reta R, Lutz VA et al (2016) Influence of oceanographic features on the spatial and seasonal patterns of mesozooplankton in the southern Patagonian shelf (Argentina, SW Atlantic). J Mar Syst 157:20–38. https://doi.org/10.1016/J.JMARSYS.2015.12.006

Sánchez RP, Ciechomski JD (1995) Spawning and nursery grounds of pelagic fish species in the sea-shelf off Argentina and adjacent areas. Sci Mar 59(3–4):455–478

Saunders RA, Collins MA, Schreeve R et al (2018) Seasonal variation in the predatory impact of myctophids on zooplankton in the Scotia Sea (Southern Ocean). Progr Ocean 168:123–144. https://doi.org/10.1016/j.pocean.2018.09.017

Schnack-Schiel SB, Niehoff B, Hagen W et al (2008) Population dynamics and life strategies of *Rhincalanus nasutus* (Copepoda) at the onset of the spring bloom in the Gulf of Aqaba (Red Sea). J Plankton Res 30(6):655–672. https://doi.org/10.1093/plankt/fbn029

Segura V, Lutz VA, Dogliotti A et al (2013) Phytoplankton types and primary production in the Argentine Sea. Mar Ecol Prog Ser 491:15–31. https://doi.org/10.3354/meps10461

Severo A, Cepeda GD, Acha EM (2024) The effects of the Patagonian shelf-break front on copepod abundance, biodiversity, and assemblages. J Mar Syst 103921:103921. https://doi.org/10.1016/j.jmarsys.2023.103921

Tarling GA, Ward P, Sheader M et al (1995) Distribution patterns of macrozooplankton assemblages in the Southwest Atlantic. Mar Ecol Prog Ser 120(1/3):29–40

Turner JT (2004) The importance of small planktonic copepods and their roles in pelagic marine food webs. Zool Stud 43(2):255–266

Verona C, Carreto JI, Ramirez F et al (1972) Plancton y condiciones ecológicas en las aguas de la plataforma bonaerense frente a Mar del Plata. Campaña "Transeccion I". Doc tec Prel Proy Des pesq 24:1–35

Weiss G (1974) Hallazgo y descripción de larvas de la polaca *Micromesistius australis* en aguas del sector patagónico argentino. Physis 33(87):537–542

Weiss G (1975) Hallazgo, descripción y distribución de las postlarvas del bacalao criollo, *Salilota australis* y del pez sable, *Lepidotus caudatus* en aguas de la plataforma Argentina. Physis 34(89):319–325

Yamaguchi A, Ikeda T, Hirakawa K (1999) Diel vertical migration, population structure and life cycle of the copepod *Scolecithricella minor* (Calanoida: Scolecitrichidae) in Toyama Bay, southern Japan Sea. Plankton Biol Ecol 46(1):54–61

Zeidler W, De Broyer C (2009) Catalogue of the Hyperiidean Amphipoda (Crustacea) of the Southern Ocean with distribution and ecological data. Bull Séances Inst Roy Colon Belge 79(1):1–104. https://doi.org/10.15468/99f6q0

Chapter 5
Nekton in the Patagonian Shelf-Break Front: Fishes and Squids

Daniela Alemany, Mauro Belleggia, Gabriel Blanco, Mariana Deli Antoni, Marcela Ivanovic, Nicolás Prandoni, Natalia Ruocco, María Luz Torres Alberto, and Anabela Zavatteri

Abstract Marine fronts affect the life of marine species at all trophic levels, determining distributional patterns. However, at higher levels such as nekton, behavior and swimming abilities become increasingly important, and the association with fronts becomes more intricate. The aim of this chapter is to provide a synthesis about nektonic species (fish and squids) at the Patagonian Shelf-break Front (PSBF), describing and compiling information on general aspects of nekton with emphasis on key species. A large number of nektonic species make use of the PSBF; more than 135 taxa of nekton are reported at this front between 1978 and 2019, during several scientific cruises and catches from commercial fleets. The PSBF is a complex adaptive system promoting suitable conditions for nektonic species and assemblages of high functional and ecological relevance. We highlight the PSBF as a key area due to its nekton biodiversity and the presence of threatened species, particularly cartilaginous fish. Moreover, key biological processes for the life cycle of fish and squids (feeding, reproduction, and/or migration) take place at the PSBF, improv-

Supplementary Information The online version contains supplementary material available at https://doi.org/10.1007/978-3-031-71190-9_5.

D. Alemany (✉) · M. Belleggia · M. L. Torres Alberto
Instituto de Investigaciones Marinas y Costeras (IIMyC, UNMdP-CONICET), Facultad de Ciencias Exactas y Naturales, Universidad Nacional de Mar del Plata (UNMdP), Consejo Nacional de Investigaciones Científicas y Técnicas (CONICET), Mar del Plata, Argentina

Instituto Nacional de Investigación y Desarrollo Pesquero (INIDEP), Mar del Plata, Argentina
e-mail: dalemany@inidep.edu.ar

G. Blanco · M. Ivanovic · N. Prandoni · N. Ruocco · A. Zavatteri
Instituto Nacional de Investigación y Desarrollo Pesquero (INIDEP), Mar del Plata, Argentina

M. Deli Antoni
Instituto de Investigaciones Marinas y Costeras (IIMyC, UNMdP-CONICET), Facultad de Ciencias Exactas y Naturales, Universidad Nacional de Mar del Plata (UNMdP), Consejo Nacional de Investigaciones Científicas y Técnicas (CONICET), Mar del Plata, Argentina

E. M. Acha et al. (eds.), *The Patagonian Shelfbreak Front*, Aquatic Ecology Series 13, https://doi.org/10.1007/978-3-031-71190-9_5

ing survival opportunities. Nektonic species identified at the PSBF, due in part to complex trophic relationships, large home ranges, as well as extensive feeding and spawning migrations, connect the frontal system with other ecosystems, redistributing matter and energy.

Acronyms and Abbreviations

BMC	Brazil–Malvinas Confluence
BNPS	Bonaerensis-northpatagonic stock
FAO	Food and Agriculture Organization of the United Nations
FTMH	Freezer-trawlers mainly targeting *Merluccius hubbsi*
INIDEP	Instituto Nacional de Investigación y Desarrollo Pesquero
ITMH	Ice-trawlers mainly targeting *Merluccius hubbsi*
IUCN	International Union for Conservation of Nature
MDS	Nonmetric multidimensional scaling
OB	On-board observers' program
PSBF	Patagonian Shelf-break Front
RV	Research vessel
SL	Standard length
SPS	South Patagonian stock

5.1 Introduction

The Patagonian Shelf-break Front (PSBF) is an area of high biological relevance, aggregating low trophic level organisms such as phytoplankton (Chap. 3) and zooplankton (Chap. 4), and in turn, this concentration of food attracts species of higher trophic levels. The influence of fronts on phytoplankton and zooplankton is well documented, but uncertainties are particularly large for nekton (Bakun 2006). Several pelagic and demersal fish are tightly coupled with transition zones or fronts (Longhurst 1998), and also squid species (early stages and adults) have evolved to exploit productive regions, migrating between spawning and feeding grounds (Rodhouse et al. 1992). Marine fronts affect the distribution of marine species at all trophic levels; however, at higher levels, behavior and swimming abilities become increasingly important (Olson 2002), determining distributional patterns. Thus, the association with fronts becomes more intricate as we move up the food web to higher trophic levels, such as nekton.

Nekton comprises organisms able to swim and move independently of currents or wind (i.e., free-swimming animals that can resist a strong current, contrary to plankton, which are passive organisms at the mercy of water masses; Aleyev 2012). They have demersal or pelagic habits and move through the water column, while others live close to the bottom. Nektonic species are heterotrophic, inhabiting coastal and oceanic environments, and many of them are important resources for marine fisheries

(FAO 2020). It is proposed that nekton may be linked to frontal zones in a complex way through their life-history strategies. In that sense, it is recognized that frontal systems play a significant role in the reproduction, foraging, and migration of nekton (Olson 2002); however, this relationship is species-specific (affecting different species differently) and also depends on life stage. As fronts are regions with high food availability and retentive properties, several nektonic species take advantage of them as spawning or feeding grounds (Bakun 2006). Particularly in the Argentine Sea, inhabiting fish and squid are highly mobile organisms with large home ranges that swim and cross frontal regions, exploring different environments. Given their complex behavior and migratory patterns, it is sometimes challenging to assess the association between them and the PSBF. Although nekton is represented by animals belonging to 12 classes (Aleyev 2012), this chapter focuses on 3 of them: Cephalopoda (squids), Osteichthyes (bony fishes), and Chondrichthyes (cartilaginous fishes).

Marine ecology and biogeography often complement each other to describe patterns of species distribution and diversity. Historically, based on faunal composition, several studies focused on the marine biogeography of the Southwest Atlantic (e.g., Balech and Ehrlich 2008; Cousseau et al. 2020; Sabadin et al. 2020). In that sense, the PSBF is located within one of the two defined biogeographic provinces: the Magellanic Province (Boschi 2000; Balech and Ehrlich 2008), characterized by cold-water species associated with the Malvinas Current. In relation to demersal fisheries (fish and squid species) in the Argentine Sea, Angelescu and Prenski (1987) defined a series of regional fishery assemblages, and the "slope deep waters assemblage" coincides with the PSBF location. But it was only recently that a biogeographic region encompassing the PSBF was described. Based on the distribution and main assemblages of cartilaginous fishes of the Southwest Atlantic, three biogeographic provinces were identified (Sabadin et al. 2020). Two of them are on the continental shelf and correspond with the Argentinean and Magellanic Provinces (Balech and Ehrlich 2008), and a third and new one called the Patagonian Slope. Within the Argentinean Province, the Patagonian Shelf-break ecoregion is described for the first time, restricted to the upper slope in two sections: one between 35 °S and 39 °S and the other longitudinally very narrow, south of 41 °S to 49 °S. In the new Patagonian Slope Province, the two ecoregions are the Southern Patagonian Slope (between 46 °S and 55 °S) and the Northern Patagonian Slope (between 38 °S and 42 °S; Sabadin et al. 2020). The Patagonian Shelf-break Front (PSBF) includes these last three ecoregions.

Within this framework, in the following sections (Biodiversity and Assemblages, Reproduction, Feeding Ecology, Migrations, and Life Cycles), we describe and compile information involving fish and squids at the PSBF, focusing on general aspects of nekton with emphasis on species of ecological or fishing relevance and/or for which information is available.

5.2 Biodiversity and Assemblages

Fish assemblages in continental slopes or shelf-breaks (in contrast to coastal fishes) show a more homogeneous distribution of species richness; however, departures from the general pattern could occur within biogeographical regions (Macpherson et al. 2009). In the Atlantic Ocean, southwards of approximately 40 °S, there are poleward decreases in species richness of pelagic organisms (Angel 1997), which coincide with well-defined transition zones (Longhurst 1998). In terms of nekton biodiversity, the Patagonian Shelf Large Marine Ecosystem has diverse frontal types exerting different effects on fish diversity and assemblages' composition (Alemany et al. 2009). The diversity–productivity relationship is an open debate, and several forms of this relationship have been proposed, sustaining that high productivity can lead to either an increase or decrease in species richness, or a combination of both (e.g., a hump-shaped distribution).

The species-energy hypothesis (Hutchinson 1959) proposes that diversity is driven by food supply, suggesting that the amount of available energy sets limits to the richness of the system (Bridges et al. 2022). In that sense, biodiversity usually peaks at fronts, where primary and secondary production are high. However, it has been reported that at permanent frontal features, such as the PSBF, fish diversity decreases when the nutrient availability is high (Alemany et al. 2009). This is potentially because, when food resources increase, fewer species become dominant. In the Southwest Atlantic, the diversity of chondrichthyans is heterogeneous; in Argentine waters, a general decreasing trend from the coast to the shelf-break was reported, but in the specific case of batoids (e.g., skates or rays), very high species richness was detected at the PSBF (Sabadin et al. 2020). Another study, although based on a more limited data set (Lucífora et al. 2012), reported several hotspots of chondrichthyans diversity along the PSBF. In that sense, Alemany et al. (2022) also found that, although nekton diversity has a heterogeneous spatial distribution, there are some areas in the PSBF of very high diversity (hotspots). Regarding squids, Rodhouse et al. (1992) reported high diversity associated with large-scale oceanographic features in the Southwestern Atlantic Ocean. In a recent comprehensive report (1978–2019, 1445 fishing hauls), encompassing part of the PSBF, nine squid species were recorded, some of which are of high commercial importance (Alemany et al. 2022).

In this chapter, to describe nekton diversity along the PSBF between 37 °S and 55 °S, we compiled information from several research cruises (Table 5.1) performed between 1978 and 2019, extracting those fishing hauls that overlapped the PSBF.

The spatial distribution of the 2279 fishing hauls analyzed to elaborate a nekton species list for the PSBF (see Appendix) is shown in Fig. 5.1. From these hauls, 140 taxa have been identified along the PSBF: 16 orders of bony fish grouped into 45 families (104 species), 4 orders of cartilaginous fish grouped into 8 families (23 species), 2 orders of squids grouped into 8 families (12 species), and 1 order of myxini (1 species). Furthermore, to evaluate fish assemblages at the PSBF, a multivariate analysis (PRIMER V6 routines) was performed on the Walther Herwig

Table 5.1 Research cruises performed between 1978 and 2019 and the total fishing hauls analyzed at the PSBF

Research cruise	Month	Year	Fishing hauls
Walther Herwig	Jan-Dec	1978	95
Evrika	August	1988	7
Kaiyo Maru	Aug-Sep	1989	6
INIDEP—squid	Febr and April	1994–2019	286
Kaiyo Maru	Sep-Oct	2005	12
INIDEP OB ITMH	Jan-Dec	2014–2018	1539
INIDEP OB FTMH	Jan-Dec	2018	334
		Total	**2279**

INIDEP Instituto Nacional de Investigación y Desarrollo Pesquero, *OB* on-board observers program, *ITMH* Ice-trawlers mainly targeting *Merluccius hubbsi*; *FTMH*: Freezer-trawlers mainly targeting *Merluccius hubbsi*

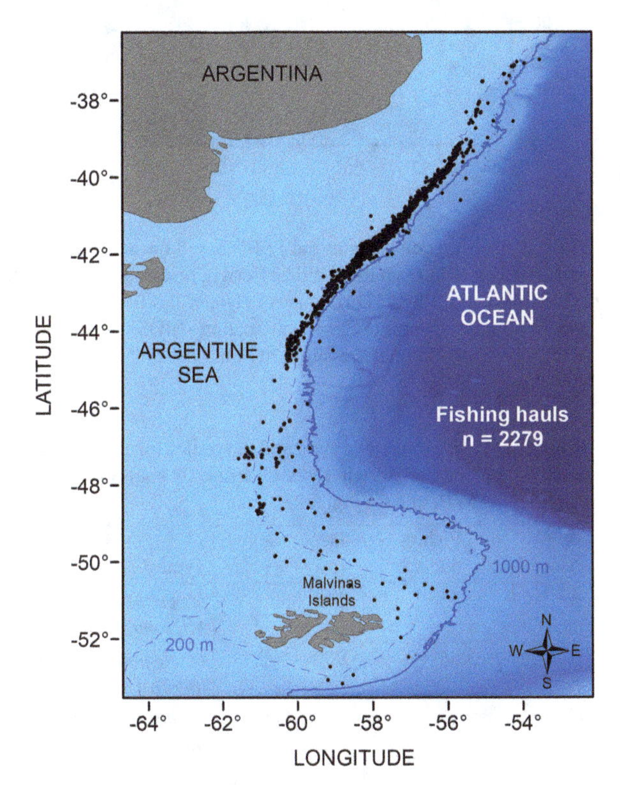

Fig. 5.1 Spatial distribution of the 2279 fishing hauls analyzed at the PSBF between 1978 and 2019, during several scientific cruises and catches from commercial fleets (Table 5.1)

cruise since its hauls ($n = 95$) covered almost the entire latitudinal extension of the PSBF. In that sense, three main assemblages were recognized, defined as A, B, and C groups (Fig. 5.2), whose differences from each other exceeded 75%. Their depth range and the most representative/typical fish species are summarized in Table 5.2. There was no latitudinal difference in the fish assemblages along the PSBF, but a

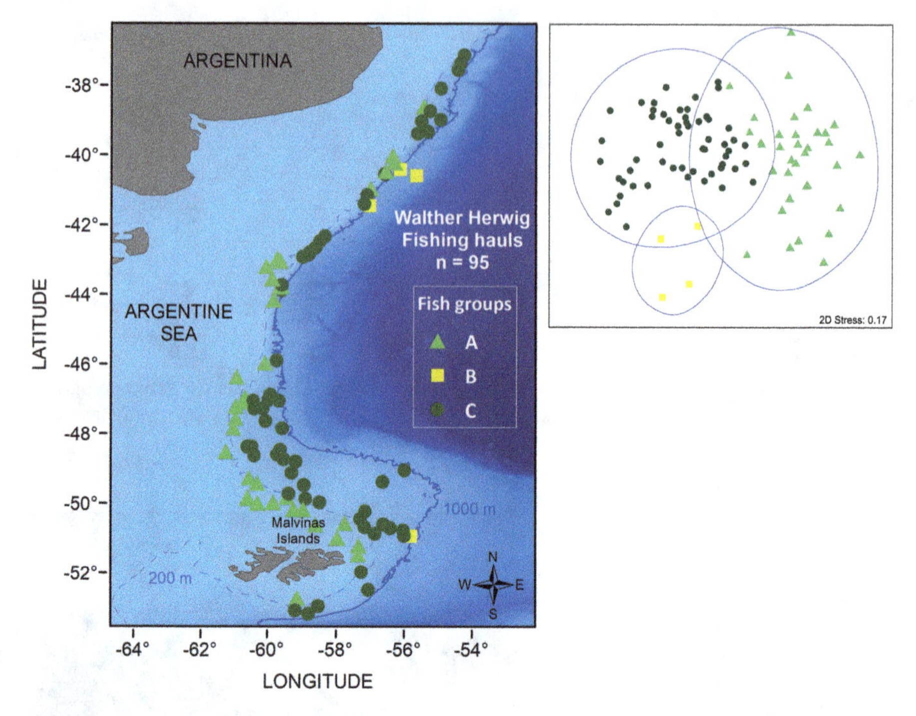

Fig. 5.2 Fish assemblages at the PSBF based on multivariate analysis of fishing hauls (n = 95) performed on data from the Walther Herwig research cruise, covering almost the entire latitudinal extension of the PSBF. Differences between the three main assemblages exceeded 75%. Right panel: Nonmetric multidimensional scaling (MDS) ordination of fish assemblages, constructed with Bray–Curtis abundance similarity matrices on fourth-root transformed data

Table 5.2 Depth range and typical fish species for the three assemblages identified at the PSBF, based on the Walther Herwig research cruise (N hauls = 95). Three main groups (assemblages) are identified as A, B, and C (Fig. 5.2)

Assemblage	Fishing hauls	Depth range (m)	Typical species
A	35	100–330	*M. hubbsi, P. ramsayi, G. nicholsi, S. australis, Micromesistius australis*
B	56	950–1500	*D. eleginoides, A. rostrata, B. griseocauda, M. holotrachys, B. brachyurops*
C	4	300–1000	*M. holotrachys, D. eleginoides, Micromesistius australis, P. ramsayi*

differentiation based on depth seems clear. In accordance, a recent study found that the structure of fish assemblages depends on depth (Cousseau et al. 2020). In that sense, it is widely recognized that marine fauna composition changes with depth, and shelf-breaks are among the steepest environmental gradients on Earth (Zintzen et al. 2012). In agreement with Bridges et al. (2022), species richness may not vary with depth, but there would be a species turnover (species replacement) at different depths, leading to different assemblages.

5.2.1 Bony Fishes

The most abundant demersal resource of the Argentine continental shelf is the Argentine hake, *Merluccius hubbsi* (Fig. 5.3). It is widely distributed across the Southwestern Atlantic Ocean from Brazil (21° 30′S) to Southern Patagonia (55 °S) (Cousseau and Perrotta 2013). It inhabits depths ranging from 50 to 500 m, with the highest abundances found between 100 and 200 m depth (Bezzi et al. 1995). The

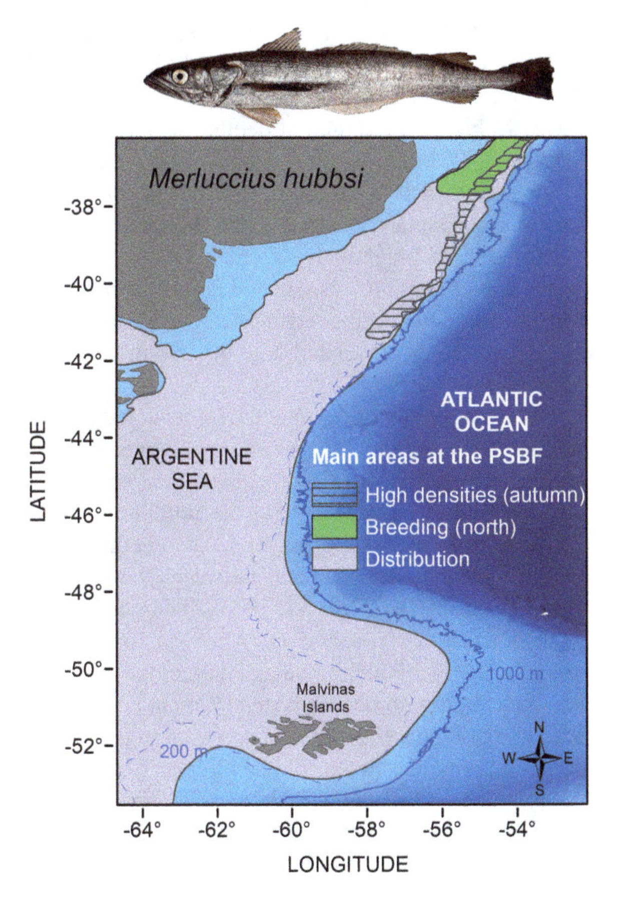

Fig. 5.3 Distribution of the most abundant demersal bony fish on the Argentine continental shelf, the Argentine hake (*Merluccius hubbsi*), and the main areas (breeding and high densities of the northern stock) at the PSBF. (Adapted from Allega et al. 2019. Photo credits: Cecilia Ravalli (INIDEP Photography laboratory))

distribution of 1- and 2-year-old hakes is restricted to 50–100 m depth, while the older age classes gradually disperse further offshore, reaching the shelf-break surroundings (Renzi et al. 2003). High abundances of the Argentine hake can often be found in the PSBF, mainly from autumn to spring, with the highest biomasses occurring between 44° 30′ and 46° 30′ (Irusta et al. 2000; Irusta et al. 2010).

South of 48 °S, the main Southern demersal fish species living in the Southwest Atlantic are the long tail hake or hoki (*Macruronus magellanicus*), the southern blue whiting (*Micromesistius australis*), the red cod (*Salilota australis*), the Patagonian toothfish (*Dissostichus eleginoides*), the southern hake (*Merluccius australis*), and the grenadiers (genders *Macrourus* and *Coelorinchus*). Their distribution is on the continental shelf and shelf-break, closely related to the cold waters of subantarctic origin (Angelescu and Gneri 1960) and to the PSBF.

The long tail hake is a demersal-pelagic species with a wide distribution around South America. In the Southwest Atlantic Ocean, it inhabits mainly the continental shelf and shelf-break between 36 °S and 56 °S (Angelescu et al. 1958) and from 20 to 600 m depths (Angelescu and Prenski 1987; Wöhler 1987; Chesheva 1995; Giussi et al. 2016). The southern blue whiting is a species of demersal-pelagic behavior (Wöhler et al. 2004a) that is distributed in the Southern cone of South America (Bellisio et al. 1979; Perrota 1982) and is associated with the continental shelf and shelf-break of the Southern region of Argentina. This species is located between 37 °S and 55 °S latitude in the Argentine shelf (Perrota 1982; Cousseau and Perrotta 2000; Wöhler et al. 2004a) at depths between 100 and 800 m (Bellisio et al. 1979; Perrota 1982; Madirolas 1999; Wöhler et al. 2004a).

The red cod is a gadiform fish with demersal habits, which is distributed in waters of subantarctic origin from 38 °S to 56 °S in the Atlantic Ocean (Perez Comas 1980; Wöhler et al. 2004b). Its bathymetric distribution extends between 30 and 900 m of depth (Cousseau and Perrota 2000). The highest concentrations of this species are found on the Argentine continental shelf, although there are records of significant yields associated with the 200 m isobath (Bellisio et al. 1979), coinciding with the PSBF.

The Patagonian toothfish is a benthic demersal species that has a wide distribution in the Southern Hemisphere (Oyarzún et al. 1988). In the Atlantic Ocean, it is related to the Malvinas Current, extending between 36° 30′ and 55 °S (Cassia and Perrotta 1996; Cousseau and Perrota 2000; Prenski and Almeyda 2000). The highest concentrations of this species are located in the northern PSBF, south and northeast of the Malvinas Islands, and between Namuncurá/Burdwood Bank and Staten Island (Prenski and Almeyda 2000). It can be found in a broad bathymetric range from 80 to 2500 m (Prenski and Almeyda 2000; Collins et al. 2010), mainly in submarine canyons (which are abundant across the shelf-break). Larger/older animals are associated with deeper waters (Cassia and Perrotta 1996), as evidenced by the presence of highest densities of juveniles in depths shallower than 600 m and adults commonly found at depths of 700–2000 m (Agnew et al. 1999; Prenski and Almeyda 2000; Barrera-Oro et al. 2005).

The southern hake is a gadiform species with demersal habits and is widely distributed around Southern South America (Bezzi et al. 1995). In particular, in the

Atlantic Ocean, its distribution follows the cold waters of the Malvinas Current between 50 °S and 56 °S (Otero and Simonazzi 1980) and depths comprised in the 100–1000 m range (García de la Rosa et al. 1997a; Wöhler 1987; Giussi et al. 2016). The species mainly inhabits the Argentinian shelf (Bianchi et al. 1982), but its presence has also been reported in the PSBF, probably related to prey availability in the area (Marí and Sánchez 2002; Sánchez and Marí 2005).

The *Macrouridae* family, commonly named as grenadiers, is very diverse, and in the Southwest Atlantic Ocean, it is represented mainly by two genera, *Macrourus* and *Coelorinchus* (Bellisio et al. 1979), which are distributed between 36 °S and 55 °S associated with the deep waters of the Malvinas Current, along the PSBF (Bellisio, et al. 1979; Cousseau and Perrota, 2000; Giussi et al. 2010). The predominant *Macrourus* species are *M. carinatus* and *M. holotrachys*, commonly referred to as large grenadiers, while the genus *Coelorinchus* is mainly represented by *C. fasciatus*, called the little grenadier (Cousseau and Perrota 2000). Species of the genus *Macrourus* inhabit waters between 300 and more than 1700 m of depth, with maximum abundance varying according to species (Cohen et al. 1990; Laptikhovsky et al. 2008; Giussi et al. 2010). *Coelorinchus fasciatus* is found at depths from 100 to 900 m (Laptikhovsky et al. 2008), with its highest concentration found between 52 °S and 55 °S, northwest of the Namuncurá/Burdwood Bank, and between 200 and 600 m (Laptikhovsky et al. 2008; Giussi et al. 2010).

Two species of the family *Ophidiidae*, the pink cuskeel *Genypterus blacodes* and the Brazilian cuskeel *G. brasiliensis*, cohabit in the Southwest Atlantic Ocean (Díaz de Astarloa and Figueroa 1993; Mabragaña et al. 2011). The former is an important commercial species, traditionally caught by bottom trawlers as by-catch of the Argentine hake fisheries and by the bottom long-line fleet in the Argentine Sea (Bertolotti et al. 1996; Villarino 1997, 1998; Nielsen et al. 1999). *Genypterus blacodes* has a wider geographic distribution, over the South Atlantic continental shelf and along the PSBF from 34 °S to 55 °S, at depths between 45 and 350 m (Bertolotti et al. 1996; Nielsen et al. 1999; Cordo 2004; Del Río-Iglesias et al. 2009; Cousseau and Rosso 2019). The highest densities of the pink cuskeel have been found between 40 °S and 48 °S, mostly over the PSBF region (Renzi 1986; Villarino 1998; Cordo 2004).

The Brazilian cuskeel is one of the most valuable species for the bottom trawl and longline fisheries of the outer shelf and shelf-break off Southern Brazil (Haimovici et al. 2008; Alvarez Perez et al. 2009). In the northern PSBF region, both species share their habitat, with some degree of segregation by temperature; *G. brasiliensis* was found up to 38 °S in warmer waters, while *G. blacodes* was distributed southwards, in colder ones (Haimovici 1997; Viera 2011). Despite the broad distribution of *G. blacodes*, seasonal and bathymetric geographic variations have been reported. During autumn and winter, two high-concentration areas have been registered: one located north of 43 °S and another one to the south, between 46 °S and 48 °S. In both regions, the individual sizes increased with depth, and the greatest densities were found mostly over the PSBF (Renzi 1986; Villarino 1998; Cordo 2004). Arkhipkin et al. (2013a) also pointed out the northwestern shelf-break of the Malvinas Islands as an area of high density for *G. blacodes*.

Myctophidae is one of the dominant and most widespread fish families of the mesopelagic zone (Gjøsaeter and Kawaguchi 1980; Salvanes and Kristoffersen 2001), distributed in all oceans except the Arctic waters (Catul et al. 2011). Its ubiquity in the world's oceans is associated with an enormous abundance, which is suggested to be underestimated (Lam and Pauly 2005; Kaartvedt et al. 2012), and a great diversity, including at least 248 species in 33 genera (Nelson 2016). These are small-sized fish (2–30 cm SL) belonging to micronekton and referred to as lanternfish due to a variety of bioluminescent organs known as "photophores," whose arrangement pattern is species-specific (Carpenter 2002; Catul et al. 2011). Myctophids are typically oceanic, offshore fishes; however, the family also includes species known as pseudoceanic, associated with continental shelf-break regions and in the neighborhood of oceanic islands (Hulley 1991). Although less abundant than at lower latitudes, lanternfishes are still numerous in the Southwest Atlantic Ocean. Around 68 myctophid species have been recorded south of the Subtropical Front (Duhamel et al. 2014), almost 40 of which have been recognized in the Southwest Atlantic Ocean between 33 °S and 56 °S (Cousseau and Rosso 2019; Figueroa 2019). A heterogeneous species composition and several pattern distributions (related to the complex hydrography) have been described (Krefft 1976; Hulley 1981; Konstantinova et al. 1994; Figueroa et al. 1998).

Information about the composition and biology of lanternfish species in shelf and PSBF waters is scarce. Konstantinova et al. (1994) recorded three species of subantarctic myctophids on the PSBF from a few sampling stations: *Protomyctophum choriodon*, *Gymnoscopelus piabilis*, and *G. nicholsi*. A most comprehensive study carried out in 1966 by the research vessel (RV) "Walther Herwig" shows a latitudinal pattern in species composition along the PSBF, associated with different water masses (Angelescu and Cousseau 1969). The northern sector of the PSBF, near the Brazil–Malvinas Confluence (BMC), shows high species richness and a remarkable mixture of tropical, subtropical, and confluence myctophids. The abundance is dominated by species of the genus *Diaphus* and *Ceratoscopelus* sp., many of which occur predominantly in tropical–subtropical water masses (Angelescu and Cousseau 1969; Hulley 1981; Konstantinova et al. 1994; Figueroa et al. 1998). Furthermore, considering larval occurrence, more lanternfish species were reported in the northern PSBF waters, which probably utilize the BMC region as a breeding site (*Hygophum hanseni*, *Lampanyctus australis*, *Lepidophanes guentheri*, *Myctophum* spp., *Notoscopelus* spp., and *Symbolophorus* spp.) (Acha et al. 2018).

South of the BMC and until 45 °S, several species and genera of *Myctophidae* were reported, such as *Protomyctophum normani*, *Hygophum hanseni*, *Lampadena* sp., and *Lampanyctus australis* (Angelescu and Cousseau 1969; research cruises Table 5.1). Among these, many are typically associated with the BMC and have a circumglobal distribution in the Southern Hemisphere (Hulley 1991; Konstantinova et al. 1994; Figueroa et al. 1998). Common species south of 45 °S are *Gymnoscopelus bolini*, *G. nicholsi*, *Electrona subaspera*, *Electrona* sp., and *Symbolophorus boops* (Angelescu and Cousseau 1969; Hulley 1981; Figueroa et al. 1998).

In oceanic waters and other worldwide slope areas, the mesopelagic fish community showed a clear stratification, with some species occupying distinct depth layers (e.g., Hulley 1992; Collins et al. 2012; Braga et al. 2014). It has been postulated that temperature is an important determinant of lanternfish distribution (Brandt 1983). According to Hulley (1992), not only the water column temperature but also the depth requirements of the individual species shape the myctophids' down-slope structure. A bathymetric zonation of the lanternfish community was also observed in the PSBF, particularly in the northern area (Angelescu and Cousseau 1969). In the upper layers (150–200 m), several species of the genus *Diaphus* and *Symbolophorus boops* are common. At greater depths, up to 400 m, species of *Diaphus*, *Gymnoscopelus*, *Lampadena*, *Lampanyctus*, *Hygophum*, and *Myctophum* have been found. With the increasing depth and the decrease in temperature (less than 6 °C), the diversity and abundance of lanternfish decline; only certain species (e.g., *Notoscopelus* spp. and *Ceratoscopelus* spp.) reach greater depths (1000–2000 m; Angelescu and Cousseau 1969; Hulley 1991).

5.2.2 Cartilaginous Fishes

The shelf-break region is characterized by a relatively high diversity of Chondrichthyes. Among the most representative chondrichthyans of the PSBF are several shark species (*Lamna nasus*, *Squalus acanthias*, *Schroederichthys bivius*, *Etmopterus* spp., *Somniosus antarcticus*). Also, 15 skate species belonging to 5 genera: *Dipturus* (*D. trachydermus*), *Zearaja* (*Z. brevicaudata*), *Bathyraja* (*B. brachyurops*, *B. macloviana*, *B. albomaculata*, *B. griseocauda*, *B. multispinis*, *B. cousseauae*, *B. scaphiops*, *B. magellanica*, *B. papilionifera*, and *B. meridionalis*), *Psammobatis* (*P. normani* and *P. rudis*), and *Amblyraja* (*A. doellojuradoi*). The high number of species found is consistent with the spatial distribution of chondrichthyan richness, which is higher on marine fronts than elsewhere in the Southwest Atlantic Ocean (Colonello et al. 2014; Sabadin et al. 2020).

Chondrichthyes are among the most threatened marine fish due to their life-history traits characterized by slow growth, late maturity, long life spans, and low fecundity (Holden 1974; Camhi et al. 1998; Smith et al. 1998), and particularly the deep-sea chondrichthyans would be the most vulnerable ones (García et al. 2008). In general, cartilaginous fishes are characterized by low biological productivity, which implies that they can only sustain low removal rates (Anderson 1990; Hoenig and Gruber 1990; Barker and Schluessel 2005).

The porbeagle shark *Lamna nasus* is probably the most important epi- and mesopelagic/neritic chondrichthyan of the PSBF region, found at depths of up to 1300 m (Waessle and Cortés 2011; Cortés and Waessle 2017; Skomal et al. 2021). The narrow-mouthed catshark *Schroederichthys bivius* is distributed from Brazil to Argentina and Chile (Bornatowski et al. 2014; Weigmann 2016). In Argentina, it occurs at depths from 50 to 350 m (Sánchez et al. 2009), but the highest abundances and frequencies of occurrence are between 50 and 200 m depth, particularly close to the 100 m isobath (Colonello et al. 2020).

The high degree of biodiversity and endemism of skates is given by their relatively conservative dorsoventrally flattened body morphology and apparent restrictive habitat preference (e.g., soft bottom substrates; Ebert and Compagno 2007). Skates are primarily marine benthic dwellers found from the intertidal down to 3000 m depth (Ebert and Compagno 2007). In the southwestern Atlantic, particularly at the eastern Patagonian Shelf and PSBF, it has been reported a high biodiversity and abundance of skates (Arkhipkin et al. 2012).

5.2.3 Squids

Seven families of the Order *Oegopsida* (*Brachioteuthidae*, *Chiroteuthidae*, *Cranchiidae*, *Gonatidae*, *Histioteuthidae*, *Ommastrephidae*, and *Onychoteuthidae*) and one family of the Order *Myopsida* (*Loliginidae*) have been identified in the PSBF (see Appendix). Squid species in the PSBF have a relevant role since they occupy intermediate levels in the marine food web, feeding on plankton and small fish, and being an important food resource for higher trophic levels such as fish, seabirds, and marine mammals. Two species are commercially exploited in the region: *Illex argentinus* (Fam. *Ommastrephidae*) and *Doryteuthis gahi* (Fam. *Loliginidae*).

The Argentine shortfin squid (*Illex argentinus*) is a neritic-oceanic species broadly distributed in the Southwest Atlantic Ocean from southern Brazil (23 °S) to southern Argentina (54 °S) at depths between 50 and 1000 m (Castellanos 1960, 1964; Hatanaka 1986; Haimovici and Álvarez Pérez 1990). It is by far the most abundant and the first most fished squid in the PSBF. Its range spans the continental shelf, the shelf-break, and adjacent oceanic areas but is limited to the region influenced by temperate-cold waters of subantarctic origin; however, it is mostly found between 35 °S and 52 °S at depths between 100 and 400 m (Brunetti 1988; Brunetti et al. 1998a, 1998b), coinciding with the PSBF (Fig. 5.4). The population dynamics are strongly related to the variability of the Malvinas Current and the BMC position (Haimovici et al. 1998; Torres Alberto et al. 2021).

The Patagonian longfin squid (*Doryteuthis gahi*) inhabits cold waters in the Eastern Pacific and Southern Atlantic (Jereb and Roper 2010). In the Southwest Atlantic Ocean, it is found in the Patagonian coastal region up to 42 °S and in the PSBF up to 36–38 °S, following the northward flow of the Malvinas Current (Pineda et al. 1998; Arkhipkin et al. 2004). It is considered a key species in the region, playing an important role in the marine ecosystem and for fisheries (Jereb and Roper 2010; Arkhipkin et al. 2013b).

Martialia hyadesi, also known as the Sevenstar flying squid, is an Antarctic circumpolar species, although the most frequent area of appearance is in the Southwest Atlantic Ocean from the Scotia Sea to the PSBF (Nigmatullin 1989). The latter is the only region where commercial catches have been reported in years with exceptional intrusion of cold water from the Malvinas Current during the autumn–winter months (Rodhouse 1991; Ivanovic et al. 1998). This squid represents a less-known

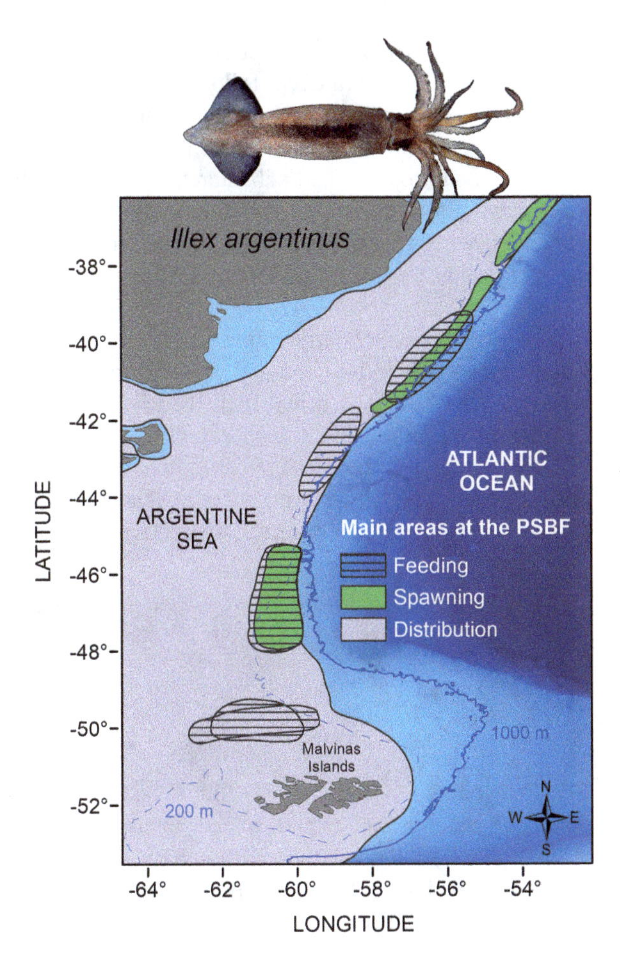

Fig. 5.4 Main areas (distribution, feeding, and spawning) of the most abundant squid species of the Argentine continental shelf, the Argentine shortfin squid (*Illex argentinus*), at the PSBF. (Adapted from Allega et al. 2019. Photo credits: Cecilia Ravalli (INIDEP Photography laboratory))

potential resource for the commercial squid-catching fleet. As mentioned, in some cold years, a considerable bycatch of *M. hyadesi* was observed in the *I. argentinus* jigging fishery in the PSBF, between 38 °S and 50 °S, from March until June (Rodhouse 1991; Ivanovic et al. 1998). However, further studies are necessary to understand the real fishery potential of this squid.

The Greater hooked squid (*Onykia ingens*) is a bathypelagic circumpolar species found in subantarctic waters. In the Southwest Atlantic Ocean, it can be found off the Patagonian continental shelf and in the PSBF, mostly at depths between 300 and 1450 m (Kubodera et al. 1998; Jereb and Roper 2010).

Gonatus antarcticus is a squid of the family *Gonatidae* with a circumpolar distribution in the Southern Hemisphere, mainly in Antarctic waters south of 40 °S (Jereb and Roper 2010). It may play an important ecological role, serving as prey for numerous seabirds breeding in the Malvinas Islands, such as penguins and albatrosses (Thompson 1994), as well as subantarctic fish and marine mammals (Jereb and Roper 2010).

The squid *Brachioteuthis riisei* is a cosmopolitan species that can be found in midwaters up to 3000 m deep, following the PSBF. It is an important food source for marine predators in the region (Jereb and Roper 2010).

Adults of *Chiroteuthis veranyi* are found in the meso- and bathypelagic realms, in tropical, subtropical, and subantarctic waters (Jereb and Roper 2010). Little is known about this species' distribution in the South Atlantic Ocean; however, individuals have been found north of the South Georgia islands between 250 and 2000 m deep (Rodhouse and Lu 1998) and sporadically in the PSBF (Ivanovic and Brunetti 2005).

The distribution of *Galiteuthis glacialis* includes the circum-polar Antarctic water masses (45 °S to 70 °S). However, it can also be found in the subantarctic waters of the Malvinas Current in the PSBF (Jereb and Roper 2010).

Two species of the genus *Histioteuthis* inhabit the PSBF: in the northern region, *H. atlantica*, and in the south, *H. eltaninae* (Rodhouse et al. 1992). Both typically occupy highly productive waters over ocean basins and continental shelf-breaks, descending from surface waters up to 2000 m deep (Jereb and Roper 2010).

5.3 Reproduction

5.3.1 Bony Fishes

The Patagonian stock of the Argentine hake (*M. hubbsi*) is related to the PSBF mainly in autumn and winter, after spawning during early summer in shallower waters (about 50 m depth) at ~43–44 °S (Macchi et al. 2007). During the cold season, Argentine hake return to deeper waters, remaining broadly dispersed throughout the continental shelf and reaching the PSBF (Macchi et al. 2007). The spawning grounds and seasons of the long tail hake (*M. magellanicus*) are not completely known. However, the presence of males and females in spawning stages reported between 200 and 400 m depths from June to November may suggest reproduction during winter–spring (Machinandiarena 1999; Machinandiarena and Ehrlich 1999; Gorini and Pájaro 2014).

The presence of gravid females, eggs, and larvae of the southern blue whiting *M. australis* (Ehrlich et al. 1999) to the south and southwest of the Malvinas Islands indicates reproductive activity in this region between 200 and 500 m deep, from August to October (Perrota 1982; Machinandiarena 1999). The highest concentrations of juveniles have been found between 48 °S and 54 °S, to the east of the Malvinas Islands, at depths greater than 200 m, which suggests a breeding area near the PSBF (Perrota 1982; Cousseau and Perrotta, 2000).

There are two spawning and breeding areas reported for *Salilota australis*. One of them is on the Patagonian inner shelf at 100 m depth, and the other is close to the southern PSBF, west of the Malvinas Islands, at 200 m depth (Perez Comas 1980; Agnew 2002; Wöhler et al. 2004a, b; Arkhipkin et al. 2010). Reproduction takes place between October and November (austral spring) (Perez Comas 1980; Cousseau and Perrota, 2000; Brickle et al. 2011).

The spawning season of the Patagonian toothfish (*D. eleginoides*) in the Atlantic occurs between July and October south of 54 °S (Agnew et al. 1999; Prenski and Almeyda 2000; Pájaro et al. 2005; Pájaro et al. 2009), and its larvae have been reported close to the Malvinas Islands, between 100 and 200 m depth in spring (Ehrlich et al. 1999). The most important spawning areas are south and west of the Burdwood Bank and south of Tierra del Fuego and Staten Island, deeper than 1000 m (Pájaro et al. 2009).

Little information is available on the reproduction of the southern hake (*M. australis*). Spawning would take place from October to November in waters surrounding the Malvinas Islands, based on the presence of larvae and postlarval stages on the southwest side of the islands and in the Patagonian shelf, between 82 and 355 m depths (Ciechomski and Weiss 1974; Ciechomski et al. 1975).

Macrourus species reproduce throughout the year in waters of the Malvinas Island south of 50 °S and Namuncurá/Burdwood Bank. The spawning of *M. carinatus* occurs between 257 and 1097 meters deep, while *M. holotrachys* takes place in deeper areas, between 789 and 1809 m (Laptikhovsky et al. 2008). The little grenadier (*C. fasciatus*) breeds between late winter and early spring in waters close to Namuncurá/Burdwood Bank, between 250 and 800 m (Ciechomski and Booman 1981; Laptikhovsky et al. 2008) at the southern PSBF.

Although *Genypterus* species have been related to the PSBF during the austral cold season, during the spawning season (austral spring), adults move coastwards (Renzi 1986; Villarino 1998; Cordo 2004).

More than 80% of the fish larvae found in the PSBF waters belong to the family *Myctophidae* (*Krefftichthys anderssoni*, *Gymnoscopelus* spp., and *Protomyctophum* spp.; Ehrlich et al. 1999; Acha et al. 2018). The front could therefore be considered a breeding area for these species. This is consistent with the findings of Sabatés and Olivar (1989) in the Southeast Atlantic, where mesopelagic fish larvae tend to be concentrated along the shelf-break, with high larvae abundances indicating that seasonal spawning took place during periods of intense upwelling. One (e.g., *Protomyctophum choriodon*) or two (e.g., *Diaphus* spp.) spawning peaks have been reported between late winter and spring, matching the spring phytoplankton bloom and the consequent increase in zooplankton abundance (Gartner 1993; Catul et al. 2011; Saunders et al. 2015). In subantarctic species, such as *Krefftichthys anderssoni* and *Ceratoscopelus* sp., the spawning season has been reported to be confined to winter, while in other species (e.g., *Lepidophanes guentheri*), spawning may continue year-round (Gartner 1993; Lourenço et al. 2017).

At the northern region of the PSBF (34–37 °S), eggs and larvae of the Argentine anchovy (*Engraulis anchoita*) were found following the 200 m isobath (Ciechomski et al. 1986). Besides, along the southern PSBF, eggs of *C. fasciatus*, *M. whitsoni*, *S. fuegensis*, *Bathylagus antarcticus*, *Idiacanthus atlanticus*, and some species of *Nototheniidae* and *Zoarcidae* have also been reported (Acha et al. 2018).

5.3.2 Cartilaginous Fishes

A nursery area of the porbeagle shark (*L. nasus*) was detected north of 41 °S in the PSBF (Waessle and Cortés 2011; Forselledo 2012; Soto and Montealegre-Quijano 2012), while the adult feeding ground is in a colder area south of 52 °S (Waessle and Cortés 2011; Cortés et al. 2017; Belleggia et al. 2021). For the rest of the cartilaginous species, information on reproduction related to the PSBF is scarce.

Oviparous skates, like those found along the PSBF region, did not show peaks of reproductive activity, probably due to the stability in the physical conditions of the deeper environment (Ruocco et al. 2006; Colonello 2009; Scenna 2011; Delpiani 2016). Egg cases of several skate species, such as *A. doellojuradoi*, *B. albomaculata*, *B. brachyurops*, *B. macloviana*, *P. rudis*, and *P. normani*, were reported near or on the PSBF, where dense beds of the Patagonian scallop, *Zygochlamys patagonica*, are found (Vazquez et al. 2016).

5.3.3 Squids

In the Argentine Sea, four spawning stocks of the Argentine shortfin squid (*I. argentinus*) have been identified based on size structure, length at maturity, and location-season of spawning (Brunetti 1988); part of the life cycle of the two largest stocks is related to the PSBF ecosystem. During the late austral summer and autumn, while they are growing and maturing, large aggregations of the South Patagonia Stock (SPS) migrate to the outer shelf and shelf-break south of 44 °S (Brunetti and Pérez Comas 1989a; Brunetti et al. 1998b, c; Haimovici et al. 1998), where they are intensively fished, sustaining one of the largest squid fisheries in the ocean world (Brunetti 1990; Brunetti et al. 1999). Between April and May, these squids mature, and by winter, adult concentrations of the SPS disappear south of 44 °S. This is presumably due to the death of the specimens, indicating the end of the semelparous life cycle of this species.

The SPS spawning ground location is controversial; it was postulated to be located either on the PSBF between 45 °S and 48 °S (Koronkiewicz 1980, 1986; Brunetti 1988) or in Southern Brazil between 27 °S and 34 °S (Waluda et al. 1999, 2001; Arkhipkin et al. 2015). In the first case, egg masses would be transported by the waters of the outer shelf and Malvinas Current to the north up to BMC, where sea temperature would be suitable to trigger the hatching and development of paralarvae, and circulation would favor their retention on the shelf, even though exportation to the oceanic region was also reported (Brunetti and Ivanovic 1992). In fact, paralarvae were found on the shelf during autumn-winter, confirming that the surroundings of BMC between 35 °S and 40 °S are the location of the hatching ground (Brunetti and Ivanovic 1992; Leta 1992; Torres Alberto et al. 2022). In the second case, mature adults would migrate along the PSBF to reach the northern spawning ground, presumably located near the surroundings of BMC. Torres

Alberto et al. (2021) tested both hypotheses by analyzing how the SPS recruitment variability would be affected by oceanographic fluctuations related to the transport of the egg masses spawned on each of the two possible spawning grounds. They found that if spawning occurs in Patagonia, 52% of the recruitment variability could be explained by circulation variability, but if it occurs in southern Brazil, it does not result in successful recruitment for the SPS, and it could be another stock.

A similar scenario is observed north of 44 °S from autumn onwards, when pre-reproductive concentrations of the Bonaerensis-Northpatagonic stock (BNPS) are migrating to the outer shelf and shelf-break (Brunetti and Pérez Comas 1989b; Brunetti et al. 1998b), while they are also caught by commercial fleets, as occurs south of 44 °S earlier in the year. By June–July, these squids reach maturity and spawn near the western border of the BMC, the same place where paralarvae hatch.

Although migrations of *D. gahi* are not as spatially extensive as other squid, juveniles migrate from the Malvinas south and eastern shelf waters to deepwater feeding grounds in winter, where they experience continuous growth and maturation (Arkhipkin et al. 2004). Therefore, they would undertake horizontal and vertical migrations during their life cycle, from inshore spawning and nursery grounds (15–50 m depth) to offshore feeding grounds (200–350 m depth) in the PSBF. Eventually, after a feeding period of 3–5 months offshore in the shelf-break, mature squids would return to the inner shelf to spawn (Hatfield and Rodhouse 1994; Arkhipkin et al. 2004, 2013b).

Rodhouse et al. (1992) reported young juveniles of *Martialia hyadesi* in October/November in the vicinity of the interface between the PSBF and the western boundary of the Malvinas Current, suggesting that at least part of the Southwest Atlantic Ocean population spawns close to the PSBF. Besides, juveniles of this species (1–8 cm mantle length) were also found in the PSBF and in the adjacent oceanic region during spring (Brunetti et al. 1998a).

Larvae and juveniles of *Gonatus antarcticus* have also been described in relation to the PSBF. The shelf/slope waters around the Malvinas Islands, with temperatures around 4 °C and depths greater than 500 m, are suggested as a possible nursery ground for the species (Rodhouse et al. 1992).

5.4 Feeding Ecology

5.4.1 Bony Fishes

In the PSBF, the Argentine hake (*M. hubbsi*) feeds mainly on fish (myctophids, grenadiers, Argentine anchovy, and Patagonian sprat), followed by cephalopods (*I. argentinus* and *D. gahi*) (Angelescu and Cousseau 1969). Myctophids are the most important fish prey, followed by the notothenid *Patagonotothen ramsayi* and other Argentine hakes (cannibalism). Crustaceans are important but in low proportion (e.g., euphausiids, Argentine red shrimp, hyperiid amphipod *T. gaudichaudii*), and their consumption decreases with depth (Angelescu and Cousseau 1969; Cordo 1981). South of 52 °S, the long tail hake (*Macruronus magellanicus*) and the southern blue whiting (*Micromesistius australis*) are important prey of the Argentine hake (Angelescu and Cousseau 1969). The consumption of pelagic organisms (crustaceans, Argentine anchovy) decreases with depth and practically disappears after 300 m depth (Cordo 1981). The euphausiids (*Euphausia* spp.), which contributed 66.5% to the diet of *M. hubbsi* in the PSBF, appear to have a key role in the regional food web by transferring the phytoplankton carbon to higher trophic levels. By contrast, a much more diversified and heterotrophic food web was observed in the continental shelf region since, instead of one principal primary consumer, there are many of them (e.g., *Euphausia* spp., *T. gaudichaudii*, *Munida* spp., *P. petrunkevitchi*, *P. muelleri*), and the Argentine hake have the choice of feeding on a variety of prey (Belleggia et al. 2014a, 2017a, 2019).

Because juvenile hake are less abundant in the PSBF, the cannibalism rate is lower than on the continental shelf (Angelescu and Cousseau 1969; Cordo 1981; Sánchez and García de la Rosa 1999; Ocampo Reinaldo et al. 2011; Belleggia et al. 2014a, 2017a, 2019). More than 95% of the cannibalism occurred upon 0- to 2-year-old hake (Angelescu and Prenski 1987). Although some studies suggested that cannibalism occurs when food is scarce (Angelescu and Prenski 1987; Sánchez and Prenski 1996; Ruiz and Fondacaro 1997; Sánchez 2009, Belleggia et al. 2014a), it was demonstrated that cannibalism increased when pelagic young-of-the-year hake remained close to the bottom, being more accessible to larger demersal hakes (Belleggia et al. 2022).

The long tail hake (*M. magellanicus*) has a generalist behavior, being mainly carcinophagous and secondarily ichthyophagous and malacophagous (Bezzi 1984; Prenski et al. 1997; Marí and Sánchez 2002; Sánchez and Marí 2005; Giussi et al. 2016; Álvarez et al. 2021). Adults feed on other juveniles or small fishes such as *Sprattus fuegensis* and *Patagonotothen* spp. (Marí and Sánchez 2002; Giussi et al. 2004). In the PSBF, myctophids are the main prey. The southern blue whiting (*M. australis*) feeds mainly on meso- and macrozooplankton, with euphausiid and amphipod crustaceans being the most important prey (Wöhler et al. 2004a, b). Between 45 °S and 54 °S, the most abundant prey in stomach contents were *Themisto gaudichaudii* and *Euphausia lucens* (Wöhler et al. 1999; Marí and Sánchez 2002), while around the Malvinas Islands, they were *Thysanoessa gregaria* and *Euphausia vallentini* (Sabatini et al. 1999).

The red cod (*S. australis*) shows vertical trophic movements, with juveniles feeding mainly on pelagic crustaceans such as euphausiids, hyperiids, and fish larvae, while adults prey on crustaceans and benthic fishes (Perez Comas 1980). *S. australis* is the main prey for commercial fish such as long tail hake, southern hake, skates, and sharks (Wöhler et al. 2004a, b). *Dissostichus eleginoides* feeds mainly on fish, cephalopods, and crustaceans (Cassia and Perrotta 1996; García de la Rosa et al. 1997b; Prenski and Almeyda 2000; Barrera-Oro et al. 2005; Troccoli et al. 2020) and shows ontogenetic changes in its diet (García de la Rosa et al. 1997b; Arkhipkin et al. 2003; Collins et al. 2007). The prey items change according to the depth and the season of the year, which is influenced by changes in the abundance and migration of prey (Arkhipkin et al. 2003). *Merluccius australis* is a macrophagous predator, being primarily ichthyophagous and secondarily malacophagous (Wöhler et al. 1999; Marí and Sánchez 2002; Sánchez and Marí 2005). Even though their main prey are related to gadiform species such as long tail hake and southern blue whiting, other fish species are also incorporated depending on their seasonal availability (Giussi et al. 2004, 2016).

Large grenadiers feed on gelatinous plankton, crustaceans, fish (mesopelagic and benthic), cephalopods, and echinoderms (Dudochkin 1988; Morley et al. 2004; Laptikhovsky 2005; Giussi et al. 2010). *M. carinatus* forages mainly at depths below 900 m, while *M. holotrachys* in depths above 1100 m (Laptikhovsky 2005). The little grenadier feeds mainly on crustaceans (amphipods, euphausiids, and isopods), while polychaetes are much less important in its diet (Cousseau and Perrota 2000; Giussi et al. 2010). Romanelli (2017) observed a specialist tendency in the feeding strategy of the three species of grenadiers (*C. fasciatus, M. carinatus*, and *M. holotrachys*) present in the PSBF and a limited amplitude of the trophic niche dependent on depth and season of the year. The principal item in autumn was amphipods, while secondary prey (e.g., polychaetes, euphausiids, isopods, or even bony fishes and echinoderms) were different for each species; nonetheless, there was an overlap between the diets of the three species, and their trophic level was four (Romanelli 2017).

Genypterus blacodes has a demersal-benthic behavior, feeding on demersal and benthic fish and crustaceans (Belleggia et al. 2023). The diet composition of *G. blacodes* in the Southwest Atlantic showed geographic variations, preying mainly upon crustaceans (isopods, stomatopods, *Munida* spp.) inshore of San Jorge and San Matías Gulf waters and upon various fish species (*P. ramsayi* and *M. hubbsi*) offshore (Renzi 1986; Sanchez and Prenski 1996; Belleggia et al. 2023). In the PSBF, the pink cuskeel feeds mostly on the nototheniid *P. ramsayi* and Argentine hake *M. hubbsi* (Renzi 1986; Bertolotti et al. 1996; Cordo 2004; Belleggia et al. 2023), followed by other fish prey such as *Salilota australis*, *Micromesistius australis*, and *Macruronus magellanicus* (Renzi 1986; Bertolotti et al. 1996; Pierce et al. 2002; Cordo 2004; Nyegaard et al. 2004). Crustaceans are the second most important prey of *G. blacodes* in the PSBF, mainly represented by benthic isopods. Cephalopods, other mollusks, and brachiopods are of minor importance in its diet (Renzi 1986; Pierce et al. 2002; Nyegaard et al. 2004; Belleggia et al. 2023). Ontogenetic shifts in the diet of the smaller pink cuskeel were also observed, which feeds mainly on crustaceans and fishes, while larger individuals are primarily ichthyophagous (Renzi 1986; Pierce et al. 2002; Nyegaard et al. 2004).

Although no trophic ecology studies have been developed for *G. brasiliensis* in the PSBF waters, a similar diet to that of *G. blacodes* was reported for the outer shelf and shelf-break region of southern Brazil. *G. brasiliensis* is included in the large ichthyophagous group of fishes in the region, which feeds on benthic and demersal fish and, to a lesser extent, on benthic macrocrustaceans and cephalopods (Haimovici et al. 1994; Castello et al. 1997). Moreover, cannibalism is also reported in this species (Eleutério 2008), as in *G. blacodes* (Belleggia et al. 2023). As with *G. blacodes*, a more important carcinophage component was observed in the

Brazilian cuskeel diets, with specimens at the shelf-break being almost exclusively ichthyophagous (Nascimento 2012).

Myctophids are mainly zooplanktivorous fishes; their larvae have daylight feeding patterns and consume eggs and larval stages of copepods (Sabatés and Saiz 2000; Contreras et al. 2015), while adults exhibit predominant nighttime feeding, mostly on copepods but also on ostracods, euphausiids, hyperiid amphipods, chaetognaths, pteropods, fish eggs, and fish larvae (Pakhomov et al. 1996; Catul et al. 2011; Olivar et al. 2019). Variations in diet between species correlated with morphological features and size have been documented for myctophids in numerous studies (Clarke 1980; Williams et al. 2001).

Generally, the main forage item of smaller species, such as those of the genus *Diaphus* (47–84 mm SL), is copepods (Kinzer and Schulz 1985; Alwis and Gjøsaeter 1988; Pakhomov et al. 1996). In species that attain larger body sizes, such as *Lampanyctus australis* (138 mm SL), *Gymnoscopelus* spp. (146–280 mm SL), and *Electrona subaspera* (127 mm SL), large prey items like amphipods, euphausiids, pteropods, decapods, and their larvae become increasingly important in their diet (Pakhomov et al. 1996; Williams et al. 2001; Saunders et al. 2015). It was also observed that the diet composition of myctophids varies geographically and/or with the season (Clarke 1980; Kinzer and Schulz 1985; Williams et al. 2001). For instance, species of *Ceratoscopelus*, which have the broadest prey diversity, including appendicularians, salps, and siphonophores, showed higher regional variability than other lanternfish species (Pusch et al. 2004).

High zooplankton abundances at the front, mainly composed of large-sized copepods, could be the main prey of myctophids in the PSBF. Euphausiids and amphipods, in lesser proportions than copepods, also occur in the area (Cepeda et al. 2018). Several species of the genera *Lampanyctus*, *Gymnoscopelus*, and *Lampadena* in the PSBF area feed primarily on cryophilic crustaceans. The main prey items were calanoid copepods, such as the medium-sized *Drepanopus forcipatus* and the largest *Calanus* spp. (Angelescu and Cousseau 1969), which are highly abundant and widely distributed in the Argentine Sea (Cepeda et al. 2018). Myctophids also consumed euphausiids and hyperiid amphipods of the genus *Themisto* (Angelescu and Cousseau 1969), which strongly dominate hyperiid abundance in the area (Cepeda et al. 2018).

5.4.2 Cartilaginous Fishes

In southern PSBF (52 °S to 56 °S), the diet of the porbeagle shark, *Lamna nasus*, consisted mainly of fish (90%), followed by cephalopods and crustaceans (Belleggia et al. 2021). The main fish consumed by the porbeagle shark were the long tail hake (*Macruronus magellanicus*) and the southern blue whiting (*Micromesistius australis*), species that constituted the main catches of the austral trawl fisheries in which porbeagle sharks were caught as bycatch (Belleggia et al. 2021). Cephalopods in the diet were represented by the families *Octopodidae*, *Ommastrephidae*, and

Onychoteuthidae, while crustaceans included lithodids, shrimps, lobsters, crabs, and euphausiids. The estimated trophic level of the porbeagle shark in this region was 4.35 (Belleggia et al. 2021).

Given that the diet of several cartilaginous species in the PSBF is unknown, it is inferred from continental shelf studies. South of 41 °S, the narrowmouth catshark *S. bivius* feeds mostly on cephalopods and crustaceans, followed by fish and polychaetes (Sánchez et al. 2009; Belleggia et al. 2017b, 2018; Villa et al. 2018). North of 41 °S, *S. bivius* feeds mainly on crustaceans, followed by polychaetes, fish, and cephalopods (Sánchez et al. 2009; Belleggia et al. 2017b, 2018; Villa et al. 2018). Diet changes in relation to predator size and season. The trophic level was estimated at 3.94 and 3.57, on the south and north of 41 °S, respectively (Belleggia et al. 2017b, 2018; Villa et al. 2018).

The diet of the spiny dogfish (*Squalus acanthias*) in the PSBF is dominated by fish (*Engraulis anchoita*, *Sprattus fueguensis*, *Merluccius hubbsi*, *Macruronus magellanicus*, *Micromesistius australis*, *Patagonotothen* spp.) and cephalopods (*Illex argentinus* and *Doryteuthis gahi*) (García de la Rosa and Sánchez 1997a, b; García de la Rosa 1998; Laptikhovsky et al. 2001; Koen Alonso et al. 2002; Belleggia et al. 2012). Benthic invertebrates are present in the stomach contents of the spiny dogfish, but in low proportions (Laptikhovsky et al. 2001; Koen Alonso et al. 2002; Belleggia et al. 2012).

Feeding habits of most skate species are poorly known, and data on their diet in the PSBF region remains limited and largely speculative. However, trophic habits can be assumed from studies carried out on the continental shelf. Skates (Rajidae) and some shark species, such as the narrow-mouthed catshark (*Schroederichthys bivius*), exhibit inferior mouth types that denote a benthic feeding specialization. Of the eight known species of *Bathyraja*, three are exclusively benthic feeders. For instance, *Bathyraja multispinis* has pavement-type dentition and feeds exclusively on crabs, mainly decapods *Peltarion spinosulum* and *Libidoclaea granaria*, followed by the isopod *Acanthoserolis* spp. (Belleggia et al. 2014b). The most important prey of *Bathyraja albomaculata* are polychaetes (*Travisia* spp.), followed by gammarid amphipods (*Ampelisca* spp.) and isopods (*Cirolana* spp.) (Sánchez and Mabragaña 2002; Brickle et al. 2003; Ruocco et al. 2009; Shimabukuro 2009). *Bathyraja macloviana* also feeds on benthic organisms, mainly polychaetes, followed by amphipods, isopods, and decapods (Sánchez and Mabragaña 2002; Mabragaña et al. 2005; Scenna et al. 2006; Barbini et al. 2013).

The remaining five species feed on teleosts and crustaceans, and in some cases show diet shifts with body size (Brickle et al. 2003; Belleggia et al. 2008, 2014b; Barbini et al. 2010). The broad-nose skate *Bathyraja brachyurops* feeds on fish, followed by crustaceans (mainly brachyuran crabs and isopods) and exhibits diet shifts with size (Belleggia et al. 2008). Similarly, the diets of *Bathyraja cousseauae* and *Bathyraja scaphiops* consist mostly of teleosts, followed by isopods and amphipods (Belleggia et al. 2014b). Although some authors consider *Bathyraja griseocauda* to be mainly ichthyophagous (Sánchez and Mabragaña 2002), the isopods *Acanthoserolis* spp. are an important component of its diet (Brickle et al. 2003; Belleggia et al. 2014b). Similarly, *Bathyraja magellanica* is primarily ichthyophagous, and secondarily, it feeds on amphipods, isopods, and decapods (Barbini et al. 2010). In the case of two characteristic skates of the PSBF, such as *Bathyraja papilionifera* and *Bathyraja meridionalis*, knowledge of their biology is null.

The yellownose skate *Zearaja brevicaudata* is mainly an ichthyophagous species that feeds on demersal fishes, primarily nototheniids *Patagonotothen ramsayi* and Argentine hake *M. hubbsi* (García de la Rosa 1998; Lucífora et al. 2000; Koen Alonso et al. 2001; Belleggia et al. 2016). Small *Z. brevicaudata* also feed on benthic prey, such as the isopod *Acanthoserolis schythei* and crabs *Peltarion spinosulum* and *Libidoclaea granaria* (Lucífora et al. 2000; Belleggia et al. 2016). Some pelagic fishes and the Argentine squid *Illex argentinus* are also consumed, but generally with cut marks indicating that they were discarded from fisheries and scavenged by *Z. brevicaudata* (Lucífora et al. 2000).

Psammobatis rudis and *P. normani* have similar trophic preferences. The most important prey for *P. normani* are crustaceans (mainly crabs and isopods), followed by polychaetes. *Psammobatis rudis* feeds almost exclusively on crustaceans (mainly isopods, crabs, and gammarids), followed by fish and polychaetes (Mabragaña and Giberto 2007).

The southern thorny skate, *Amblyraja doellojuradoi*, feeds mainly on crabs, and occasionally on polychaetes, teleosts, isopods, and mollusks (Sánchez and Mabragaña 2002; Delpiani et al. 2013). Among the crabs, *Libidoclea granaria* and *Peltarion spinulosum* were the most consumed species (Delpiani et al. 2013).

5.4.3 Squids

The PSBF plays a main role as a feeding ground for *I. argentinus*, a voracious and opportunistic predator that feeds mainly on zooplanktonic species (hyperiid amphipods and euphausiids), small fishes (myctophids), and squids (Koronkiewicz 1980; Ivanovic and Brunetti 1994; Ivanovic 2000, 2010; Mouat et al. 2001). The Argentine shortfin squid feeds mostly during daylight, near the seabed, starting in the morning and reaching a maximum in the afternoon (Ivanovic and Brunetti 1994). Besides, it plays an important role in the PSBF food web, serving as prey for diverse species such as the Argentine hake (*Merluccius hubbsi*), the long tail hake (*Macruronus magellanicus*), petrels, albatrosses, and marine mammals (Ivanovic 2000; Jereb and Roper 2010).

Adult concentrations of *D. gahi*, feeding on euphausiids and other zooplankton, have been reported mostly around the 200 m isobath to the south and northeast of the Malvinas Islands (Agnew et al. 2005), between 51 °S and 52 °S, coinciding with the southern part of the PSBF. To the north of the PSBF, it preys mostly on zooplanktonic crustaceans such as euphausiids and amphipods, but decapods and small squids may also be ingested by larger individuals. Among its main predators are the Argentine hake (*Merluccius hubbsi*), the southern blue whiting (*Micromesistius australis*), the long tail hake (*Macruronus magellanicus*), and some species of squids (Pineda et al. 1996, 1998; Brunetti et al. 1999).

Adults of the Sevenstar flying squid (*M. hyadesi*) in the PSBF prey mainly on fish (43.9% of the total, with 90% being myctophids), squid (36.6%, mostly juveniles of the same species), and zooplankton (9.5%) (Ivanovic et al. 1998).

Little is known about the ecological role of the greater hooked squid in the PSBF; however, like most deep-cold water squids, *O. ingens* feeds mainly on deep-water fish (e.g., myctophids) and crustaceans. It also serves as prey for large oceanic predators, such as birds (e.g., penguins, petrels, albatrosses) and marine mammals (Phillips et al. 2003; Jereb and Roper 2010; Rosas-Luis et al. 2014). Therefore, it functions as a link between smaller deep-water species and large top predators in subantarctic oceanic waters.

5.5 Migrations and Life Cycles

5.5.1 *Bony Fishes*

The southern blue whiting *Micromesistius australis* is a highly migratory species, which tends to aggregate in schools, usually performing daily vertical migrations. During the day, it concentrates in deep areas, rising and dispersing in the water column during the night (Madirolas 1999). The inferred migratory pattern in the Atlantic states that adults spend the winter around the De los Estados Island and on the Patagonian shelf; then, in spring, they make a displacement toward the breeding area southwest of Malvinas, through the channel formed by De los Estados Island and the Burdwood Bank (Agnew 2002; Wöhler et al. 2004a, b). After spawning during September and October (Macchi et al. 2007), adults disperse to feed on the southern Argentine shelf, in the Scotia Sea (Wöhler et al. 2004a, b), or northwest of the islands (McKeown et al. 2017).

In terms of the Argentine anchovy (*E. anchoita*), a small pelagic fish, three stocks have been defined, and the one that is distributed between 28 °S and 41 °S performs an extensive migratory cycle. During late fall, the shoals distribute further from the coastal regions of the southeast of Buenos Aires Province, reaching the external shelf and the PSBF between 33 °S and 37 °S, their main place of occurrence during winter (May–July; Hansen 2004). Moreover, as part of their reproductive migration, Argentine anchovies are located in the northern region of the PSBF (north of 40 °S), where spawning has been recorded between April and August (Hansen 2004).

Regarding myctophids, juveniles and adults perform extensive diel migrations, from mesopelagic depths during the day into the epipelagic zone at night, constituting migrating sound scattering layers in the ocean (Salvanes and Kristoffersen 2001; Gordeeva 2011). Most of the species cited for the PSBF region (Sect. 5.2.1) are nyctoepipelagic at the surface, reaching depths of 100 or 200 m at night. However, many of them are found below 200 m at night, such as *Diaphus*, *Gymnoscopelus bolini*, and *Lampanyctus australis* (Hulley 1991).

Myctophids are thought to undergo vertical migration to avoid predation while the upper layers are well-lighted and feed on the surface following the nightly ascension of zooplankton (Catul et al. 2011; Olivar et al. 2019). It has been estimated that mesopelagic fauna captures about 90% of organic carbon fixed in the euphotic zone (Robinson et al. 2010), and therefore lanternfishes play an important role in transferring organic matter between depths through their main behavioral trait (Catul et al. 2011). Also, myctophids are considered to be an essential component in the mesopelagic ecosystem (Pusch et al. 2004; Olivar et al. 2019), acting as an important link between zooplankton and predators of higher trophic levels, such as seabirds, squids, fishes, and marine mammals (e.g., Angelescu and Cousseau 1969; Jackson et al. 2000; Ivanovic et al. 1998; Cherel et al. 2002; Laptikhovsky 2005). In spite of their vertical migration capability, lanternfish distributions appear to be primarily affected by sea currents, and they permanently drift with moving water (Gjøsaeter and Kawaguchi 1980; McGinnis 1982; Suneetha and Salvanes 2001; Gordeeva 2011). Seasonal horizontal migration is yet to be clarified for mesopelagic species, as knowledge of their feeding or spawning grounds is limited (Gjøsaeter and Kawaguchi 1980).

In the Southwest Atlantic Ocean, the majority of the subantarctic myctophid species are sexually mature throughout their geographic range, and no feeding concentrations have been reported (Hulley 1981). However, a latitudinal trend in mean size and age has been found for many species that also occur in Antarctic waters, suggesting massive migrations and expatriate populations to higher latitudes (Saunders et al. 2017). A progressive increase in mean body size with latitude has been described for several species in the region (*Protomyctophum* spp., *Gymnoscopelus nicholsi*, *Krefftichthys anderssoni*). Also, expatriate, large-sized adult cohorts were found farther south in the Antarctic Polar Front, which are probably lost to the reproductive part of the population (Gjøsaeter and Kawaguchi 1980; Saunders et al. 2015, 2017; Lourenço et al. 2017).

5.5.2 Cartilaginous Fishes

Adult females of the porbeagle shark *Lamna nasus* show a seasonal migratory pattern between breeding and foraging grounds. The seasonal pattern and total length frequency distributions of *L. nasus* registered in commercial catches indicate that the nursery areas are located in northern and temperate regions (Waessle and Cortés 2011; Forselledo 2012; Soto and Montealegre-Quijano 2012), while the adult feeding grounds are in southern (52–56 °S) and colder areas (Waessle and Cortés 2011; Cortés et al. 2017; Belleggia et al. 2021).

The majority of the skate species inhabiting the PSBF, although limited, show migratory movements between habitats, moving north in winter to feed in warmer areas and returning south in summer, presumably to lay their eggs on the shelf area (Arkhipkin et al. 2012; Winter et al. 2015; Cortés et al. 2021). There are some common patterns in the seasonal movements of skates in the region. For instance, during the cold season, *A. doellojuradoi*, *B. albomaculata*, *B. brachyurops*, *B. griseocauda*, *B. multispinis*, and *B. scaphiops* remain deeper at the shelf-break between ca.

49 °S and 51 °S, but during the warm season, they migrate to shallower waters in the outer continental shelf of the Malvinas Islands, between 50 °S and 51 °S (Arkhipkin et al. 2012).

5.5.3 Squids

Seasonal migratory movements of the two main stocks of the Argentine shortfin squid include a phase of feeding, maturing, and spawning that takes place in the PSBF; thus, interannual variations in this frontal region play a major role in recruitment regulation (Brunetti 1988; Waluda et al. 2001; Acha et al. 2004; Torres Alberto et al. 2021). The species also performs a daily vertical migration, scattering in the water column during the night and going close to the bottom during the day (Brunetti 1988). This behavior was confirmed for the PSBF (125–850 m) by submarine observations (Moiseev 1991).

The Patagonian longfin squid, *D. gahi*, is a nektonic species that moves from nursery grounds located in shallow inshore waters to offshore feeding grounds in the PSBF when reaching maturity. Thus, its main niche is at depths of 50–350 m associated with cold waters between 5.5 and 8.5 °C (Hatfield and Rodhouse 1994; Pineda et al. 1998).

5.6 Discussion

The PSBF is a highly productive area of the Southwestern Atlantic Ocean, recognized as an *oasis* (sensu Acha et al. 2015) immersed in less productive waters. It is one of the largest fronts in the world ocean, and nektonic species take advantage of this permanent feature in different ways. More than 135 taxa of nekton were reported for the PSBF between 1978 and 2019, during several scientific cruises and catches from commercial fleets. However, the PSBF is still an area with very low sampling effort, especially in the deepest regions; therefore, it is expected that richness will be even greater than reported. Although we did not compare the diversity of the front and the Argentine continental shelf, we can conclude that a large number of nekton species make use of the PSBF. Most of the species mentioned throughout the text have large home ranges, so their distributions are not restricted exclusively to the front. However, nektonic species at the PSBF evidently find suitable conditions, some to feed in this profitable foraging ground, others to reproduce, or even on the way to another area (migration).

The PSBF exhibits a wide latitudinal range, and at an oceanographic scale, this large region is connected by the Malvinas Current, which links nekton from the southernmost area of Burdwood Bank with northern submarine canyons at ca. 38 °S. Several nektonic species, given their optimal temperature ranges, show an affinity with the Malvinas Current and thus with the entire extent of the PSBF. There are many species of subantarctic origin distributed along the front, and several of them are commercially exploited and exported to other countries (FAO 2020), such as *Merluccius hubbsi*, *Genypterus blacodes*, *Macruronus magellanicus*, *Dissostichus eleginoides*, *Macrourus carinatus*, *Coelorinchus fasciatus*, *Micromesistius australis*, *Salilota australis*, *Patagonotothen ramsayi*, among others.

It is widely recognized that the PSBF is a key region in the life cycles of several species, including the most important fishery resources, such as the Argentine shortfin squid (*I. argentinus*) and the Argentine hake (*M. hubbsi*). These commercially valuable species spend at least part of their life cycles in close association with the front. In that sense, the relationship between the life history of the Argentine squid and the PSBF is the most documented; the highest squid concentrations, their reproduction, and migration are associated with this dynamic feature. The spawned egg masses of the Argentine shortfin squid (South Patagonian Stock), which reproduces during autumn, are transported by the Malvinas Current along the PSBF from south to north in order to reach favorable temperatures for embryonic development and hatching.

The relationship of bony and cartilaginous fishes with the PSBF is more complex, but several commercially important fish stocks are associated with the frontal area during part or the whole life cycle (Bertolotti et al. 1996). Myctophids deserve special attention; their larvae are present throughout the year along the entire PSBF, and they are also the most abundant ichthyoplankton component, with adults being the most abundant mesopelagic fish in the region. Myctophids play a paramount role in linking lower trophic levels of the food web with higher ones, as they feed on zooplankton and, at the same time, are prey for other nekton species and top predators. Other of the most abundant fish species on the Argentine continental shelf include the Argentine anchovy and the Argentine hake, whose important spawning areas during austral spring are close to the coast. However, both are also reported to be adapted to spawn along the PSBF, particularly in autumn and winter.

As mentioned before, despite the fact that the association of nekton with frontal features is somehow difficult to assess given their high mobility and complex behavior, most nektonic species (bony fish and squids) start out life as part of the plankton (planktonic larvae) and gradually incorporate into nekton as they grow. As pointed out in Chap. 4, ichthyoplankton (eggs and larvae) of demersal and mesopelagic fish are found in the PSBF (e.g., *Salilota australis*, *Micromesistius australis*, *Sprattus fuegensis*, myctophids, and grenadiers). These fish eggs and larvae, in addition to other zooplanktonic prey, are in turn food for adult fish and squids that distribute along the front. Thus, key biological processes (feeding and/or reproduction) take place at the PSBF, improving nektonic survival opportunities.

Although considerable scientific information is available on some species of the greatest ecological and economic interest at the PSBF, knowledge about other nektonic species, such as cartilaginous fish, is still limited. Particularly, given their specific life-history traits, they are highly vulnerable to overexploitation and population depletion. Particularly, deep-water chondrichthyans, like those inhabiting the PSBF, are somehow more vulnerable to depletion than shelf-water species (García et al. 2008). Several species that use the PSBF are endemic to the Southwest Atlantic and are listed by the IUCN (2023) as endangered (EN; *Bathyraja brachyurops*, *Bathyraja griseocauda*, *Dipturus trachydermus*, *Zearaja brevicaudata*) or vulnerable (VU; *Bathyraja albomaculata*, *Lamna nasus*, *Schroederichthys bivius*, *Squalus acanthias*).

Even though there are several coincident biogeographical studies in the Argentine Sea that set the more conspicuous patterns of nekton distribution, all the regions identified are interconnected and far from being static. The possible divisions that have been proposed are dynamic, even more so in the ocean, where physical processes (e.g., dynamics and spatial-temporal heterogeneity of the water column) influence the distribution of species (Bakun 2006). The PSBF could be considered an ecotone (sensu Longhurst 1998) or a transitional zone between the neritic and the oceanic realms, where overlapping, replacement, and partial segregation of species occur. The ecotone region has more species shared with adjacent areas than typical ones. In that sense, nektonic species identified in the PSBF, due in part to complex trophic relationships, large home ranges, as well as extensive feeding and spawning migrations, connect the frontal system with other ecosystems, some of which are at great distances from the front, redistributing matter and energy.

5.7 Conclusion

This chapter brings together all available, dispersed, and difficult-to-access information and provides a synthesis of nektonic species at the PSBF, compiling numerous scientific publications and unpublished data. It is important to highlight that despite the information summarized in this chapter, the shelf-break and deeper waters have a very low sampling effort; thus, it is expected that the number of nektonic species that make use of the PSBF would be even greater. This chapter highlighted the role of the PSBF in the life history (reproduction, foraging, and migration) of several nektonic species. The PSBF is a complex adaptive system that is not only one of the areas with the highest productivity in the Southwestern Atlantic, but also, on a regional scale, promotes suitable conditions for the presence of key species and assemblages of high functional and ecological relevance. In this chapter, the PSBF is emphasized as a key area due to its nekton biodiversity and the presence of threatened species, particularly cartilaginous fish.

Acknowledgments We especially thank the people who participated on board the research vessels during the several cruises analyzed in this chapter and the On-board Observer Program at INIDEP for the acquisition, processing, and storage of the primary data. This is INIDEP contribution No. 2310.

References

Acha EM, Mianzan HW, Guerrero RA, Favero M, Bava J (2004) Marine fronts at the continental shelves of austral South America: physical and ecological processes. J Marine Syst 44:83–105
Acha EM, Piola AR, Iribarne O, Mianzan H (2015) Ecological processes at marine fronts. In: Oases in the ocean, Springer briefs in environmental science. Springer International Publishing, Cham
Acha EM, Ehrlich MD, Muelbert JH, Bruno D, Machinandiarena L, Cadaveira M (2018) Ichthyoplankton associated to the frontal regions of the Southwestern Atlantic. In: Hoffmeyer M, Sabatini ME, Brandini F et al (eds) Plankton ecology of the Southwestern Atlantic, from subtropical to the Subantarctic realm. Springer, Berlin, pp 219–246
Agnew DJ (2002) Critical aspects of The Falkland Islands pelagic ecosystem: distribution, spawning and migration of pelagic animals in relation to oil exploration. Aquatic Conserv: Mar Freshw Ecosyst 12(1):39–50
Agnew DJ, Heaps L, Jones C, Watson A, Berkieta K, Pearce J (1999) Depth distribution and spawning pattern of *Dissostichus eleginoides* at South Georgia. CCAMLR Sci 6:19–36
Agnew DJ, Hill SL, Beddington JR, Purchase LV, Wakeford RC (2005) Sustainability and management of Southwest Atlantic Squid Fisheries. Bull Mar Sci 76(2):579–594
Alemany D, Acha EM, Iribarne O (2009) The relationship between marine fronts and fish diversity in the Patagonian Shelf Large Marine Ecosystem. J Biogeogr 36:2111–2124
Alemany D, Prandoni N, Ivanovic M, Acha EM (2022) Diversidad de peces y calamares en el área denominada Agujero Azul y zonas adyacentes. Inf Ases Transf 65:20
Aleyev YG (2012) Nekton. Springer Science & Business Media, p 441
Allega L, Braverman M, Cabreira A, Campodónico S, Colonello JH et al (2019) Estado del conocimiento biológico pesquero de los principales recursos vivos y su ambiente, con relación a la exploración hidrocarburífera en la Zona Económica Exclusiva Argentina y adyacencias. Instituto Nacional de Investigación y Desarrollo Pesquero (INIDEP), Mar del Plata, Argentina
Alvarez Perez JA, Pezzuto PR, Wahrlich R et al (2009) Deep-water fisheries in Brazil: history, status and perspectives. Lat Am J Aquat Res 37(3):513–541
Álvarez D, Giussi AR, Botto F (2021) Superposición y partición de nichos isotópicos entre especies de peces demersales de interés comercial en el ecosistema austral. Inf Invest INIDEP 41:17
Alwis A, Gjøsaeter J (1988) Feeding behaviour of *Diaphus dumerilii* in NW Africa with notes on its relation to other myctophyds in the area. Flødevigen rapportser. Havforskningsinstituttet, Tromsø, p 55–69
Anderson ED (1990) Fishery models an applied to elasmobranch fisheries. NOAA Tech Rep NMFS 90:473–484
Angel MV (1997) Pelagic biodiversity. In: Ormond RFG, Gage JD, Angel MV (eds) Marine biodiversity: patterns and processes. Cambridge University Press, pp 35–68
Angelescu V, Cousseau MB (1969) Alimentación de la merluza en la región del talud continental argentino, época invernal (Merluccidae *Merluccius hubbsi*). Bol Inst Biol Mar 19:5–84
Angelescu V, Gneri F (1960). Contribución al conocimiento bioecólogico de la merluza de cola (*Macruronus magellanicus* Lönnberg). *Actas y trabajos del Primer Congreso Sudamericano de Zoología* (La Plata) 12–24/10/59 Comisión de Investigaciones Científicas (CIC). 59 y Comisión de Investigaciones Científicas y Técnicas (CNICT) I(I). Ecología 3–18
Angelescu V, Prenski L. (1987) Ecología trófica de la merluza común del Mar Argentino (Merlucciidae, *Merluccius hubbsi*). Parte 2. Dinámica de la alimentación, analizada sobre la base de las condiciones ambientales, la estructura y las evaluaciones de los efectivos en su área de distribución. *Contribución del Instituto Nacional de Investigación y Desarrollo Pesquero.* Argentina. 561, p 205
Angelescu V, Gneri F, Nani A (1958) La merluza del Mar Argentino. Biología y taxonomía. Servicio de Hidrografía Naval de Buenos Aires. Publicación. *H*: 104, p 224
Arkhipkin AI, Brickle P, Laptikhovsky V (2003) Variation in the diet of the Patagonian toothfish with size, depth and season around The Falkland Islands. J Fish Biol 63(2):428–441

Arkhipkin AI, Grzebielec R, Sirota AM et al (2004) The influence of seasonal environmental changes on ontogenetic migrations of the squid *Loligo gahi* on the Falkland shelf. Fish Oceanogr 13:1–9

Arkhipkin AI, Brickle P, Laptikhovsky V (2010) The use of Island water dynamics by spawning red cod, *Salilota australis* (Pisces: Moridae) on the Patagonian Shelf (Southwest Atlantic). Fish Res 105:156–162

Arkhipkin A, Brickle P, Laptikhovsky V, Pompert J, Winter A (2012) Skate assemblage on the eastern Patagonian shelf and Slope: structure, diversity and abundance. J Fish Biol 80:1704–1726

Arkhipkin AI, Brickle P, Laptikhovsky V (2013a) Links between marine fauna and oceanic fronts on the Patagonian Shelf and Slope. Arquipelago-Life and Marine Sciences 30:19–37

Arkhipkin AI, Hatfield EMC, Rodhouse PGK (2013b) *Doryteuthis gahi*, Patagonian long-finned squid. In: Rosa R, Odor R, Pierce G (eds) Advances in squid biology, ecology and fisheries. Nova Science Publishers, Inc., pp 123–157

Arkhipkin AI, Graz M, Blake A (2015) Water density pathways for shelf/slope migrations of squid *Illex argentinus* in the Southwest Atlantic. Fish Res 172:234–242

Bakun A (2006) Fronts and eddies as key structures in the habitat of marine fish larvae: opportunity, adaptive response and competitive advantage. Scientia Marina 70(S2):105–122

Balech E, Ehrlich MD (2008) Esquema biogeográfico del Mar Argentino. Rev Invest Des Pesq 19:45–75

Barbini SA, Scenna LB, Figueroa DE, Cousseau MB, Díaz de Astarloa JM (2010) Feeding habits of the Magellan skate: effects of sex, maturity stage and body size on diet. Hydrobiologia 641:275–286

Barbini SA, Scenna LB, Figueroa DE, Díaz De Astarloa JM (2013) Effects of intrinsic and extrinsic factors on the diet of Bathyraja macloviana, a benthophagous skate. J Fish Bio 83:156–169

Barker MJ, Schluessel V (2005) Managing global shark fisheries: suggestions for prioritizing management strategies. Aquat Conserv 15:325–347

Barrera-Oro ER, Casaux RJ, Marschoff ER (2005) Dietary composition of juvenile *Dissostichus eleginoides* (Pisces, Nototheniidae) around Shag Rocks and South Georgia. Antarctica Polar Biol 28(8):637–641

Belleggia M, Mabragaña E, Figueroa DE, Scenna LB, Barbini SA, Díaz De Astarloa JM (2008) Food habits of the broad nose skate, *Bathyraja brachyurops* (Chondrichthyes, Rajidae), in the south-west Atlantic. Sci Mar 72:701–710

Belleggia M, Figueroa DE, Sánchez F, Bremec C (2012) Long-term changes in the spiny dogfish (*Squalus acanthias*) trophic role in the southwestern Atlantic. Hydrobiologia 684:57–67

Belleggia M, Figueroa DE, Irusta G, Bremec C (2014a) Spatio-temporal and ontogenetic changes on the diet of the Argentine hake *Merluccius hubbsi* (Marini, 1933). J Mar Biol Assoc UK 94(08):1701–1710

Belleggia M, Scenna LB, Barbini SA, Figueroa DE, Díaz De Astarloa JM (2014b) The diets of four *Bathyraja* skates (Elasmobranchii, Rajidae) from the Southwest Atlantic. Cybium 38(4):314–318

Belleggia M, Andrada N, Paglieri S et al (2016) Trophic ecology of yellownose skate *Zearaja chilensis* (Guichenot, 1848) (Elasmobranchii: Rajidae), a top predator in the southwestern Atlantic. J Fish Biol 88:1070–1087

Belleggia M, Giberto D, Bremec C (2017a) Adaptation of diet in a changed environment: increased consumption of the lobster krill *Munida gregaria* (Fabricius, 1793) by Argentine hake. Mar Ecol 38:e12445

Belleggia M, Villa A, Colonello J et al (2017b) The diet of the Narrowmouthed Catshark *Schroederichthys bivius*, from the Patagonian continental shelf. Joint Meeting of Ichthyologists and Herpetologists. American Elasmobranch Society. Austin, Texas, EEUU. July 12–16, 2017

Belleggia M, Zenoni-Lufrano M, Villa A et al (2018) The diet of the narrowmouthed catshark *Schroederichthys bivius* in Argentine sea. Sharks International Conference. João Pessoa, PB, Brazil. June 03–08, 2018

Belleggia M, Alves NM, Leyton MM et al (2019) Are hakes truly opportunistic feeders? A case of prey selection by the Argentine hake *Merluccius hubbsi* off southwestern Atlantic. Fish Res 214:166–174

Belleggia M, Colonello J, Cortés F, Figueroa DE (2021) Eating catch of the day: the diet of porbeagle shark *Lamna nasus* (Bonnaterre 1788) based on stomach content analysis, and the interaction with trawl fisheries in the southwestern Atlantic (52°S–56°S). J Fish Biol 99:1591–1601

Belleggia M, Alvarez Colombo GL, Santos B, Castelletta M, Mattera B (2022) Who let the YOY up? The pelagic habitat as a nursery and feeding ground area for the young-of-the-year (YOY) Argentine hake, Merluccius hubbsi Marini, 1933. J Fish Biol 100:378–389

Belleggia M, Álvarez CD, Pisani E, Descalzo M, Zuazquita E (2023) Prey contribution to the diet of pink cusk-eel *Genypterus blacodes* (Forster, 1801) revealed by stomach content and stable isotopic analyses in the southwestern Atlantic. Fish Res 262:106660

Bellisio NB, López BR, Torno A (1979) Peces Marinos Patagónicos, Ministerio de Economía, Secretaría de Estado de Intereses Marítimos, Subsecretaría de Pesca. Buenos Aires, p 279

Bertolotti MI, Brunetti NE, Carreto JI, Prenski LB, Sanchez RP (1996) Influence of shelf-break fronts on shellfish and fish stocks off Argentina. International Council for the Exploration of the Sea CM 1996/S:41 (Theme Session S), p 24

Bezzi S (1984) Aspectos biológico pesqueros de la merluza de cola del Atlántico Sudoccidental. Rev Invest Des Pesq 4:63–80

Bezzi SI, Verazay GA, Dato CV (1995) Biology and fisheries of Argentine hakes (*M. hubbsi and M. australis*). In: Alheit J, Pitcher TJ (eds) Hake: biology, fisheries and markets. Chapman & Hall, London, pp 229–267

Bianchi A, Massoneau M, Olivera RM (1982) Análisis estadístico de las características T-S del sector austral de la plataforma continental Argentina. Acta Oceanographica Argentina 3:93–118

Bornatowski H, Santos L, De Castro RM, Weiser PA (2014) Occurrence of the narrowmouth catshark *Schroederichthys bivius* (Chondrichthyes: Scyliorhinidae) in southern Brazil. Mar Biodivers Rec 7:1–3

Boschi EE (2000) Species of decapod crustaceans and their distribution in the American marine zoogeographic provinces. Rev Invest Des Pesq 13:7–136

Braga AC, Costa PAS, Martins AS et al (2014) Lanternfish (Myctophidae) from eastern Brazil, Southwest Atlantic Ocean. Lat Am J Aquat Res 42:245–257

Brandt SB (1983) Temporal and spatial patterns of lanternfish (family Myctophidae) communities associated with a warm-core eddy. Mar Biol 74:231–244

Brickle P, Laptikhovsky V, Pompert J, Bishop A (2003) Ontogenic changes in the feeding habits and dietary overlap between three abundant rajid species on The Falkland Islands shelf. J Mar Biol Assoc UK 83:1119–1125

Brickle P, Laptikhovsky V, Arkhipkin A (2011) The reproductive biology of a shallow water morid (*Salilota australis* Günther, 1878), around The Falkland Islands. Est Coast Shelf Sci 94:102–110

Bridges AEH, Barnes DKA, Bell JB, Ross RE, Howell KL (2022) Depth and latitudinal gradients of diversity in seamount benthic communities. J Biogeogr 49:904–915

Brunetti NE (1988) Contribución al conocimiento biológico-pesquero del calamar argentino (Cephalopoda: Ommastrephidae: *Illex argentinus*). Doctoral dissertation, Universidad Nacional de La Plata

Brunetti NE (1990) Evolución de la pesquería de *Illex argentinus* (Castellanos, 1960). Inf Téc Inv Pesq 155:3–19

Brunetti NE, Ivanovic ML (1992) Distribution and abundance of early life stages of squid (*Illex argentinus*) in the south-west Atlantic. ICES J Mar Sci 49:175–183

Brunetti NE, Pérez Comas JA (1989a) Abundancia, distribución y estructura poblacional del calamar (*Illex argentinus*) en aguas de la plataforma patagónica en diciembre de 1986 y enero-febrero de 1987. Frente Marítimo 5(A):61–70

Brunetti NE, Pérez Comas JA (1989b) Abundancia, distribución y estructura poblacional del recurso calamar (*Illex argentinus*) en aguas uruguayo-bonaerenses en mayo, setiembre y noviembre de 1986 y en marzo y mayo de 1987. Frente Marítimo 5(A):39–59

Brunetti NE, Rossi GR (1990) Informe preliminar sobre la campaña argentino-soviética en el B/I "Evrika" (agosto-octubre 1988). Inf Téc Int INIDEP 1:35

Brunetti NE, Ivanovic ML, Elena B (1998a) Calamares omastréfidos (Cephalopoda, Ommastrephidae). In: Boschi EE (ed) El Mar Argentino y sus recursos pesqueros, Tomo 2: Los moluscos de interés pesquero. Cultivos y estrategias reproductivas de bivalvos y equinoideos. INIDEP, Mar del Plata, pp 37–68

Brunetti NE, Ivanovic ML, Rossi GR et al (1998b) Fishery biology and life history of *Illex argentinus*. In: Okutani T (ed) Large pelagic squid. Japan marine fishery resources center (JAMARC) special publication. Tanaka Printing Co. Ltd, Tokio, pp 216–231

Brunetti NE, Elena B, Rossi GR et al (1998c) Summer distribution, abundance and population structure of *Illex argentinus* on the Argentine shelf in relation to environmental features. S Afr J Marine Sci 20:175–186

Brunetti NE, Ivanovic ML, Sakai M (1999) Calamares de importancia comercial en Argentina. Biología, distribución, pesquerías y muestreo biológico. Contrib INIDEP 1121

Camhi M, Fowler S, Musick J et al (1998) Sharks and their relatives: ecology and conservation. Occas Pap IUCN Species Surviv Comm 20:1–39

Carpenter KE (ed) (2002) The living marine resources of the Western Central Atlantic. Volume 2: Bony fishes part 1 (Acipenseridae to Grammatidae). FAO, Rome, p 601–1374

Cassia MC, Perrotta R (1996) Distribución, estructura de tallas, alimentación y pesca de la merluza negra (*Dissostichus eleginoides* Smith, 1898) en un sector del Atlántico Sudoccidental. INIDEP Inf Téc 9:19

Castellanos ZA (1960) Una nueva especie de calamar argentino, *Ommastrephes argentinus sp. nov.* (Mollusca, Cephalopoda). Neotropica 6(20):55–58

Castellanos ZA (1964) Contribución al conocimiento biológico del calamar *Illex argentinus*. Bol Inst Biol Mar 8:4–34

Castellanos ZA, Menni R (1968) Los cefalópodos de la expedición "Walter Herwig". Com Inv Cient Prov Bs As 6(2):1–31

Castello JP, Haimovici M, Odebrecht C, Vooren CM (1997) The continental shelf and slope. In: Seeliger U, Odebrecht C, Castello JP (eds) Subtropical convergence environments. The coast and sea in the Southwestern Atlantic, 1st edn. Springer-Verlag, Berlin, pp 171–178

Catul V, Gauns M, Karuppasamy PK (2011) A review on mesopelagic fishes belonging to family Myctophidae. Rev Fish Biol Fish 21:339–354

Cepeda GD, Temperoni B, Sabatini ME et al (2018) Zooplankton communities of the Argentine continental shelf (SW Atlantic, ca. 34°–55°S), an overview. In: Hoffmeyer M, Sabatini M, Brandini F et al (eds) Plankton ecology of the Southwestern Atlantic. Springer, New York, pp 171–199

Cherel Y, Pütz K, Hobson KA (2002) Summer diet of king penguins (*Aptenodytes patagonicus*) at The Falkland Islands, southern Atlantic Ocean. Polar Biol 25:898–906

Cheseva ZA (1995) The biology of Magellan hake (*Macruronus magellanicus*) from the Southwest Atlantic. J Ichthyol 35(3):29–39

Ciechomski JD, Booman CI (1981) Descripción de embriones y de áreas de reproducción de los granaderos *Macrourus whitsoni* y *Coelorhynchus fasciatus*, de la polaca *Micromesistius australis* y del bacalao austral *Salilota australis* en zona Patagonica y Fuegina del Atlantico Sudoccidental. Physis 40(98):5–14

Ciechomski JD, Weiss G (1974) Características del desarrollo embrionario y larval de las merluzas *Merluccius hubbsi y Merluccius polylepis*. Physis (A) 33(87):527–536

Ciechomski JD, Cassia M, Weiss G (1975) Distribución de huevos, larvas y juveniles de peces en los sectores bonaerenses, patagónicos y fueguinos del mar epicontinental argentino en relación con las condiciones ambientales, en noviembre 1973 – enero 1974. Ecosur 2(4):219–248

Ciechomski JD, Ehrlich MD, Lasta CA, Sánchez RP (1979) Campañas realizadas por el buque de investigación "Walther Herwig" en el Mar Argentino, desde mayo hasta noviembre de 1978. Organización y reseña de datos obtenidos. Contrib INIDEP 374, p 313

Ciechomski JD, Ehrlich MD, Lasta CA, Sánchez RP (1981) Distribución de huevos y larvas de peces en el Mar Argentino y evaluación de los efectivos de desovantes de anchoíta y merluza. Contrib INIDEP 383:59–79

Ciechomski JD, Sánchez RP, Lasta CA (1986) Evaluación de la biomasa de adultos desovantes, distribución vertical y variación cuantitativa de la intensidad de los desoves de la anchoíta (*Engraulis anchoita*) durante la primavera de 1982. Rev Invest Desarr Pesq 5:30–48

Clarke TA (1980) Diets of fourteen species of vertically migrating mesopelagic fishes in Hawaiian waters. Fish Bull 78:619–640

Cohen DM, Inada T, Iwamoto T et al (1990) FAO species catalogue.Vol. 10. Gadiform fishes of the world (Order Gadiformes). An annotated and illustrated catalogue of cods, hakes, grenadiers and other gadiform fishes known to date. FAO Fish Synop N° 125(10) Rome FAO, p 442

Collins MA, Ross KA, Belchier M et al (2007) Distribution and diet of juvenile Patagonian toothfish on the South Georgia and Shag Rocks shelves (Southern Ocean). Mar Biol 152(1):135–147

Collins MA, Brickle P, Brown J et al (2010) The Patagonian Toothfish. Adv Mar Biol:227–300

Collins MA, Stowasser G, Fielding S et al (2012) Latitudinal and bathymetric patterns in the distribution and abundance of mesopelagic fish in the Scotia Sea. Deep-Sea Res II Top Stud Oceanogr 59-60:189–198

Colonello JH (2009) Ecología reproductiva de tres batoideos (Chondrichthyes): *Atlantoraja castelnaui* (Rajidae), *Rioraja agassizii* (Rajidae) y *Zapteryx brevirostris* (Rhinobatidae). Implicancias de distintas estrategias adaptativas en un escenario de explotación intensiva. PhD Thesis, Universidad Nacional de La Plata, Argentina

Colonello JH, Cortés F, Massa AM (2014) Species richness and reproductive modes of chondrichthyans in relation to temperature and fishing effort in the Southwestern Atlantic Shelf (34–54° S). Fish Res 160:8–17

Colonello JH, Cortés F, Belleggia M (2020) Male-biased sexual size dimorphism in sharks: the narrowmouth catshark *Schroederichthys bivius* as case study. Hydrobiologia 847:1873–1886

Contreras T, Olivar MP, Bernal A, Sabatés A (2015) Comparative feeding patterns of early stages of mesopelagic fishes with vertical habitat partitioning. Mar Biol 162:2265–2277

Cordo HD (1981) Resultados sobre la alimentación de la merluza del mar epicontinental argentino (*Merluccius hubbsi*). Análisis biológico y estadístico de los datos obtenidos de las campañas de B/I "Shinkai Maru" y "Walther Herwig" (1978–1979). In Angelescu V (ed) Campañas de investigación pesquera realizadas en el Mar Argentino por los B/I "Shinkai Maru" y "Walther Herwig", años1978 y 1979. Resultados de la parte argentina. INIDEP, Mar del Plata, Argentina p 299–312

Cordo HD (2004) Abadejo (*Genypterus blacodes*). Caracterización biológica y estado del recurso. In: Sánchez RP, Bezzi SI (eds) El Mar Argentino y sus Recursos Pesqueros. Tomo 4. Los Peces Marinos de Interés Pesquero. Caracterización Biológica y Evaluación del Estado de Explotación. Instituto Nacional de Investigación y Desarrollo Pesquero (INIDEP), Mar del Plata, p 237–253

Cortés F, Waessle JA (2017) Hotspots for porbeagle shark (Lamna nasus) bycatch in the southwestern Atlantic (51°S–57°S). Can J Fish Aquat Sci 74(7):1100–1110

Cortés F, Waessle JA, Massa AM, Hoyle SD (2017) Aspects of porbeagle shark bycatch in the Argentinean surimi fleet operating in the Southwestern Atlantic Ocean (50–57°S) during 2006–2014. Western and Central Pacific Fisheries Commission. Rarotonga, Cook Islands. 9–17 August, 2017. Available at https://meetings.wcpfc.int/index.php/meetings/sc13

Cortés F, Colonello J, Hozbor NM, Perez MA, Belleggia M (2021) Áreas de agregación y tendencias en la abundancia de rayas de altura en el Atlántico Sudoccidental (34°-48°S). Inf Invest INIDEP 37:22

Cousseau MB, Perrotta RG (2000) Peces marinos de Argentina: biología, distribución, pesca. 4a. ed. Mar del Plata: INIDEP, p 193

Cousseau MB, Perrotta RG (2013) Peces marinos de Argentina: biología, distribución, pesca. 4a. ed. Mar del Plata: Instituto Nacional de Investigación y Desarrollo Pesquero INIDEP, p 193

Cousseau MB, Rosso JJ (2019) Fishes of Argentina: marine waters, 1st edn. Vázquez Mazzini Editores, Buenos Aires, p 152

Cousseau MB, Hansen JE, Gru DL (1979) Campañas realizadas por el buque de investigación "Shinkai Maru" en el Mar Argentino desde abril de 1978 hasta abril de 1979. Organización y reseña de datos básicos obtenidos. Contr INIDEP, p 373

Cousseau MB, Pequeño G, Mabragaña E, Lucífora LO, Martínez P, Giussi A (2020) The Magellanic Province and its fish fauna (South America): several provinces or one? J Biogeogr 47:220–234

García de la Rosa SB, Sanchez F, Prenski B (2004) Caracterización biológica y estado de explotación de la raya (*Dipturus chilensis*). In: Sánchez R, Bezzi S (eds) El Mar Argentino y sus recursos pesqueros. Tomo 4. Biología y evaluación del estado de explotación. INIDEP, Mar del Plata, p 53–66

Del Río-Iglesias JL, Martínez-Portela J, Patrocinio T (2009) Informe de la Campaña de Investigación Pesquera ATLANTIS 2009, 24 de febrero al 1 de abril, B/O Miguel Oliver. Dpto. Pesquerías Lejanas, Instituto Español de Oceanografía, Vigo, p 158

Delpiani GE (2016) Reproductive biology of the southern thorny skate Amblyraja doellojuradoi (Chondrichthyes, Rajidae). J Fish Biol 88(4):1413–1429

Delpiani GE, Spath MC, Figueroa DE (2013) Feeding ecology of the southern thorny skate, *Amblyraja doellojuradoi* on the Argentine continental Shelf. J Mar Biol Assoc UK 93(8):2207–2216

Díaz de Astarloa JM, Figueroa DE (1993) Las especies del género Genypterus (Pisces, Ophidiiformes) presentes en aguas argentinas. Facultad de Ciencias del Mar, Universidad Católica del Norte, Coquimbo, Chile, Serie Ocasional 2:47–56

Dudochkin AS (1988) The food of the grenadier, *Macrourus holotrachys*, in the south western Atlantic. J Ichthyol 28(2):72–76

Duhamel G, Hulley PA, Causse R, Koubbi P, Vacchi M, Pruvost P, Vigetta S, Irisson JO, Mormède S, Belchier M, Dettai A, Detrich HW, Gutt J, Jones CD, Kock KH, Lopez Abellan LJ, Van de Putte AP (2014) Chapter 7. Biogeographic patterns of fish. In: De Broyer C, Koubbi P, Griffiths HJ et al (eds) Biogeographic atlas of the Southern Ocean. Scientific Committee on Antarctic Research, Cambridge, pp 328–362

Ebert D, Compagno L (2007) Biodiversity and systematics of skates (Chondrichthyes: Rajiformes: Rajoidei). Environ Biol Fish 80:111–124

Ehrlich MD, Sánchez RP, Ciechomski JD, Machinandiarena L, Pájaro M (1999) Ichthyoplankton composition, distribution and abundance on the southern patagonian shelf and adjacent waters. INIDEP Doc Cient 5:37–65

Eleutério CLT (2008) Crescimento, idade e mortalidade do congro rosa *Genypterus brasiliensis* (Regan, 1903) da região Sudeste e Sul do Brasil. Dissertação de Pós-graduação, Instituto de Pesca São Paulo, Brazil, p 60

FAO (2020) Worldwide review of bottom fisheries in the high seas in 2016. FAO Fisheries and Aquaculture Technical Paper vol 657. Rome

Figueroa DE (2019) Clave de peces marinos del Atlántico Sudoccidental entre los 35°S y 56°S. INIDEP, Mar del Plata, p 365

Figueroa DE, Díaz de Astarloa JM, Martos P (1998) Mesopelagic fish distribution in the Southwest Atlantic in relation to water masses. Deep-Sea Res I Oceanogr Res Pap 45:317–332

Filippova JA (1972) New data on the squids (Cephalopoda: Oegopsida) from the Scotia Sea (Antarctic). Malacologia 11(2):391–406

Forselledo R (2012) Distribución, estructura poblacional y aspectos reproductivos del tiburón pinocho *Lamna nasus* (Bonaterre, 1788) en el Atlántico Sudoccidental. M.Sc. thesis, Universidad de la República de Uruguay

Gabbanelli V, Díaz de Astarloa JM, Gonzalez-Castro M, Vazquez DM, Mabragaña E (2018) Almost a century of oblivion: integrative taxonomy allows the resurrection of the longnose skate *Zearaja brevicaudata* (Marini, 1933) (Rajiformes; Rajidae). C R Biol 341(9–10):454–470

García de la Rosa SB (1998) Estudios de las interrelaciones tróficas de dos elasmobranquios de la plataforma continental del Mar Argentino, en relación con las variaciones espacio-temporales y ambientales. *Squalus acanthias* (Squalidae) y *Raja flavirostris* (Rajidae). PhD Thesis, Universidad Nacional de Mar del Plata, Argentina

García de la Rosa SB, Sanchez F (1997) Alimentación de *Squalus acanthias* y predación sobre merluza *Merluccius hubbsi* en el Mar Argentino entre 34°47°–47°S. Rev Invest Des Pesq 11:119–133

García de la Rosa SB, Giussi AR, Sánchez F (1997a) Distribución, estructura de tallas y alimentación de la merluza austral (*Merluccius australis*) en el Mar Argentino. Resúmenes expandidos del VII COLACMAR I:350–352

García de la Rosa SB, Sánchez F, Figueroa D (1997b) Comparative feeding ecology of Patagonian toothfish (*Dissostichus eleginoides*) in the southwestern Atlantic. CCAMLR Sci 4:105–124

García VB, Lucífora LO, Myers RA (2008) The importance of habitat and life history to extinction risk in sharks, skates, rays and chimaeras. Proc Royal Soc B 275(1630):83–89

Gartner JV Jr (1993) Patterns of reproduction in the dominant lanternfish species (pisces: Myctophidae) of the eastern Gulf of Mexico, with a review of reproduction among tropical-subtropical Myctophidae. Bull Mar Sci 52(2):721–750

Giussi AR, Hansen JE, Wöhler OC (2004). Biología y pesquería de la merluza de cola (Pisces: Macruronidae, *Macruronus magellanicus*). En: R.P. Sánchez & S.I. Bezzi (ed) Los peces marinos de interés pesquero. Caracterización biológica y evaluación del estado de explotación. El Mar Argentino y sus recursos pesqueros. Tomo 4. *Publicaciones especiales INIDEP*, Mar del Plata, p 321–346

Giussi AR, Sanchéz F, Wöhler OC et al (2010) Grenadiers (Pisces: Macrouridae) of the Southwest Atlantic Ocean: biologic and fishery aspects. Rev Invest y Desarr Pesq 20:19–33

Giussi AR, Gorini F, Di Marco E, Zavatteri A, Marí N (2016) Biology and fishery of the southern hake (*Merluccius australis*) in the Southwest Atlantic Ocean. Rev Invest Des Pesq 28:37–53

Gjøsaeter J, Kawaguchi K (1980) A review of the world's resources of mesopelagic fish. FAO Fish Tech Pap 193:1–151

Gordeeva NV (2011) On structure of species in pelagic fish: the results of Populational–genetic analysis of four species of Lanternfish (Myctophidae) from the Southern Atlantic. J Ichthyol 51(2):152–165

Gorini F, Pájaro M (2014) Características reproductivas y longitud de primera madurez de la merluza de cola (*Macruronus magellanicus*) en el Atlántico Sudoccidental. Período 2003-2010. Rev Invest Des Pesq 25:5–19

Haimovici M (1997) Demersal and benthic Teleosts. In: Seeliger U, Odebrecht C, Castello JP (eds) Subtropical convergence environments. The coast and sea in the Southwestern Atlantic, 1st edn. Springer-Verlag, Berlin, pp 129–136

Haimovici M, Álvarez Pérez JJ (1990) Distribución y maduración sexual del calamar argentino, *Illex argentinus* (Castellanos, 1960) (Cephalopoda, Ommastrephidae), en el sur de Brazil. Scient Mar 54(2):179–185

Haimovici M, Martins AS, Figueiredo JL, Vieira PC (1994) Demersal bony fishes of the outer shelf and uper slope of the southern Brazil subtropical convergence ecosystem. Mar Ecol Prog Ser 108:59–77

Haimovici M, Brunetti NE, Rodhouse PG et al (1998) *Illex argentinus* in Squid recruitment dynamics: the genus *Illex* as a model, the commercial Illex species and influence on variability (vol 376). Food & Agriculture Org

Haimovici M, Rossi-Wongstchowski CLDB, Bernardes RA et al (2008) Prospecção pesqueira de espécies demersais com rede de arrasto-de-fundo na Região Sudeste-Sul do Brasil. Série documentos Revizee, Score Sul. Instituto Oceanográfico, USP, São Paulo, p 183

Hansen JE (2004) Anchoíta (*Engraulis anchoita*). In: Sánchez R, Bezzi SI (eds) El Mar Argentino y sus recursos pesqueros, vol 4. Los peces marinos de interés pesquero. Caracterización biológica y evaluación del estado de explotación. Instituto Nacional de Investigación y Desarrollo Pesquero, Mar del Plata, Argentina, p 101–115

Hatanaka H (1986) Growth and life span of the short-finned squid *Illex argentinus* in the waters off Argentina. Bull Jap Soc Sci Fish 52(1):11–17
Hatfield EMC, Rodhouse PG (1994) Distribution and abundance of juvenile *Loligo gahi* in Falkland Island waters. Mar Biol 121:267–272
Hoenig JM, Gruber SH (1990) Life-history patterns in the elasmobranchs: implications for fisheries management. In: Pratt HL, Gruber SH, Taniuchi T (eds) Elasmobranchs as Living Resources: Advances in the Biology, Ecology, Systematics, and the Status of the Fisheries, NOAA Technical Report, pp 1–16
Holden MJ (1974) Ray migrations – Do bigger eggs mean better dispersal? Proc Challenger Soc 4:1–215
Hulley PA (1981) Results of the research cruises of FRV "Walther Herwig" to South America. LVIII. Family Myctophidae (Osteichthyes, Myctophiformes). Arch Fisch 31(1):1–300
Hulley PA (1991) Order myctophiformes. In: Smith MM, Heemstra PC (eds) Smith's sea fishes, 1st edn. Southern Book Publishers, Johannesburg, pp 282–321
Hulley PA (1992) Upper-slope distributions of oceanic lanternfishes (family: Myctophidae). Mar Biol 114:365–383
Hutchinson GE (1959) Homage to Santa Rosalia or why are there so many kinds of animals? Am Nat 93:145–159
Irusta CG, Pérez M, Simonazzi M, Castrucci R (2000) Aspectos biológicos pesqueros de la merluza en el sector del Mar Argentino comprendido entre 170 y 200 millas, entre 41°S – 48°S. Inf Téc INIDEP 87:11
Irusta CG, D'atri LL, Castrucci R (2010) Análisis y estimación de la CPUE de merluza (*Merluccius hubbsi*) correspondiente al efectivo patagónico localizado entre 41° S y 48° S entre los años 1986-2009. Inf Inv INIDEP 56:33
IUCN (2023) *The IUCN Red List of Threatened Species. Version 2023-1.* https://www.iucnredlist.org. Accessed on 25 Feb 2023
Ivanovic ML (2000) Alimentación y relaciones tróficas del calamar *Illex argentinus* en el ecosistema pesquero. PhD Thesis, Facultad de Ciencias Exactas y Naturales, Universidad Nacional de Mar del Plata, p 251
Ivanovic ML (2010) Alimentación del calamar *Illex argentinus* en la región patagónica durante el verano de los años 2006, 2007 y 2008. Rev Invest Des Pesq 20:51–63
Ivanovic ML, Brunetti NE (1994) Food and feeding of *Illex argentinus*. Antarct Sci 6:185–193
Ivanovic ML, Brunetti NE (2005) Informe preliminar campaña Kaiyo Maru 2005. Crucero conjunto argentino – japonés para el estudio de los juveniles del calamar argentino (*Illex argentinus*). Inf Téc INIDEP 94
Ivanovic ML, Brunetti NE, Elena B, Rossi GR (1998) A contribution to the biology of the ommastrephid squid *Martialia hyadesi* (Rochebrune and Mabille, 1889) from the SouthWest Atlantic. Afr J Mar Sci 20:73–79
Jackson GD, Buxton NG, George MJA (2000) The diet of the southern *Opah Lampris immaculatus* on the Patagonian Shelf; the significance of the squid *Moroteuthis ingens* and anthropogenic plastic. Mar Ecol Prog Ser 206:261–271
Jereb P, Roper CFE (2010) Cephalopods of the world, an annotated and illustrated catalogue of cephalopod species known to date. Volume 2. Myopsid and Oegopsid squids. FAO Species Catalogue for Fishery Purposes, 4, p 605
Kaartvedt S, Staby A, Aksnes DL (2012) Efficient trawl avoidance by mesopelagic fishes causes large underestimation of their biomass. Mar Ecol Prog Ser 456:1–6
Kinzer J, Schulz K (1985) Vertical distribution and feeding patterns of midwater fish in the central equatorial Atlantic. Mar Biol 85:313–322
Koen Alonso M, Crespo EA, Garcia NA et al (2001) Food habits of *Dipturus chilensis* (Pisces: Rajidae) off Patagonia, Argentina. ICES J Mar Sci 58:288–297
Koen Alonso M, Crespo EA, Garcia NA et al (2002) Fishery and ontogenetic driven changes in the diet of the spiny dogfish, *Squalus acanthias*, in Patagonian waters, Argentina. Environ Biol Fish 63:193–202

Konstantinova MP, Remeslo AV, Fedulov PP (1994) The distribution of Myctophíds (Myctophidae) in the Southwest Atlantic in relation to water structure and dynamics. J Ichthyol 34(7):151–160
Koronkiewicz A (1980) Size, maturity, growth and food of squid *Illex argentinus* (Castellanos, 1960). Int Coun Exp Sea, CM 18:18
Koronkiewicz A (1986) Growth and life cycle of squid *Illex argentinus* from Patagonian and Falkland Shelf and polish fishery of squid for this region, 1978–1985. ICES C. M. 1986/K: 27 Shellfish Commitee, p 16
Krefft G (1976) Distribution patterns of oceanic fishes in the Atlantic Ocean. Selected problems. Rev Trav Inst Pêches Marit 40:439–460
Kubodera T, Piatkowski U, Okutani T et al (1998) Taxonomy and zoogeography of the family Onychoteuthidae. Sm C Zoology 586:277–291
Lam V, Pauly D (2005) Mapping the global biomass of mesopelagic fishes. Sea Around Us Project Newsl 30:4
Laptikhovsky VV (2005) A trophic ecology of two grenadier species (Macrouridae, Pisces) in deep waters of the Southwest Atlantic. Deep-Sea Res I Oceanogr Res Pap 52(8):1502–1514
Laptikhovsky VV, Arkhipkin AI, Henderson AC (2001) Feeding habits and dietary overlap in spiny dogfish *Squalus acanthias* (Squalidae) and narrowmouth catshark *Schroederichthys bivius* (Scyliorhinidae). J Mar Biol Assoc UK 81:1015–1018
Laptikhovsky VV, Arkhipkin AI, Brickle P (2008) Biology and distribution of grenadiers of the family Macrouridae around the Falkland Islands. In Grenadiers of the World Oceans: Biology, Stock Assessment, and Fisheries Am Fish Soc Symp 63, p 24
Lattuca ME, Llompart F, Avigliano E, Renzi M, De Leva I, Boy CC, de Albuquerque CQ (2020) First insights into the growth and population structure of *Cottoperca trigloides* (Perciformes, Bovichtidae) from the Southwestern Atlantic Ocean. Front Mar Sci 7:421
Leta HR (1992) Abundance and distribution of Rhynchoteuthion larvae of *Illex argentinus* (Cephalopoda: Ommastrephidae) in the south-western Atlantic. S Afr J Marine Scie 12(1):927–941
Longhurst A (1998) Ecological geography of the sea. Academic Press, New York
Lourenço S, Saunders RA, Collins MA et al (2017) Life cycle, distribution and trophodynamics of the lanternfish *Krefftichthys anderssoni* (Lönnberg, 1905) in the Scotia Sea. Polar Biol 40:1229–1245
Lucífora LO, Valero JL, Bremec CS, Lasta ML (2000) Feeding habits and prey selection by the skate *Dipturus chilensis* (Elasmobranchii: Rajidae) from the south-western Atlantic. J Mar Biol Assoc UK 80:953–954
Lucífora LO, García VB, Menni RC, Worm B (2012) Spatial patterns in the diversity of sharks, rays, and chimaeras (Chondrichthyes) in the Southwest Atlantic. Biodivers Conserv 21:407–419
Mabragaña E, Giberto DA (2007) Feeding ecology and abundance of two sympatric skates, the shortfin sand skate *Psammobatis normani* McEachran, and the smallthorn sand skate *P. rudis* Günther (Chondrichthyes, Rajidae), in the Southwest Atlantic. ICES J Mar Sci 64:1017–1027
Mabragaña E, Giberto DA, Bremec CS (2005) Feeding ecology of *Bathyraja macloviana* (Rajiformes: Arhynchobatidae): a polychaete-feeding skate from the south-west Atlantic. Sci Mar 69:405–413
Mabragaña E, Díaz de Astarloa JM, Hanner R et al (2011) DNA barcoding identifies Argentine fishes from marine and brackish waters. PLoS One 6(12):e28655
Macchi GJ, Pájaro M, Dato C (2007) Spatial variations of the Argentine hake (*Merluccius hubbsi* (Marini, 1933)) spawning shoals in the Patagonian area during a reproductive season. Rev Biol Mar Oceanog 42(3):345–356
Machinandiarena L (1999) Biología reproductiva de polaca (*Micromesistius australis*), merluza de cola (*Macruronus magellanicus*) y abadejo (*Genypterus blacodes*). Avances en métodos y tecnología aplicados a la investigación pesquera. Seminario Final Proyecto INIDEP-JICA sobre evaluación y monitoreo de recursos pesqueros, 1994–1999 Sección 1, p 61–64
Machinandiarena L, Ehrlich M (1999) Detección de un área de cría de la merluza de cola (*Macruronus magellanicus*) en el Mar Argentino. Rev Invest Des Pesq 12:45–50

Macpherson E, Hastings PA, Robertson DR (2009) Macroecological patterns among marine fishes. In: Witman JD, Kaustuv R (eds) Marine macroecology. University of Chicago Press, London, pp 122–152

Madirolas A (1999) Acoustic surveys on the southern blue whiting (*Micromesistius australis*) Mar del Plata: Instituto Nacional de Investigación y Desarrollo Pesquero INIDEP Doc Cient 5:81–93. http://hdl.handle.net/1834/2567

Marí NR, Sánchez F (2002) Espectros tróficos específicos de varias especies de peces demersales de la región austral y sus variaciones anuales entre 1994 y 2000. Inf Téc Int INIDEP 88:9

McGinnis RF (1982) Biogeography of lanternfishes (Myctophidae) south of 30° S. Amercian Geophysical Union, Washington

McKeown NJ, Arkhipkin AI, Shaw PW (2017) Regional genetic population structure and fine scale genetic cohesion in the southern blue whiting *Micromesistius australis*. Fish Res 185:176–184

Menni RC, Ringuelet RA, Aramburu RH (1984) Peces marinos de la Argentina y Uruguay. Editorial Hemisferio Sur S.A, Buenos, Aires, p 359

Moiseev SI (1991) Observation of the vertical distribution and behaviour of nektonic squids using manned submersibles. Bull Mar Sci 49:446–456

Morley SA, Mulvey T, Dickson J et al (2004) The biology of the bigeye grenadier at South Georgia. J Fish Biol 64(6):1514–1529

Mouat B, Collins MA, Pompert J (2001) Patterns in the diet of *Illex argentinus* (Cephalopoda: Ommastrephidae) from the Falklands Islands jigging fishery. Fish Res 52:41–49

Nascimento MC (2012) Alimentação e relações tróficas de peixes demersais marinhos da região sudeste e sul do Brasil. Tese de doutorado, Universidade Estadual de Campinas, Brazil, p 145

Nelson JS (2016) Fishes of the world, 5th edn. Wiley, New Jersey, p 707

Nesis KN (1987) Cephalopods of the world. In: Burgess LA (ed) Squids, cuttlefishes, octopuses and allies. TFH Publications Inc. Ltd, London, p 351

Nielsen JG, Cohen DM, Markie DF, Robins CR (1999) FAO species catalogue. Ophidiiform fishes of the world. (order Ophidiiformes). FAO Fish Synop 18:1–178

Nigmatullin CM (1989) Las especies de calamar más abundantes del Atlántico Sudoeste y sinopsis sobre la ecología del calamar (*Illex argentinus*). Frente Marítimo 5(A):71–82

Nyegaard M, Arkhipkin A, Brickle P (2004) Variation in the diet of *Genypterus blacodes* (Ophidiidae) around The Falkland Islands. J Fish Biol 65:666–682

Ocampo Reinaldo M, González R, Romero MA (2011) Feeding strategy and cannibalism of the Argentine hake *Merluccius hubbsi*. J Fish Biol 79:1795–1814

Olivar MP, Bode A, López-Pérez C et al (2019) Trophic position of lanternfishes (Pisces: Myctophidae) of the tropical and equatorial Atlantic estimated using stable isotopes. ICES J M Sci 76(3):649–661

Olson DB (2002) Biophysical dynamics of ocean fronts. In: Robinson AR, McCarthy JJ, Rothschild BJ (eds) The sea biological-physical interactions in the sea, vol 12. Wiley, New York, pp 187–218

Otero HO, Simonazzi MA (1980) Los recursos pesqueros demersales del Mar Argentino. Parte I. Evaluación de la biomasa (*standing stock*) de la merluza común (*Merluccius hubbsi*) y de la merluza austral (*Merluccius polylepis*) en el área de distribución estival. Rev Invest Desarr Pesq 2:5–12

Oyarzún C, Campos P, Valeria HR (1988) Adaptaciones para la flotabilidad en *Dissostichus eleginoides* Smitt, 1898 (Pisces, Perciformes, Nototheniidae). Invest Pesq (Spain) 52(4):455–466

Pájaro M, Macchi GJ, Gorini F (2005) Detección de un área de puesta de merluza negra (*Dissostichus eleginoides*) sobre la base del análisis histológico. Inf Tec Int INIDEP 87(05):8

Pájaro M, Macchi GJ, Martínez PA et al (2009) Características reproductivas de dos agregaciones de merluza negra (*Dissostichus eleginoides*) del Atlántico Sudoccidental. Inf Tec Int INIDEP 49(09):16

Pakhomov EA, Perissinotto R, McQuaid CD (1996) Prey composition and daily rations of myctophid fishes in the Southern Ocean. Mar Ecol Prog Ser 134:1–14

Perez Comas JA (1980) Distribución, áreas de concentración y estructura de la población del bacalao austral (*Salilota australis*, Gunther 1887) del Atlántico Sudoccidental. Rev Invest Desarr Pesq INIDEP 2:23–37

Perrotta RG (1982) Distribución y estructura poblacional de la polaca (*Micromesistius australis*). Rev Invest Desarr Pesq 3:35–50

Phillips KL, Nichols PD, Jackson GD (2003) Dietary variation of the squid *Moroteuthis ingens* at four sites in the Southern Ocean: stomach contents, lipid, and fatty acid profiles. J Mar Biol Assoc UK 83:523–534

Pierce GJ, Santos MB, Bishop A et al (2002) The trophic relationships of several commercial finfish species from the southwest Atlantic. ICES CM 2002/N:13 (Theme Session on Environmental Influences on Trophic Interactions), p 21

Pineda SE, Aubone A, Brunetti NE (1996) Identificación y morfometría de las mandibulas de *Loligo gahi* y *Loligo sanpaulensis* (Cephalopoda, Loliginidae) del Atlántico Sudoccidental. Rev Invest Des Pesq 10:85–99

Pineda SE, Brunetti NE, Scarlato NA (1998) Calamares loligínidos (Cephalopoda, Loliginidae). El Mar Argentino y sus Recursos Pesqueros 2:13–36

Prenski LB, Almeyda S (2000) Some biological aspect relevant to Patagonian toothfish (*Dissostichus eleginoides*) exploitation in the Argentine exclusive economic zone and adjacent ocean sector. Frente Marit 18(A):103–124

Prenski LB, Ehrhardt NM, Legault C (1997) Evaluación del estado de explotación de la merluza de cola (*Macruronus magellanicus*) en la plataforma sudpatagónica Argentina. Rev Invest Des Pesq 11:5–17

Pusch C, Hulley PA, Kock KH (2004) Community structure and feeding ecology of mesopelagic fishes in the slope waters of King George Island (South Shetland Islands, Antarctica). Deep-Sea Res 51:1685–1708

Renzi M (1986) Aspectos biológico-pesqueros del abadejo (*Genypterus blacodes*). Rev Invest Desarr Pesq 6:5–19

Renzi M, Santos B, Simonazzi M (2003) Estructura por edad y sexo de la población de merluza. Aportes para la evaluación del recurso merluza (*Merluccius hubbsi*) al sur de los 41° S. Año 1999. Ser Inf Téc INIDEP 51:57–76

Robinson C, Steinberg DK, Anderson TR et al (2010) Mesopelagic zone ecology and biogeochemistry – a synthesis. Deep-Sea Res II Top Stud Oceanogr 57:1504–1518

Rodhouse PG (1991) Population structure of *Marlialia hyadesi* (Cephalopoda: Ommastrephidae) at the Antartic Polar Front and the Patagonian Shelf, South Atlantic. Bulletin Mar Sci 49(1–2):404–418

Rodhouse PG, Lu CC (1998) *Chiroteuthis veranyi* from the Atlantic sector of the Southern Ocean (Cephalopoda: Chiroteuthidae). S Afr J Marine Sci 20:311–322

Rodhouse PG, Symon C, Hatfield EMC (1992) Early life cycle of cephalopods in relation to the major oceanographic features of the Southwest Atlantic Ocean. Mar Ecol Prog Ser 89:183–195

Romanelli PJ (2017) Ecología trófica de tres especies de granaderos *Coelorhinchus fasciatus*, *Macrourus carinatus* y *Macrourus holotrachys*, presentes en el Océano Atlántico Sudoccidental. Thesis, Universidad Nacional de Mar del Plata

Rosas-Luis R, Sánchez P, Portela JM et al (2014) Feeding habits and trophic interactions of *Doryteuthis gahi*, *Illex argentinus* and *Onykia ingens* in the marine ecosystem off the Patagonian Shelf. Fish Res 152:37–44

Ruiz AE, Fondacaro RR (1997) Diet of hake (*Merluccius hubbsi* Marini) in a spawning and nursery area within Patagonian shelf waters. Fish Res 30:157–160

Ruocco NL, Lucífora LO, Díaz de Astarloa JM, Wöhler O (2006) Reproductive biology and abundance of the white–dotted skate, *Bathyraja albomaculata*, in the Southwest Atlantic. ICES J Mar Sci 63:105–116

Ruocco NL, Lucífora LO, Díaz de Astarloa JM, Bremec C (2009) Diet of the white-dotted skate, *Bathyraja albomaculata*, in waters of Argentina. J Appl Ichthyol 25:94–97

Sabadin DE, Lucífora LO, Barbini SA, Figueroa DE, Kittlein M (2020) Towards regionalization of the chondrichthyan fauna of the Southwest Atlantic: a spatial framework for conservation planning. ICES J Mar Sci 77:1893–1905
Sabatés A, Olivar MP (1989) Comparative spawning strategies of mesopelagic fishes in two marine systems with different productivity. Rapports des Procès Verbaux du Conseil International pour l'Exploration de la Mer 191:27–33
Sabatés A, Saiz E (2000) Intra-and interspecific variability in prey size and niche breadth of myctophiform larvae. Mar Ecol Prog Ser 201:261–271
Sabatini M, Alvarez Colombo GL, Ramirez FC (1999) Zooplankton biomass in the reproductive area of the southern blue whiting (*Micromesistius australis*). INIDEP Doc Cient 5:23–35
Salvanes AG, Kristoffersen JB (2001) Mesopelagic fishes. In: Steele J (ed) Encyclopedia of ocean sciences, 3rd edn. Academic Press, London, pp 1711–1717
Sánchez F (2009) Alimentación de la merluza (*Merluccius hubbsi*) en el Golfo San Jorge y aguas adyacentes. Inf Tec INIDEP 75:1–21
Sánchez F, García de La Rosa SB (1999) Alimentación de *Merluccius hubbsi* e impacto del canibalismo en la región comprendida entre 34°50'S – 47°S del Atlántico Sudoccidental. Rev Invest Des Pesq 12:77–93
Sánchez F, Mabragaña E (2002) Características biológicas de algunas rayas de la región sud patagónica. Inf Tec INIDEP 48:1–15
Sánchez F, Marí NR (2005) Interacciones tróficas entre especies de peces demersales en la región austral entre 45°S y 54°S. Inf Téc Int INIDEP 91:9
Sánchez F, Prenski LB (1996) Ecología trófica de peces demersales en el Golfo San Jorge. Rev Invest Des Pesq 10:57–71
Sánchez F, Marí NR, Bernardele JC (2009) Distribución, abundancia y alimentación de pintarroja *Schroederichthys bivius* Müller & Henle, 1838 en el Océano Atlántico sudoccidental. Rev Biol Mar Oceanog 44:453–466
Saunders RA, Collins MA, Ward P et al (2015) Trophodynamics of *Protomyctophum* (Myctophidae) in the Scotia Sea (Southern Ocean). J Fish Biol 87:1031–1058
Saunders RA, Collins MA, Stowasser G, Tarling GA (2017) Southern Ocean mesopelagic fish communities in the Scotia Sea are sustained by mass immigration. Mar Ecol Prog Ser 569:173–185
Scenna LB (2011) Biología y ecología reproductiva de las especies del género *Bathyraja* (Elasmobranchii: Rajidae) en la Plataforma Continental Argentina. PhD Thesis, Universidad Nacional de Mar del Plata, Argentina
Scenna LB, García de la Rosa SB, Díaz De Astarloa JM (2006) Trophic ecology of the Patagonian skate, *Bathyraja macloviana*, on the Argentine continental shelf. ICES J Mar Sci 63:867–874
Shimabukuro V (2009) Hábitos alimentarios y dentición de *Bathyraja albomaculata* (Norman, 1937) (Chondrichthyes: Rajidae). M.Sc. thesis, Universidad Nacional de Mar del Plata, Argentina
Skomal G, Marshall H, Galuardi B et al (2021) Horizontal and vertical movement patterns and habitat use of juvenile porbeagles (*Lamna nasus*) in the Western North Atlantic. Front Mar Sci 8:624158
Smith SE, Au DW, Show C (1998) Intrinsic rebound potentials of 26 species of Pacific sharks. Mar Freshw Res 41:663–678
Soto JMR, Montealegre-Quijano S (2012) Elevación de Río Grande, una importante área de cría del tiburón sardinero *Lamna nasus* en el Atlántico Sur. II Simposio Iberoamericano de ecología reproductiva, reclutamiento y pesquerías. Mar del Plata, Argentina. November 19–22, 2012
Suneetha KB, Salvanes AGV (2001) Population genetic structure of the glacier lanternfish, *Benthosema glaciale* (Myctophidae) in Norwegian waters. Sarsia 86:203–212
Thompson KR (1994) Predation on *Gonatus antarcticus* by Falkland Islands seabirds. Antarct Sci 6:269–274
Torres Alberto ML, Bodnariuk N, Ivanovic M, Saraceno M, Acha EM (2021) Dynamics of the confluence of Brazil and Malvinas currents, and a southern Patagonian spawning ground, explain recruitment fluctuations of the main stock of *Illex argentinus*. Fish Oceanogr 2020:1–15

Torres Alberto ML, Saraceno M, Ivanovic M, Acha EM (2022) Habitat of Argentine squid (*Illex argentinus*) paralarvae in the southwestern Atlantic. Mar Ecol Prog Ser 688:69–82
Troccoli GH, Aguilar E, Martínez PA et al (2020) The diet of the Patagonian toothfish *Dissostichus eleginoides*, a deep-sea top predator off Southwest Atlantic Ocean. Polar Biol 43:1595
Vazquez DM, Mabragaña E, Gabbanelli V, Díaz De Astarloa JM (2016) Exploring nursery sites for oviparous chondrichthyans in the Southwest Atlantic (36°S–41°S). Mar Biol Res 12:715–725
Viera M (2011) Características ecomorfométricas de los otolitos sagitta de *Genypterus Blacodes* y *Genypterus Brasiliensis* provenientes de la zona común de pesca Argentino-Uruguaya. Tesis de grado, Universidad de la República, Uruguay, p 30
Villa A, Colonello J, Belleggia M (2018) Ecología trófica de *Schroederichthys bivius* en el Golfo San Jorge y aguas adyacentes. Jornadas Nacionales de Ciencias del Mar. Buenos Aires, Argentina. 30 July-3 August, 2018
Villarino MF (1997) Evolución de las capturas de abadejo (*Genypterus blacodes*) versus merluza común (*Merluccius hubbsi*) por mes y área de pesca durante los años 1987-1990. INIDEP Inf Téc 12:16
Villarino MF (1998) Distribución estacional y estructura de tallas del abadejo (*Genypterus blacodes*) en el Mar Argentino. INIDEP Inf Téc 18:25
Waessle JA, Cortés F (2011) Captura incidental, distribución y estructura de tallas de *Lamna nasus* en aguas argentinas (período 2006—2010). Inf Invest INIDEP 84:1–11
Waluda CM, Trathan PN, Rodhouse PG (1999) Influence of oceanographic variability on recruitment in the *Illex argentinus* (Cephalopoda: Ommastrephidae) fishery in the South Atlantic. Mar Ecol Progr Ser 183:159–167
Waluda C, Rodhouse P, Podestá G et al (2001) Surface oceanography of the inferred hatching grounds of *Illex argentinus* (Cephalopoda: Ommastrephidae) and influences on recruitment variability. Mar Biol 139:671–679
Weigmann S (2016) Annotated checklist of the living sharks, batoids and chimaeras (Chondrichthyes) of the world, with a focus on biogeographical diversity. J Fish Biol 88(3):837–1037
Weiss G (1974) Hallazgo y descripción de larvas de polaca (*Micromesistius australis*) en aguas del sector patagónico (Pisces, Gadidae). Physis Secc A 33:537–542
Williams A, Koslow JA, Terauds A, Haskard K (2001) Feeding ecology of five fishes from the mid-slope micronekton community off southern Tasmania, Australia. Mar Biol 139:1177–1192
Winter A, Pompert J, Arkhipkin A, Brewin PE (2015) Interannual variability in the skate assemblage on the South Patagonian shelf and slope. J Fish Biol 87:1449–1468
Wöhler OC (1987) Contribución al estudio de la distribución batimétrica de algunas especies de peces demersales y calamares del Mar Argentino. Tesis de Licenciatura, Facultad de Ciencias Exactas y Naturales, Universidad Nacional de Mar del Plata, p 78
Wöhler OC, Giussi AR, García de La Rosa S et al (1999) Resultados de la Campaña de Evaluación de Peces Demersales Australes efectuada en el verano de 1997. Inf Téc INIDEP 24:1–70
Wöhler OC, Cassia MC, Hansen JE (2004a) Caracterización biológica y evaluación del estado de explotación de la polaca (*Micromesistius australis*). En Sanchéz RP, Bezzi SI (eds) El Mar Argentino y sus Recursos Pesqueros Publicaciones Especiales INIDEP Mar del Plata 4:283–305
Wöhler OC, Cassia MC, Hansen JE (2004b) Biología y pesquería del bacalao austral (*Salilota australis*). En Sánchez RP, Bezzi SI (eds) El Mar Argentino y sus Recursos Pesqueros, Publicaciones Especiales INIDEP Mar del Plata 4:347–359
Zintzen V, Anderson MJ, Roberts CD, Harvey ES, Stewart AL, Struthers CD (2012) Diversity and composition of demersal fishes along a depth gradient assessed by baited remote underwater stereo-video. PLoS One 7(10):e48522

Chapter 6
Benthic Assemblages and Biodiversity Patterns of the Shelf-Break Front

Diego A. Giberto, Laura Schejter, María Virginia Romero, Mauro Belleggia, and C. S. Bremec

Abstract Although the shelf-break frontal region covers a distance of around 2000 kilometers, past studies of the benthic organisms mostly included information based on qualitative surveys with a narrow geographic scope. The body of knowledge has been expanding over the past few decades, leading to current attempts towards an all-encompassing ecosystemic approach. In this chapter, we provide a summary of the benthic assemblages and diversity trends along the frontal ecosystem, from 36 °S to 50 °S. In particular, we describe the megabenthic assemblages using data from surveys of the Patagonian scallop *Zygochlamys patagonica*, the Argentine hake *Merluccius hubbsi,* and the Argentine squid *Illex argentinus* fisheries. With the use of this information, the shelf-break front's most characteristic species—a total of about 150 taxa—are presented. Some diversity trends with latitude and depth are also shown. Information on deep habitats, including submarine canyons, is also compiled. Therein, a huge number of new species were recently described.

Supplementary Information The online version contains supplementary material available at https://doi.org/10.1007/978-3-031-71190-9_6.

D. A. Giberto (✉) · L. Schejter · M. Belleggia
Consejo Nacional de Investigaciones Científicas y Técnicas (CONICET), Buenos Aires, Argentina

Laboratorio de Bentos, Instituto Nacional de Investigación y Desarrollo Pesquero (INIDEP), Mar del Plata, Argentina

Instituto de Investigaciones Marinas y Costeras (IIMyC), Mar del Plata, Argentina
e-mail: diegogiberto@inidep.edu.ar

M. V. Romero
Consejo Nacional de Investigaciones Científicas y Técnicas (CONICET), Buenos Aires, Argentina

Instituto de Investigaciones Marinas y Costeras (IIMyC), Mar del Plata, Argentina

C. S. Bremec
Consejo Nacional de Investigaciones Científicas y Técnicas (CONICET), Buenos Aires, Argentina

E. M. Acha et al. (eds.), *The Patagonian Shelfbreak Front*, Aquatic Ecology Series 13, https://doi.org/10.1007/978-3-031-71190-9_6

The importance of biodiversity boosters, such as sclerobionts, is highlighted. Several Vulnerable Marine Ecosystems (VMEs) were recorded beyond 200 m deep, where coral gardens, coral reefs, rocky environments, submarine canyons, and sponge beds may act as hotspots of biodiversity. Due to the challenges of sampling in deep environments, studying these kinds of ecosystems remains a challenging task. In conclusion, the biological patterns discussed in this chapter call attention to the need for conservation efforts in order to preserve typical regions of the system's entire benthic diversity.

6.1 Introduction

The oceans of the world are subject to human influence, with several stressors having an effect on practically the whole ocean. The many stressors associated with climate change dominate humanity's footprint on the open ocean (anomaly high sea surface temperatures, ocean acidification, and increasing ultraviolet radiation), but commercial fishing and shipping also cover large areas of the oceans and contribute significantly to overall impact. The least impacted areas are mostly near the poles, although they also encompass relatively extensive areas such as temperate ecoregions around Argentina (Halpern et al. 2015). However, as marine ecosystems are influenced by global and regional processes, standardized information on community structure has become crucial for assessing broad-scale responses to natural and anthropogenic disturbances. Extensive bottom regions present numerous theoretical and methodological challenges for understanding community patterns on a macroecological scale. In particular, the shelf-break front is composed of a complex system of soft sediments and punctual areas of hard bottoms, with contrasting histories and geophysical-chemical environments.

The Argentinean continental shelf is extensive, its width reaches nearly 550 km at 35 °S (Río de la Plata) and more than 2000 km at 52 °S, where Malvinas Islands are located (Parker et al. 1997). The continental slope is associated with the 200 m isobath, although it varies between 110 and 165 m depth in many areas. Sediments are mainly sandy / muddy at the slope, although shells, gravel, or rocks are also found. Deep submarine canyons break through the continental margin and slope (Lonardi and Ewing 1971; Parker et al. 1997; Bozzano et al. 2017, see Chap. 2 of this book). Water masses in the shelf and slope area develop a thermo-haline shelf-break front, including the development of strong thermo- and pycnoclines (30–40 m depth). Bottom water temperature in 100 m water depth is about 6–8 °C during this season. Thermo- and pycnoclines are broken by vertical mixing due to convective processes during autumn-winter, thus leading to about 1–2 °C higher bottom water temperatures (Guerrero and Piola 1997).

The shelf-break frontal area in the Argentine Sea is one of the most productive ecosystems in the southwest (SW) Atlantic Ocean (Acha et al. 2004, 2015). Regarding commercial resources, the importance of the shelf-break front is well known, at least for several fish and invertebrate species. The front plays an important role in the northward feeding migration of the hake (*Merluccius hubbsi*), and the anchovy (*Engraulis anchoita*), as the fish are closely associated with it for 5–6 months of the year. Myctophiids are the most abundant small pelagic fish in the area, and their larvae occurred throughout the year. Dense concentrations of the short fin squid *Illex argentinus* and the Patagonian scallop *Zygochlamys patagonica* occur along the shelf-break (Acha et al. 2004). Increased phytoplankton densities along the front intensify carbon exports to the bottoms with a direct effect on benthic-pelagic interactions, positively impacting on the benthic communities (Marrari et al. 2017). Indeed, dense beds of the Patagonian scallop are located in coincidence with the shelf-break front in depths between 90 and 150 m. This spatial pattern is attributed to the high productivity of the front (Lasta and Bremec, 1998).

Despite the large extent of the study region (around 2000 kilometers long), most studies of benthic communities have been historically limited to qualitative surveys and/or have been geographically restricted (mainly fisheries focused). Fortunately, present knowledge about benthic communities of the shelf-break front has been growing during the last decades, starting with studies of limited spatial coverage (basically using by-catch samples linked to the development of the scallop fishery) and essentially describing the presence or distribution of different faunal groups (Lasta and Bremec, 1998, 1999) to present-day attempts of advancing to a comprehensive ecosystemic approach (e.g., Acha et al. 2004; Schejter et al. 2017). In order to link biodiversity patterns with key ecosystem processes and to develop the ecological basis for the management and conservation of the shelf-break communities, we present a synoptic view of benthic assemblages and diversity patterns along the frontal ecosystem, from 50 °S to 36 °S. We also identify the major gaps in the knowledge of benthic diversity of the region.

6.2 Patagonian Scallop Fishery Assemblages

6.2.1 Diversity and Assemblages

The bottom is characterized by the presence of extensive beds of the Patagonian scallop *Zygochlamys patagonica*, a pectinid species exploited by scallopers since 1996 and distributed between 37 °S and 47 °S along the 100 m isobath (Lasta and Bremec 1998; Schejter et al. 2017) (Fig. 6.1). This commercial resource and the invertebrate by-catch are yearly monitored as part of a scallop management plan that defines different management units (Schejter et al. 2017; Campodónico et al. 2020). Early research on composition and structure of the by-catch in spatial and temporal scales was developed (Bremec and Lasta 2002; Escolar et al. 2011; Mauna et al. 2011; Schejter et al. 2012a, 2017) as well as studies on production and trophic interactions (Bremec et al. 2000; Schejter et al. 2002; Botto et al. 2006; Souto 2009).

The total number of taxa sampled during the monitoring of the Patagonian scallop fishing grounds usually reaches around 90 (Schejter et al. 2014). However, research on different taxonomical groups gives information on richness of sponges

Fig. 6.1 (**a**) Typical catch at the Patagonian scallop fishing grounds (scale bar = 200 mm). B, C. Shell of the Patagonian scallop, *Zygochlamys patagonica*, without (**b**) and with (**c**) epibiotic organisms (polychaetes, mainly serpulids, and brachiopods) (scale bar = 10 mm). D-L. Common invertebrates from the bycatch of the fishery: (**d**). Sea urchin *Sterechinus agassizii* (scale bar = 20 mm). (**e**). Basket star *Gorgonocephalus chilensis* (scale bar = 60 mm). (**f**). The volutid snail *Adelomelon ancilla* (scale bar = 30 mm). (**g**). Ophiuroid *Ophiacantha vivipara* (scale bar = 10 mm). (**h**). The spider crab *Libidoclaea granaria* (scale bar = 20 mm). (**i**). The sea star *Diplopteraster clarki* (scale bar = 25 mm). (**j**). The hairy snail *Fusitriton magellanicus*, with several epibiotic organisms in the apex (sponges and polychaetes) (scale bar = 12 mm). (**k**). The sea urchin *Austrocidaris canaliculata* with several epibionts on the spines (hydrozoans and the lepadomorph *Weltnerium gibberum*) ((scale bar = 30 mm). (**l**). The brachiopod *Magellania venosa* with epibiotic sea anemones (i.e., the red anemone *Isotealia antarctica*) (scale bar = 30 mm)

(Schejter et al. 2006, 2008a, 2011a), echinoderms (Escolar and Bremec 2015), hydroids (Genzano et al. 2009), bryozoans (López Gappa and Landoni 2009), infaunal species (Sánchez et al. 2011), demersal and benthic frequent fish (Schejter et al. 2012a), and sponges's endobionts (Schejter et al. 2012b) in those areas. As a result, the known benthic invertebrate's richness reaches nearly 250 species (Schejter et al. 2013), including more than 50 scallop epibionts (Schejter and Bremec 2007a, 2009, López Gappa and Landoni 2009) (Fig. 6.1). Epibiotic associations highly contribute to benthic richness, as organisms provide settlement substrate in absence of hard bottoms (see *Epibiosis and bioerosion: biodiversity enhancers* below).

The Patagonian scallop behaves as an ecosystem engineer and provides substrate and refuge to a variety of associated organisms (Schejter and Bremec 2007a, 2009). The main assemblage of benthic invertebrates associated to the scallops in fishing grounds is composed by the sponge *Tedania* sp., the anemone *Actinostola crassicornis* and the echinoderms *Ophiactis asperula*, *Ophiacantha vivipara*, *Ophiura lymani*, *Sterechinus agassizii*, *Diplasterias brandti*, *Ctenodiscus australis*, *Psolus patagonicus,* and *Pseudocnus dubiosus* (Bremec and Lasta 2002, Bremec et al. 2003) (Fig. 6.1), with some variations along different areas of the front. Some fishing areas between 38 °S and 40 °S exhibit high density of the tubiculous worm *Chaetopterus antarcticus* and the frequent echinoderms *Labidiaster radiosus* and *Gorgonocephalus chilensis*. Other common taxa found are the sea stars *Ctenodiscus australis* and *Diplasterias brandti*, the urchin *Austrocidaris canaliculata* and the ophiuroid *Ophiactis asperula*.

Patches with high density of the hermit crab *Sympagurus dimorphus* and the urchin *Sterechinus agassizii* were also sampled (Schejter and Bremec 2007b; Schejter and Mantelatto 2015; Escolar 2010). Southern scallop beds (40 °S to 41° 30′ S) are less dense and exhibit higher species richness than the heavily exploited northern ones, as well as dominance of sponges in the catches that reach 90% of wet biomass (Schejter and Bremec 2013). On the other side, the fishing grounds located nearly at 42 °S and southwards show very high biomass of ophiuroids, mainly *Ophiactis asperula* and *Ophiacantha vivipara* and patches of the coral *Flabellum* cf. *curvatum* and the urchin *Sterechinus agassizii* (Escolar 2010).

6.2.2 *Trophic Links and Productivity*

Hydrodynamic conditions, seasonality of the shelf-break front, variations in species recruitment, trophic interactions, and fishing trawling influence the abundance and distribution patterns of the different species that inhabit the areas where Patagonian scallop is exploited (Mauna et al. 2011; Bremec et al. 2015). The high benthic production along the shelf break front, in areas with dense aggregations of pectinids (Bremec et al. 2000; Bogazzi et al. 2005 Souto 2009), is sustained by food supply from the photic zone (Schejter et al. 2002 Acha et al. 2004, 2015). Scallops contribute nearly 2/3 of carbon flux in the environment (Bremec et al. 2000), as they feed on phytoplankton from the surface that reaches the bottom (Schejter 2000; Schejter et al. 2002). Studies on epifaunal benthic secondary production can be found in Bremec et al. (2000) and Souto (2009), where conversion models were used to estimate the productivity of the community associated with *Zygochlamys patagonica* in the 39 °S. The total community values found in these studies varied between 45.22 $KJ.m^{-2}.year^{-1}$ and 67.57 $KJ.m^{-2}.year^{-1}$. These numbers were lower than those estimated in the infaunal communities of the San Jorge Gulf, with an average community production value of 87 $KJ.m^{-2}.year^{-1}$ (up to 152 $KJ.m^{-2}.year^{-1}$) (Suby et al. 2019a). On the other hand, the productivity values found for some sectors of the Patagonian coast (between 40 and 56 KJ.m^{-2}.year^{-1}) were within the range of the values estimated for the *Z. patagonica* beds (Suby et al. 2019a, b).

Secondary production patterns can be explained by differences in the oceanographic conditions at each ecosystem (high temperatures at coastal areas), differences in the community composition (polychaetes dominance vs. bivalves dominance), and differences in the energy efficiency with which different parts of the benthic community (epibenthic or infauna) take advantage of the resources available. Stable isotopes analysis indicates that scallops feed on a different energy source (plankton, organic matter) than other filtering or suspensivore benthic organisms, like ascidians, sponges, or worms (Botto et al. 2006). Scallop predators are the gastropods *Fusitriton magellanicus*, *Odontocymbiola magellanica*, *Adelomelon ancilla* and the sea star *Labidiaster radiosus* (Fig. 6.1), while the star *Diplopteraster clarki* predates on the mentioned snails. However, these environments are subjected to commercial exploitation by trawlers that discard both non-commercial parts of pectinids (valves, gonads, mantle) and by-caught species sometimes seriously damaged (Escolar et al. 2017), so it must be considered that organisms can benefit from these available food sources as shown in other regions (Link and Almeida 2002; Jenkins et al. 2004).

6.2.3 Disturbance by Fisheries

These areas are particularly sensitive to exploitation, what affects both commercial and non-commercial organisms. No general loss of species was registered in the whole fishing area, although some taxa varied its relative density (Schejter et al. 2008b, 2016; Bremec et al. 2015). The presence of sessile and fragile species diminished in areas with high fishing pressure (Bremec et al. 2000), while the presence of predators and opportunists like asteroids and gastropods increased, showing a general decrease of benthic biomass (Escolar et al. 2009, 2011, 2015; Schejter et al. 2008b, 2016). The recovery of biomass of sessile taxa, like sponges, tunicates, and others, was observed in areas not subjected to trawling for more than 4 years (Bremec et al. 2015; Escolar et al. 2015) and species richness was always higher in areas with low than with high disturbance (Schejter and Bremec 2013). It is well known that trawling affects the bottom environment by means of homogenizing the substrate and damaging organisms and associated species (e.g., Bradshaw et al. 2002; Hinz et al. 2009). In this scallop fishery, damage of non-commercial species also occurs during the mechanical selection process on board the factory vessels. The urchin *Sterechinus agassizii* and the snail *Fusitriton magellanicus* are most heavily damaged, other organisms show different degree of destruction and others remain retained during the selection process (Escolar et al. 2017; Schwartz et al. 2016).

In the case of the "parchment worm" *Chaetopterus antarcticus* (Moore et al. 2017), its high frequency and abundance, between 37 °S and 40 °S, is associated with the commercial scallop. This polychaete was usually collected within its free "U tubes," typical of infaunal habit (Rouse 2001; Nishi et al. 2009). However, the yearly assessment of invertebrate by-catch of the scallop fishery showed notorious settlement of parchment worms on *Z. patagonica* (Bremec et al. 2008). Previous works on *Z. patagonica* epibionts (Walossek 1991; Bremec and Lasta 2002; Bremec et al. 2003) mentioned organisms like calcareous algae, poriferans, hydrozoans, anthozoans, polychaetes, molluscs, cirripeds, foraminiferans, ascidians, and bryozoans, without records of *Chaetopterus*. The variability of life habits of *C. antarcticus* was not commonly registered previously in the area; epibiotic behavior could be the result of intensive soft sediment disturbance due to trawling and hence, the selection of other available primary settlement substrate (Bremec and Schejter 2019).

6.3 Benthic Characterization Associated to Other Fisheries

6.3.1 The Argentine Hake Fishery

The composition of the benthic community in the Argentine hake *Merluccius hubbsi* fishery is usually carried out through the analysis of the by-catch of the fishery, which develops part of its activity between 100 and 200 m depth. Since benthic diversity patterns depend directly on the sampled area and the sampler type, this information has made it possible to analyze the shelf-break communities from a different point of view than the usual scallops by-catch, collecting benthic organisms through the evaluation of hake juveniles and adults (Bremec et al. 2011, 2012; Gaitán et al. 2013, 2014; Giberto et al. 2015, 2017a; Gaitán and Souto 2020). The species that usually dominate the benthic assemblages of the shelf-break region off the Río de la Plata (~35 °S, northern waters) are *Ophiactis asperula*, *Ophiura lymani*, *Pseudechinus magellanicus*, *Zygochlamys patagonica*, *Chaetopterus antarcticus*, *Ophiacantha vivipara*, *Libidoclaea granaria*, *Isotealia antarctica*, *Diplopteraster clarki*, and *Paramolgula gregaria*, among others.

At the shelf-break southern regions, between 42° and 48 °S, associations dominated by different sponges' species are usually found, with other characteristic species like *Zygochlamys patagonica*, *Sterechinus agassizii*, *Actinostola crassicornis*, *Adelomelon ancilla*, *Fusitriton magellanicus*, *Diplasterias brandti*, *Arbacia dufresnii*, *Libidoclaea granaria*, and *Magellania venosa* (Fig. 6.1), among others. The Patagonian scallop is present throughout the whole hake fishery grounds, but its biomass values are only slightly greater in the Patagonian region. However, the overall benthic species composition is quite similar, with main differences in dominance levels. These differences could be a reflection of the samplers used: while hake fishing nets only accidentally catch megabenthic species, scallop fishing nets have been designed to catch Patagonian scallops. On the other hand, given that the Argentinean hake fishery does not expressly search for areas in which Patagonian scallops are dominant, differences in the regions hauled may also play a role in the differences usually found between both fisheries by-catches.

6.3.2 *The Argentine Squid Fishery*

The Argentine squid *Illex argentinus* fishery is carried out along the southern region of the slope front. This neritic-oceanic species extends between 23 °S and 54 °S in the Southwest Atlantic (Brunetti et al. 1999). There are several stocks, but possibly the most closely related to the shelf-break front region is the South Patagonian stock. There are two typical modes of squid fishing, the most traditional and oldest is fishing with bottom trawls, used exclusively until 1987, and fishing with automatic jigging machines used almost exclusively today (only a few vessels that trawl in the area for hake incidentally catch the squid). The main fishing zone in terms of volume is carried out in the frontal region between 44 °S and 48 °S (Brunetti et al. 1999). Adults carry out daily vertical trophic migrations, in which they disperse to mid-water at night and drop to levels close to the bottom during the day (Brunetti 1988; Haimovici et al. 1998).

As part of a comprehensive study of the fishery ecosystem, studies on the benthic communities associated with the squid fishery have recently begun. The focus is on both the infaunal and epifaunal communities captured during regular stock assessment expeditions between 36 and 61 °S and the isobaths 90–400 m. Dominant species at this region are the tunicates *Paramolgula gregaria* and other Polyclinidae, the sponges *Tedania mucosa* and *Tedania* sp., the mollusk *Zygochlamys patagonica*, the echinoderms *Labidiaster radiosus*, *Sterechinus agassizii*, *Diplasterias brandti*, *Ctenodiscus australis* and *Gorgonocephalus chilensis*, the crustaceans *Libidoclaea granaria* and *Grimothea gregaria*, and the cnidarian *Actinostola crassicornis* (Authors' unpublished data). The typical species are very similar to those found at the scallop fishery, but the shelf break bottoms presented an unusually high dominance of sponges and tunicates, in a similar way to those found at some shallower sectors of the shelf (Schejter and Giberto 2022). This is probably an indication that some sectors of the Argentine *Illex* fishery are relatively preserved from intense bottom fishing and may constitute a potential refugee for juveniles of species of commercial and non-commercial interest, as suggested for some regions of the San Jorge Gulf (Giberto et al. 2015).

6.4 Benthic Communities at the Slope and Beyond: >200 M Depth

6.4.1 *Submarine Canyons*

In terms of complexity, instability, material processing, and hydrodynamics, underwater canyons are unique environments. The macrobenthic diversity and abundance within canyons are somewhat influenced by the physical disturbance regime as well as the rate and volume of organic matter deposition (Portela et al. 2015). In Argentina, the continental margin shows deep submarine canyons variable in size

(Parker et al. 1997) and based on geomorphology and location were classified in four systems: "Río de la Plata" (35°–38 °S), "Colorado-Río Negro" (39°–42 °S), "Ameghino" (42°–46 °S), and "Patagonia" (46°–49 °S) (Lonardi and Ewing 1971). Lastras et al. (2011), del Río et al. (2012), and Muñoz et al. (2013) give information about their topography, sedimentology and origin. Benthic samples from a submarine canyon (43°35′S–59°33′W) between 325 and 360 m depth were taken after a hydroacoustic prospection of "Ameghino" system (Madirolas et al. 2005). The benthic assemblage, composed by 86 taxa, was similar to the scallop assemblage distributed in the adjacent fishing grounds (Bremec and Schejter 2010) and the inventory of epibionts registered on the scallops added 25 taxa to the already known (Schejter et al. 2014). It is important to notice that the Antarctic sponge *Guitarra dendyi* was recorded in this deep location, representing the northernmost distribution limit for the species and that two new species of sponges were also described in this canyon (*Stelodoryx argentinae* and *Tedania* (*Tedaniopsis*) *sarai*) (Bertolino et al. 2007). Worn mollusk shells were also registered and corresponded to species distributed both in coastal and shelf areas (e.g., *Eurhomalea exalbida*, *Buccinanops* spp., *Petricola dactylus*, and Mactridae), indicating transport of sediments and water masses to deep zones.

The Mar del Plata Canyon, located at about 38 °S, has been studied from three Argentinean Research Cruises developed during the past decade and devoted to the study of the deep biodiversity. In this particular environment, reaching depths of up to 3500 m, several new species have been discovered and many distribution updates have been recorded among invertebrates. This is still an on-going task and many samples continue to be studied. Among cnidarians, the new species *Errina argentina* (Bernal et al. 2018), *Umbellula pomona* (Risaro et al. 2020), and *Armadillogorgia albertoi* (Cerino and Lauretta 2013) were described, and new distributional records for stylasterids (Bernal et al. 2021) and the black coral *Dendrobathypathes grandis* (Lauretta and Penchaszadeh 2017) were recorded. Among crustaceans, peracarids were highly diverse (e.g., Doti et al. 2020) and several new species were described: cumaceans, *Holostylis unirramosa* and *Platytyphlops sarahae* (Roccatagliata and Alberico 2015, Roccatagliata 2020); isopods, *Neasellus argentinensis*, *Edotia abyssalis*, *Xiphoarcturus kussakini*, *X. carinatus*, and *Pseudione chiesai* (Doti 2016; Pereira and Doti 2017; Pereira et al. 2019, 2020, 2021).

New distributional records for several species, genera, and families were also noted in the literature. For decapod crustaceans, new distributional and/or bathymetric records were also reported, such as *Stereomastis suhmi* (Farias et al. 2015), *Ethusina abyssicola* (Ocampo et al. 2014), and *Paralomis spinosissima* (Olguin et al. 2015). Among echinoderms, the following new species have been described: holothuroidea, *Benthodytes violeta* (Martinez et al. 2014), *Psolus lawrencei* (Martinez and Penchazsadeh 2017) and *Paleopatides shumel* (Martinez et al. 2019), echinoidea, *Corparva lyrida* (Flores et al. 2021) and the asteroidean *Bernasconiaster pipi* (Rivadeneira et al. 2020). Regarding mollusks, the new species *Scaphander meridionalis* (Siegwald et al. 2020) was described, while new distributional records of the species *Laubierina peregrinator* (Pastorino 2016) and *Theta lyronuclea*

(Sánchez and Pastorino 2020) suggest water masses connections from the deep environments. Among ascidians, two new species were described: *Aplidium marplatensis* and *Aplidium solitarium* (Maggioni et al. 2018). Reproductive aspects of several invertebrate species inhabiting this deep habitat revealed a protective strategy (e.g., Berecoechea et al. 2017; Lauretta et al. 2020; Martinez and Penchaszadeh 2017; Martinez et al. 2020; Teso et al. 2020).

6.4.2 Fragile Habitats and Key Sensitive Species

Oceans around the world are home to fragile organisms like deep-water corals and sponges. As a result, habitat-structuring organisms are important not only in shallow water but also in the shelf-break ecosystems, hydrothermal vents, seamounts, and even in the deep-sea basins that were long thought to be constant and uniform. These ecosystems are home to sensitive deep-water corals that are endangered by human activity, particularly fishing and oil prospecting (Portela et al. 2015). A giant cold-water coral mound province called "Northern Argentine Mound Province" is located near Mar del Plata canyon, at about 38° to 39 °S, in deep waters of about 1000 m; it is linked to a contourite depositional system that covers at least 2000 km^2 (Steinmann et al. 2020). These ecosystems are characterized by particular conditions that include the availability of suspended food and sediment particles in combination with a sufficient bottom-current strength, keeping material in suspension while preventing destructive erosion and excessive sedimentation. Schejter et al. (2021) recorded cold-water corals in this area (*Bathelia candida*, *Verticillata castellviae*, *Plumarella* sp., Primnoidae, sponges and some echinoderms, Fig. 6.2) as by catch of traps devoted to the study of the crab *Chaceon notialis* at ~950 m. The first finding of chemosynthetic ecosystems in the Argentine deep sea was also recently reported (Bravo et al. 2024).

Other deep areas near 41 °S and 400 m depth showed dominance of echinoderms, Stylasteridae, Pennatulacea and Primnoidea (Fig. 6.2) (Schejter et al. 2017). In particular, the sea pens *Anthoptilum grandiflorum* and *Balticina africana* were recorded in some muddy patches, associated with the sea-anemone *Hormathia pectinata* (Schejter et al. 2018). Information from Vulnerable Marine Ecosystems (VME) in SW Atlantic deep waters from about 42° to 47° 30′S report the presence of sponge beds including the presence of carnivorous sponges (e.g., *Asbestopluma* sp., *Euchelipluma* sp., *Chondrocladia* sp.) and glass sponges (hexactinellids), cold-water coral gardens (dominated by Primnoidae and Isididae) and cold-water coral reefs between 250 and 1300 m depth, in accordance to the different strata composition (Portela et al. 2012, 2015). A rich and diverse associated fauna was also recorded in the different habitats. In particular, the record of the reef builder coral *Bathelia candida* and large hydrocorals (*Errina* spp., *Cheiloporidion pulvinatum*, *Sporadopora* sp., and *Stylaster densicaulis*) is remarkable (Durán Muñoz et al. 2012; Portela et al. 2012).

Fig. 6.2 Invertebrates recorded at >200 m depth. (**a**). The sea fan *Verticillata castellviae* (scale bar = 40 mm), (**b**). The sea pen *Balticina africana* with an epibiotic sea anemone *Hormathia pectinate* (white arrow) (scale bar = 30 mm), (**c**). The sea fan *Thouarella* sp. (Primnoidae) (scale bar = 40 mm), (**d**). The cold water coral and reef builder *Bathelia candida* (scale bar = 25 mm), (**e**). The Hexactinellid sponge *Rossella* sp. (scale bar = 45 mm), the ophiuroid *Astrotoma agassizii* (scale bar = 20 mm)

A fragile benthic habitat known as "Blue Hole" (*Agujero Azul* in Spanish) can be found in this region. It is a particular area of 6000 km^2 located 500 km east of San Jorge Gulf (45 °S to 47 °S, 60 °W to 61 °W) that has been pointed out as a potential conservation area, since the deeper bottoms are hosting a rich and fragile benthic community (Portela et al. 2012, 2015; Cairns and Polonio 2013). For this reason, currently there have been several efforts tending to encourage studies of these VME in particular. Preliminary results from an Argentinian Expedition undertaken in 2021 onboard RV "Víctor Angelescu" (INIDEP) to this area recorded at least 52 epibenthic species, including various species of sponges (i.e., *Tedania* spp. and Hexactinellida species), the echinoderms *Sterechinus agassizii* and *Diplopteraster clarki*, the cnidarians *Anthoptilum grandiflorum*, Pennatulacea, and Primnoidea. In overall, preliminary diversity estimators suggest values up to 100 epibenthic species in the "Blue Hole," with a similar community composition to those of the shelf (dominants species belong to Cnidaria and Echinodermata) (Authors, personal observations).

6.5 Epibiosis and Bioerosion: Biodiversity Enhancers

6.5.1 *Living on Others: Sclerobiosis and Beyond*

Studying associations between dead/living organic or inorganic hard substrates and colonizers, i.e., sclerobiosis (Romero et al. 2022), is an important aspect to be taken into account in the shelf-break front. This approach is a useful tool for characterizing structure of benthic assemblages, even to know ethological patterns, aspects of life cycles, or habitat requirements (Romero et al. 2017). There are sessile organisms that may be present in the environment and may go unnoticed if the colonizer-substrate associations are not studied, such as organisms that specifically settle on inorganic or hard dead substrates (e.g., Bremec et al. 2003; Schejter and Bremec 2007a; Schejter et al. 2011a) as well as colonizing specific hosts (e.g., Cook et al. 1998; Schejter et al. 2011b).

About 70% of the Argentine continental shelf is dominated by sandy and muddy bottoms with minor quantities of shells (Parker et al. 1997; Razik et al. 2015) and gravel to cobble sized rocks in several morphological depressions (Wilckens et al. 2021). In the absence of widespread hard substrates in this region, many sessile species depend on epibiotic relationships to survive (Schejter and Bremec 2008; Schejter et al. 2010). It is precisely for this prevalence of soft bottoms that the benthic fauna, mainly considering species of commercial interest in the Argentine sea, is a key component as a source of available substrates for the settlement of invertebrates. Shell species offer a high degree of structural complexity and add physic heterogeneity to the bottom; it is advantageous for the colonization, the development of different relationships between encrusters, bioeroders, and even motile fauna that may temporarily attach or associate with substrates, resulting in its highlighted contribution to the increase of local species richness. Therefore, epibiotic relationships increase the benthic species richness in the shelf-break front (Schejter and Bremec 2009).

6.5.2 *Increasing the Biodiversity in Marine Communities*

Epibionts and associated fauna are responsible for almost 50% of the total species richness in Patagonian scallop fishing areas. A wide variety of organisms colonizing the pectinid scallop include calcareous algae, poriferans, hydrozoans, anthozoans, polychaetes, mollusks, barnacles, foraminifera, ascidiaceans, and bryozoans (Schejter and Bremec 2007a; Bremec and Schejter 2019). The sponge *Iophon proximum* was the most common and abundant epibiont in all study areas and generally few or no other epibionts were recorded when this sponge was present. Individuals and empty tubes of Serpulidae (Fig. 6.1) are common in both valves and in all the areas studied, as well as colonies of Bryozoa (Schejter and Bremec 2007a; Lopez Gappa and Landoni 2009). In addition, there are cases of secondary epibiosis and sclerobiosis by hydroids and Foraminifera (on polychaete tubes and barnacles) (Schejter and Bremec 2006), the acorn barnacle *Weltnerium gibberum* (Schejter and Bremec 2007a), egg capsules of *Fusitriton magellanicus*, *Odontocymbiola magellanica* and *Adelomelon ancilla* (Bremec et al. 2003), and sponges and bryozoans on dead shells of the Patagonian scallop (Schejter and Bremec 2007a). Furthermore, some species of epibionts on *Zygochlamys patagonica* act as basibionts and consequently favor its recruitment. The hydroid *Symplectoscyphus subdichotomus* is settled on polychaete tubes, sponges, Patagonian scallops, and other hydroid colonies and, at the same time, they provide the primary settlement substrate for *Z. patagonica* spats. Encrusting sponge *Iophon proximum*, amphipod tubes and polychaete tubes on the same adult scallop species are also colonized by recruits of *Z. patagonica* (Bremec et al. 2008).

Higher species richness, compared to the neighboring scallop fishing ground at the shelf- break frontal area, was also noted in the submarine canyon of Argentine Sea located at 43°35′S to 59°33′W, 325 m depth. Patagonian scallops collected in such submarine canyon represented near 20% in biomass of the total catch and recorded a high richness of epibionts (Schejter et al. 2014). Unlike commercial fishing grounds, bryozoans and polychaetes were the most frequent epibionts, and sponges were uncommon on shells of the Patagonian scallop in the submarine canyon. Of the 53 epibiotic taxa, Bryozoa was the most diverse group (34 species) while Polychaeta was the most abundant group (e.g., *Serpula narconensis* and *Idanthyrsus macropaleus*), recorded on 94% of the scallops. The cnidarians Stylasteridae and Clavulariidae and the sponge *Tedania* (*Tedaniopsis*) *infundibuliformis* were also recorded as epibionts on *Z. patagonica*.

Other conspicuous species provide substrates and refuge for many invertebrates (Fig. 6.1). The hairy snail *Fusitriton magellanicus* is the second species most colonized by epibiont organisms (Schejter et al. 2011a, b). The greatest richness of sponges in the area is hosted on empty, pagurized and living specimens of the hairy snail. Polychaetes are the most frequent (e.g., *Potamilla antarctica* and sandy tubes of Polynoidae, Terebellidae), followed by sponges (e.g., *Hymedesmia* sp., *Clathria* spp., *Tedania (Tedania)* spp. and *Dictyonella* spp.) and ascideaceans. At the same time, the mentioned polychaete tubes promote the settlement of the byssate bivalve *Hiatella umbonata* and create refuges for amphipods, isopods, and nematodes. The gastropod *Adelomelon ancilla* is used as a substrate by two species of anemones, while empty valves of *A. ancilla* are also occupied by the crustaceans *Pagurus comptus* and *Propagurus gaudichaudii* (Schejter and Escolar 2013).

Polychaetes (empty tubes of *Chaetopterus antarcticus*), Rajoidea (empty capsules), crabs, corals, and brachiopods have been recorded as colonized substrates (Schejter and Bremec 2009; Bremec and Schejter 2019). The tube-worm *C. antarcticus* is also settled on chondrichthyes capsules and snails (Bremec and Schejter, 2019). Other epibiont organisms were also recorded on the stone coral *Desmophyllum dianthus* (Schejter et al. 2015). Last, boring organisms are also recorded in *F. magellanicus*. The more common traces were assigned to the ichnogenus *Entobia* -being clionid bioeroding sponge species identified as trace makers-, traces attributable to spionid polychaetes and probably assigned to *Maendropolydora* sp., and other unidentified marks (Authors, personal observations). These results support that epibiotic relationships and, more broadly, sclerobiosis are responsible for an increase in benthic species richness in the submarine canyon and in shelf areas, where *Z. patagonica* is the dominant basibiont species. The presence of soft bottoms and the dominance in abundance and biomass of the Patagonian scallop contribute to the increase in the richness of benthic species in the shelf break front.

6.6 Some Insights About General Biodiversity Trends

6.6.1 Biodiversity in Context

The epifaunal benthic diversity of the slope front has been studied mainly through the by-catch of the three principal Argentinean fisheries (Patagonian scallop, Argentine squid, and Argentine hake). In general terms, the total biodiversity numbers and patterns at the shelf-break front usually refer to "megabenthic species captured by fishing trawlers," since the infaunal and rocky deep communities are scarcely studied (Bremec and Giberto 2017). The typical annual data obtained during research surveys carried out by the INIDEP cover bottoms from 50 °S to 36 °S and between 100 and 400 m (unpublished data from Benthic Laboratory of INIDEP) (Fig. 6.3a). The research effort upon these bottoms varies according to the fishery considered (Patagonian scallop, for example, is the one with the most regular record), but is useful to explore some general diversity trends. For example, the number of samples needed to study the total by-catch benthic diversity of the region (the plateau in the species accumulation curve) would be around 400–500 sites when we pool the three fisheries together (see Fig. 6.3b). The average species number per site of benthic by-catch is usually around 14 taxa, with values between 1 and 34 taxa, with about 150 typical species in the whole area (Authors, unpublished data, see Supplementary Material). Using the classical diversity estimators of Gotelli and Colwell (2011), it is assumed that the shelf-break by-catch community would be composed of 160–190 species (Fig. 6.3c), which is not far from the values usually recorded in fishing research cruises.

The data mentioned in the previous paragraph reflect the broad-scale of the frontal system and define the lowest effort needed to describe megabenthic biodiversity at best. Main contributors to the diversity of the front are Echinodermata (40 taxa), Mollusca (30 taxa), Cnidaria (20 taxa), Crustacea (20 taxa), Annelida (10 taxa), Porifera (10 taxa), and Tunicata (10 taxa). The last two groups usually contribute with significant biomasses at several regions of the front. An additional issue to consider is the difference in the available taxonomic knowledge of benthic species, which will likely determine changes in diversity in the near future. The low diversity of some groups is partially due to the taxonomic complexity of some groups (species level identification is a time consuming task) and to the lack of specialists. The sponges are a particular example of this case, in which their diversity has been increasing over the years due to improved taxonomic knowledge (e.g., Schejter et al. 2006, 2008a, 2011a; Bertolino et al. 2007) and not because there have been

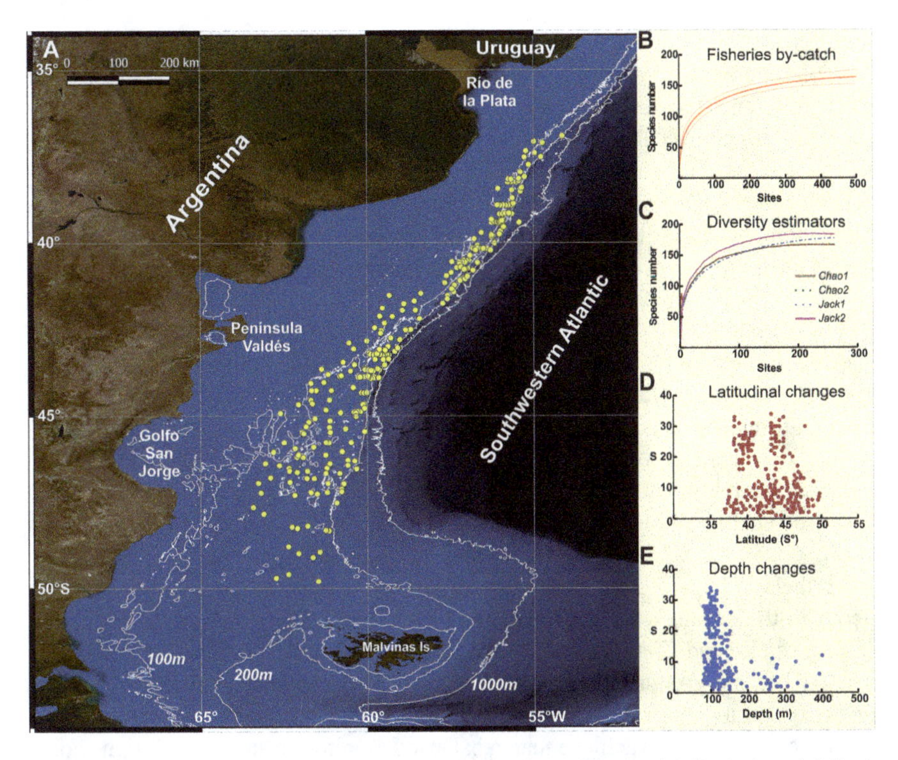

Fig. 6.3 Benthic biodiversity patterns at the shelf-break frontal system. (**a**). Typical spatial distribution of the annual monitoring of megabenthic by-catch sites (yellow circles, annual database = 262 sampling sites) surveyed by the Benthic Laboratory (National Institute for Fisheries Research and Development, Argentina) (100 m, 200 m, and 1000 m isobaths are also indicated). (**b**). Extrapolation of the species accumulation curves standardized by samples (species density) for the annual benthic monitoring exemplified in A (n = 262). (**c**). Species richness asymptotic estimators (Chao 1, Chao 2, Jacknife 1 y Jacknife 2) for the exemplified annual database. (**d**). Total site species number vs. latitude range for the annual database. (**e**). Total site species number vs. depth (m) for the annual benthic monitoring exemplified in A. (Satellite image modified from Stöckli et al. 2007)

obvious changes in the frontal ecosystem. A similar situation can be exemplified by the asteroids in the Patagonian Argentinian deep sea, where a recent review found 41 species, including 7 new records (Hurtado-García and Manjón-Cabeza 2022).

6.6.2 *Biodiversity Trends*

Biodiversity gradients are usually related to latitudinal change in the marine realm (Costello and Chaudhary 2017). As geographical variables are used as proxies to reflect meaningful environmental gradients (e.g., water temperature, primary production, and so on), it is expected to be relevant at larger scales or when significant environmental changes exist. This is not the case at the shelf break front, where by-catch benthic diversity trends suggest no discernible changes with latitude, or at least a meaningful one (Fig. 6.3d). This is not a surprise, since the oceanographic variables are quite constant along the front (the Malvinas current is the main water mass along the study area). Sediments and bottom topography may be some of the causes of biological changes in the environment on the shelf-break front, but it is not strictly related to latitudinal gradients. Nonetheless, many species-rich locations are found between 38 °S and 45 °S, while species-poor sites are located at the edges of the geographical gradient. This pattern can point to a connection or variable that hasn't been thought of before (for example, bottom current speed, primary productivity in the water column, among others).

On the other hand, a potential depth trend can be found (Fig. 6.3e), with most species-rich locations occurring at less than 150 m. This pattern should be interpreted with caution due to significant differences in sampling efforts at deep bottoms. Information about other invertebrate groups (i.e., bryozoans and hydrozoans) in the Southwestern Atlantic suggests a similar trend; in general, the species richness reflects the intensity of sampling effort expended on a determined area more than true ecological shifts with depth or other environmental traits (López Gappa 2000; Giberto 2003; López Gappa and Landoni 2005; Genzano et al. 2009). Nevertheless, large scale studies suggest a sharp decline in species richness with depth in the ocean, with great endemicity in coastal than offshore environments (e.g., Costello and Chaudhary 2017; Costello et al. 2017). Further research on the smaller benthic groups is required to finally assess regional gradients in benthic diversity and its relationships with geographical and environmental variables at the shelf break. In order to achieve that, a balanced monitoring sampling design should be applied across the main gradients of the frontal region.

6.7 Trophic Interactions: Fishes' Benthic Preys

Overall, benthic communities play a key role in energy flow of marine ecosystems by connecting different biotic and abiotic components and transferring energy between trophic levels. In turn, benthofagous fish contribute to the trophic web by serving as essential links between the benthic and pelagic ecosystems. In particular, there are several species inhabiting in close association with bottom substrates that fed upon benthic invertebrates in the shelf break front (Giberto et al. 2017b). Among the most important elasmobranch benthic dwellers in the shelf break are skate species: *Zearaja brevicaudata*, *Bathyraja brachyurops*, *Bathyraja macloviana*, *Bathyraja albomaculata*, *Bathyraja griseocauda*, *Bathyraja multispinis*, *Bathyraja cousseauae*, *Bathyraja scaphiops*, *Bathyraja magellanica*, *Psammobatis normani*, *Psammobatis rudis*, *Amblyraja doellojuradoi*, and shark species such as narrowmouthed catshark *Schroederichthys bivius* that exhibited inferior mouth type that denote a benthic feeder specialization. Also, some bony fish such as the pink cusk-eel *Genypterus blacodes* and grenadiers (*Coelorinchus fasciatus*, *Macrourus carinatus*, *Macrourus holotrachys*) fed upon benthic invertebrates in the shelf break region.

Three of the eight known *Bathyraja* species are exclusively benthic feeders. *B. multispinis* has pavement like dentition and fed exclusively in crabs, mainly decapods *Peltarion spinosulus* and *Libidoclaea granaria*, followed by the isopods *Acanthoserolis* spp. and *Arcturus* spp., and the Nephropidae *Thymops birsteini* and Munididae *Curtonida spinosa* (Belleggia et al. 2014). The most important prey of *Bathyraja albomaculata* are polychaetes (*Travisia* spp.), followed by gammarid amphipods (*Ampelisca* spp.) and isopods (*Cirolana* spp.) (Sánchez and Mabragaña 2002; Brickle et al. 2003; Ruocco et al. 2009; Shimabukuro 2009). *B. macloviana* also fed mainly on polychaetes, followed by amphipods, isopods, and decapods (Sánchez and Mabragaña 2002; Mabragaña et al. 2005; Scenna et al. 2006; Barbini et al. 2013). The broad nose skate *B. brachyurops* fed on fish, followed by brachyuran crabs, isopods (Serolidae, Cirolanidae and Antarcturidae), and gastropods (e.g., Volutidae *Odontocymbiola magellanica*) (Belleggia et al. 2008). Likewise, the diet of *B. cousseauae* and *B. scaphiops* consisted mostly of teleosts followed by isopods and amphipods (Lysianassidae *Erikus* spp.) (Belleggia et al. 2014). On the other hand, *B. griseocauda* is mainly ichthyophagous (Sánchez and Mabragaña 2002), but the isopods *Acanthoserolis* spp. and *Arcturus* spp. are an important component of their diet (Brickle et al. 2003; Belleggia et al. 2014). Similarly, *B. magellanica* is primarily ichthyophagous, and secondarily it feeds upon amphipods, isopods, and decapods (Barbini et al. 2010).

The Yellownose skate *Zearaja brevicaudata* is mainly an ichthyophagous species that fed on demersal fishes, primarily notothenids *Patagonotothen ramsayi* and Argentine hake *M. hubbsi* (García de la Rosa 1998; Lucifora et al. 2000; Koen Alonso et al. 2001; Belleggia et al. 2016). However, small *Z. brevicaudata* also fed on benthic prey, such as the isopods Serolidae and Cirolanidae, the crabs *Peltarion spinosulus* and *Libidoclaea granaria* (Lucifora et al. 2000; Belleggia et al. 2016). The skates *Psammobatis rudis* and *P. normani* have similar trophic preferences. The most important prey for *P. normani* are crustaceans (mainly crabs and isopods), followed by polychaetes, while *P. rudis* feeds almost exclusively on crustaceans (isopods, crabs, and amphipods), followed by fish and polychaetes (Mabragaña and Giberto 2007). The southern thorny skate *Amblyraja doellojuradoi* fed mainly on crabs *Libidoclaea granaria* and *Peltarion spinosulus* (Sánchez and Mabragaña 2002; Delpiani et al. 2013), followed by polychaetes, teleosts, isopods (Serolidae, Cirolanidae, and Antarcturidae), and Caridea (*Campylonotus vagans*) (Delpiani et al. 2013). The narrowmouth catshark *Schroederichthys bivius* feeds mainly on crustaceans, followed by polychaetes, fish, and cephalopods (Sánchez et al. 2009; Belleggia et al. 2017, 2018; Villa et al. 2018).

Among bony fish in the shelf break region, the pink cusk-eel *Genypterus blacodes* feeds mainly on fish (the notothenioid *Patagonotothen ramsayi* and the Argentine hake *Merluccius hubbsi*) and crustaceans (Belleggia et al. 2022). The most important crustaceans consumed are the lobster krill *Grimothea gregaria*, isopods (*Cirolana* spp., Serolidae), stomatopods (*Pterygosquilla armata*), and gammarid amphipods (Belleggia et al. 2022). The grenadiers (*Coelorinchus fasciatus, Macrourus carinatus, Macrourus holotrachys*) feed almost exclusively upon gammarid amphipods (Romanelli 2017). Finally, *Patagonotothen ramsayi*, a typical catch at the shelf break front, is reported as a generalist fish of demersal-benthic feeding habits, feeding on polychaetes, amphipods, and other crustaceans (Fischer et al. 2022).

6.8 Final Remarks

The shelf-break front benthic assemblages seem to be quite similar all over the region, particularly those of soft bottoms up to 200 m depth (see typical species in Table 1). These assemblages are rather undistinguishable when based on species richness patterns and community composition and are not related to changes in geographic factors (diversity is quite constant at any latitude along the front) but to the Malvinas Current oceanographic conditions. Echinoderms, mollusks, and cnidarians are the most megabenthic diverse groups, but the diversity of sponges and crustaceans is also significant. On the other hand, knowledge has been focused on particular sectors of the front and mainly on megabenthic species, generally as by-catch of the main fisheries of the shelf-break. When infaunal species are taken into account, benthic diversity significantly increases (e.g., Sánchez et al. 2011). The typical megabenthic by-catch species comprised around 150 taxa, which usually dominates the benthic communities in terms of abundance or biomass values. This final figure is constrained by sampling restrictions (fishing nets are not designed to catch invertebrate species) and abundance patterns (low density species are hard to catch), but also by bottom topography (fishing vessels typically trawl on soft flat bottoms, avoiding rocky irregular bottoms), among other factors.

In order to improve our comprehension and management of the benthic communities of the shelf-break, a few closing thoughts must be taken into account:

(a) The spatial distribution of scientific knowledge is heterogeneous and is not properly balanced across the bottom topography. The most studied region is that with depths up to 200 m and north of 47 °S, while the deepest shelf-break bottoms and regions of higher latitudes received much less attention. In addition, there is an obvious difference in the samplers used to collect benthic species, for example, most sites are usually sampled by means of epibenthic dredges or by bottom trawlers (by-catch). Due to the absence of well-balanced designs, the patterns obtained should be considered as hypotheses for benthic diversity research in the region. Future improvements should be directed to the meio- and macrobenthic infaunal communities, largely neglected, and to study the benthic invertebrates inhabiting deep soft and rocky bottoms.

(b) Species richness patterns are related not only to the variability of the frontal pelagic production but also to the substrate complexity, with average richness values located in bottoms with gravel and sandy sediments, and hot-spots of diversity located at rocky bottoms (diversity is also boosted by many cases of sclerobiosis). The homogenous sandy bottoms usually present low diversity values. The soft bottoms dominated by the scallop *Zygochlamys patagonica* are characterized by high invertebrate biomass and sustain a valuable commercial fishery, with a significant benthic community secondary production. This species is widespread throughout the frontal region, with the exception of the region known as "Agujero Azul," where it does not reach abundances profitable for commercial exploitation. Thus, in order to preserve essential ecosystem processes of the shelf-break front, not only high diversity regions but also productive ones should be taken into account in future management and conservation actions.

(c) The last but not the least important, the shelf-break frontal area is subjected to many human activities that have direct effects on benthic communities and are usually more intensive at depths lower than 200 m, such as contamination of the sediments by waste of commercial vessels and the presence of many bottom trawl fisheries. Recently, new oil explorations have been carried out at deep bottoms (~1500 m), potentially threatening vulnerable benthic ecosystems (VME). Therefore, basic knowledge of integral benthic diversity and ecological processes are urgently needed to design effective environmental policies, including shallow and deepest communities. All these anthropic factors should be considered in research and modeling of the frontal benthic diversity, since they add complexity and variability to the ecological processes, constraining the inferences of studies based only on natural processes.

The biological patterns discussed in this chapter suggest the necessity of conservation efforts across all distinct ecosystems (sandy and rocky bottoms, shallow and deep communities) to protect the entirety of this large frontal ecosystem's biodiversity: species composition varied among depths and sediment types, so it is necessary to make conservation efforts to preserve representative samples of the system's total benthic diversity.

Acknowledgments We would like to thank all the members of the scientific and technical staff, but also the crew of the expeditions from which the data in this chapter come from. The suggestions and comments of two anonymous reviewers also contributed to the improvement of the preliminary version of this chapter. This is an INIDEP contribution n° 2371. This was partially financed by PICT 2019-4233 to LS.

References

Acha EM, Mianzan HW, Guerrero RA, Favero M, Bava J (2004) Marine fronts at the continental shelves of austral South America. Physical and ecological processes. J Mar Syst 44:83–105

Acha EM, Piola A, Iribarne O, Mianzan H (2015) Ecological processes at marine fronts: Oases in the ocean. Springer International Publishing, p 68

Barbini SA, Scenna LB, Figueroa DE, Cousseau MB, Díaz de Astarloa JM (2010) Feeding habits of the Magellan skate: effects of sex, maturity stage and body size on diet. Hydrobiologia 641:275–286

Barbini SA, Scenna LB, Figueroa DE, Díaz De Astarloa JM (2013) Effects of intrinsic and extrinsic factors on the diet of *Bathyraja macloviana*, a benthophagous skate. J Fish Bio 83:156–169

Belleggia M, Mabragaña E, Figueroa DE, Scenna LB, Barbini SA, Díaz de Astarloa JM (2008) Food habits of the broad nose skate, *Bathyraja brachyurops* (Chondrichthyes, Rajidae), in the south-west Atlantic. Sci Mar 72:701–710

Belleggia M, Scenna LB, Barbini SA, Figueroa DE, Díaz de Astarloa JM (2014) The diets of four *Bathyraja* skates (Elasmobranchii, Rajidae) from the Southwest Atlantic. Cybium 38(4):314–318

Belleggia M, Andrada N, Paglieri S, Cortés F, Massa A, Figueroa DE, Bremec (2016) Trophic ecology of yellownose skate *Zearaja chilensis* (Guichenot, 1848) (Elasmobranchii: Rajidae), a top predator in the southwestern Atlantic. J Fish Biol 88:1070–1087

Belleggia M, Villa A, Colonello J, Figueroa DE, Massa A, Giberto D, Bremec C (2017) The diet of the Narrowmouthed Catshark *Schroederichthys bivius*, from the Patagonian continental shelf. Joint Meeting of Ichthyologists and Herpetologists. American Elasmobranch Society. Austin, Texas, EEUU. July 12–16, 2017

Belleggia M, Zenoni-Lufrano M, Villa A, Colonello J, Figueroa DE, Massa A, Giberto D, Bremec C (2018) The diet of the narrowmouthed catshark *Schroederichthys bivius* in Argentine sea. Sharks International Conference. João Pessoa, PB, Brazil. June 03–08, 2018

Belleggia M, Alvarez CD, Pisani E (2022) Ecología trófica y contribución de las presas en la dieta del abadejo *Genypterus blacodes* (Forster, 1801). Análisis de contenidos estomacales y de isótopos estables. Technical Report INIDEP N° 78, p 16

Berecoechea JJ, Brogger MI, Penchaszadeh PE (2017) New evidence of brooding in the deep-sea brittle star *Astrotoma agassizii* Lyman, 1876 from a South Western Atlantic Canyon. Deep Sea Research, Part I 127:105–110

Bernal MC, Cairns SD, Penchaszadeh P, Lauretta D (2018) *Errina Argentina* sp. nov., a new stylasterid (hydrozoa: Stylasteridae) from Mar del Plata submarine canyon (Southwest Atlantic). Mar Biodivers 49:833. https://doi.org/10.1007/s12526-018-0861-1

Bernal C, Cairns S, Penchaszadeh PE, Lauretta D (2021) Stylasterids (hydrozoa: Stylasteridae) from Mar del Plata submarine canyon and adjacent area (southwestern Atlantic), with a key to the species off Argentina. Zootaxa 4969:401–452

Bertolino M, Schejter L, Calcinai B, Cerrano C, Bremec C (2007) Sponges from a submarine canyon of the Argentine sea. In: Custódio MR, Hajdu E, Lôbo-Hajdu G, Muricy G (eds) Porifera research: biodiversity, innovation, sustainability. Museu Nacional, Rio de Janeiro, pp 189–201

Bogazzi E, Baldoni A, Rivas A, Martos P, Reta R, Orensanz JM, Lasta M, Dell'arciprete P, Werner F (2005) Spatial correspondence between areas of concentration of Patagonian scallop (*Zygochlamys patagonica*) and frontal systems in the southwestern Atlantic. Fish Oceanogr 14(5):359–376

Botto F, Bremec C, Marecos M, Schejter L, Lasta M, Iribarne O (2006) Identifying predators of the SW Atlantic Patagonian scallop *Zygochlamys patagonica* using stable isotopes. Fish Res 81:45–50

Bozzano G, Martin J, Spoltore DV, Violante RA (2017) Los cañones submarinos del margen continental argentino: una síntesis sobre su génesis y dinámica sedimentaria. Latin American Journal of Sedimentology and Basin Analysis 24(1):85–101

Bradshaw C, Veale LO, Brand AR (2002) The role of scallop-dredge disturbance in long-term changes in Irish Sea benthic communities: a re-analysis of an historical dataset. J Sea Research 47:161–184

Bravo ME, Principi S, Levin LA, Ormazabal JP, Ferronato C, Palma F, Isola J, Tassone AA (2024) Discovery of deep-sea cold seeps from Argentina host singular trophic linkages and biodiversity, Deep-Sea Res PT I, 211:104361. ISSN 0967-0637. https://doi.org/10.1016/j.dsr.2024.104361

Bremec CS, Giberto DA (Eds) (2017) Comunidades bentónicas en regiones de interés pesquero de la Argentina. Publicación Especial INIDEP. Mar del Plata, Argentina, p 129

Bremec C, Lasta ML (2002) Epibenthic assemblage associated with scallop (*Zygochlamys patagonica*) beds in the Argentine shelf. Bull Marine Sciences 70(1):89–105

Bremec C, Schejter L (2010) Benthic diversity in a submarine canyon in the Argentine sea. Rev Chil Hist Nat 83:453–457

Bremec CS, Schejter L (2019) *Chaetopterus antarcticus* (Polychaeta: Chaetopteridae) in Argentinian shelf scallop beds: from infaunal to epifaunal life habits. Rev Biol Trop 67:39–50

Bremec C, Brey T, Lasta M, Valero J, Lucifora L (2000) *Zygochlamys patagonica* beds on the Argentinian shelf. Part I: energy flow through the scallop bed community. Arch Fish Mar Res 48(3):295–303

Bremec C, Marecos A, Schejter L, Lasta M (2003) Guía técnica para la identificación de Invertebrados epibentónicos asociados a bancos de vieira patagónica (*Zygochlamys patagonica*) en el Mar Argentino. Publicaciones Especiales INIDEP, Mar del Plata, p 28

Bremec C, Escolar M, Schejter L, Genzano G (2008) Primary settlement substrate of scallop *Zygochlamys patagonica* (King and Broderip, 1832) (Mollusca: Pectinidae) in fishing grounds in the Argentine Sea. J Shellfish Res 27(2):273–280

Bremec C, Souto V, Escolar M, Giberto D (2011) Fauna bentónica asociada a prerreclutas de merluza en la Zona Común de Pesca Argentino-Uruguaya. Resultados de la campaña CC-12/09. Inf. Invest. INIDEP N°5/2011, p 15

Bremec C, Souto V, Escolar M, Giberto D (2012) Fauna bentónica asociada a prerreclutas de merluza en la zona patagónica entre 44° y 47°S. Resultados de la campaña OB-07/11. Inf. Invest. INIDEP N°28/2012, p 9

Bremec C, Schejter L, Giberto D (2015) Synoptic post fishery structure of invertebrate bycatch associated to *Zygochlamys patagonica* fishing grounds at the Southwest Atlantic shelf-break front (39°S, Argentina). J Shellfish Res 34(3):729–736

Brickle P, Laptikhovsky V, Pompert J, Bishop A (2003) Ontogenic changes in the feeding habits and dietary overlap between three abundant rajid species on The Falkland Islands shelf. J Mar Biol Assoc UK 83:1119–1125

Brunetti NE (1988) Contribución al conocimiento biológico-pesquero del calamar argentino (Cephalopoda, Ommastrephidae, *Illex argentinus*). PhD Thesis, Universidad de La Plata, p 135

Brunetti NE, Ivanovic ML, Sakai M (1999) Calamares de importancia comercial en la Argentina. Biología, distribución, pesquerías, muestreo biológico. Instituto Nacional de Investigación y Desarrollo Pesquero (INIDEP), Mar del Plata, Argentina, p 45

Cairns SD, Polonio V (2013) New records of deep-water Scleractinia off Argentina and The Falkland Islands. Zootaxa 3691:58–86

Campodónico S, Escolar M, García J, Aubone A (2020) Historical overview and current status of the Patagonian scallop *Zygochlamys patagonica* (King 1832) fishery in Argentina. Biology, stock assessment and management. Marine and Fishery Sciences 32(2):125–148. https://doi.org/10.47193/mafis.3222019121904

Cerino N, Lauretta D (2013) *Armadillogorgia albertoi* sp. nov.: new primnoid from Argentinean deep sea. Zootaxa 3741:369–376

Cook JA, Chubb JC, Veltkamp CJ (1998) Epibionts of *Asellus aquaticus* (L.) (Crustacea, Isopoda): an SEM study. Freshw Biol 39(3):423–438

Costello MJ, Chaudhary C (2017) Marine biodiversity, biogeography, Deep-Sea gradients, and conservation. Curr Biol 27(11):R511–R527

Costello MJ, Tsai P, Wong PS, Cheung AKL, Basher Z, Chaudhary C (2017) Marine biogeographic realms and species endemicity. Nat Commun 8(1):1057

Del Río JL, Acosta J, Cristobo J, Portela J, Parra-Descalzo S, Tel E, Viñas L, Pereiro-Muñoz JA, Vilela R, Elvira E, Patrocinio T, Ríos P, Almón-Pazos B, Blanco R, Murillo FJ, Polonio-Povedano V, Fernández J, Cabanas-López JM, Gago J, González-Nuevo G, Cabrero-Rodríguez AH, Besada V, Schultze F, Franco-Hernández MÁ, Bargiela J, Blanco-García X (2012) Estudio de los Ecosistemas Marinos Vulnerables en aguas internacionales del Atlántico Sudoccidental. In: Instituto Español de Oceanografía, Ministerio de Economía y Competitividad (Ed). Temas de Oceanografía, vol. 6, p 238

Delpiani GE, Spath MC, Figueroa DE (2013) Feeding ecology of the southern thorny skate, *Amblyraja doellojuradoi* on the Argentine continental Shelf. J Mar Biol Assoc UK 93(8):2207–2216

Doti BL (2016) Three new parammunids (Isopoda: Asellota: Parammunidae) from the Argentine Sea, south-west Atlantic. J Mar Biol Assoc U K:1–15. https://doi.org/10.1017/S0025315416001016

Doti BL, Chiesa I, Roccatagliata D (2020) Biodiversity of the deep-sea isopods, cumaceans and amphipods (Crustacea: Peracarida) recorded off the Argentine coast. In: Hendrickx M (ed) Deep-Sea Pycnogonids and crustaceans of the Americas. Editorial Springer Latin American, pp 157–191

Durán Muñoz P, Sayago-Gil M, Murillo FJ, Del Río JL, López-Abellán LJ, Sacau M, Sarralde R (2012) Actions taken by fishing nations towards identification and protection of vulnerable marine ecosystems in the high seas: the Spanish case (Atlantic Ocean). Mar Policy 36:536–543

Escolar M (2010) Variaciones espacio-temporales en la comunidad de invertebrados bentónicos asociada al frente de talud. Equinodermos como caso de estudio. PhD Thesis, Facultad de Ciencias Exactas y Naturales, Universidad de Buenos Aires, p 189

Escolar M, Bremec C (2015) Comunidad de equinodermos en bancos de vieira patagónica asociados al frente de talud. Revista de Investigación y Desarrollo Pesquero 26:23–36

Escolar M, Diez M, Hernández D, Marecos A, Campodónico S, Bremec C (2009) Invertebrate bycatch in Patagonian scallop fishing grounds: a study case with data obtained by the on board observers program. Rev Biol Mar Oceanografía 44:369–377

Escolar M, Schejter L, Bremec C (2011) Bancos de *Zygochlamys patagonica* en el frente de talud: el efecto del esfuerzo pesquero sobre la fauna asociada. In: VIII Congreso Latinoamericano de Malacología, Puerto Madryn, Argentina, Resúmenes, p 149

Escolar M, Campodónico C, Marecos A, Schejter L (2015) Efecto del arrastre pesquero en la comunidad bentónica asociada a la vieira patagónica. Inf. Invest. INIDP N°84/2015, p 23

Escolar M, Schwartz M, Marecos A, Herrera S, Díaz R, Schejter L, Campodónico S, Bremec CS (2017) Daño en invertebrados bentónicos en la captura incidental de la pesquería de vieira patagónica. Revista de Investigación y Desarrollo Pesquero 30:53–73

Farias N, Ocampo E, Luppi T (2015) On the presence of the deep-sea blind lobster *Stereomastis suhmi* (Decapoda: Polychelidae) in Southwestern Atlantic waters and its circum-Antarctic distribution. N Z J Zool 42(2):119–125

Fischer L, Covatti Ale M, Deli Antoni M, Díaz de Astarloa JM, Delpiani G (2022) Feeding ecology of the longtail southern cod, *Patagonotothen ramsayi* (Regan, 1913) (Notothenioidei) in the marine protected area Namuncurá-Burdwood Bank, Argentina. Polar Biol 45(9):1483–1494

Flores JN, Penchaszadeh PE, Brogger MI (2021) Heart urchins from the depths: *Corparva lyrida* gen. et sp. nov. (Palaeotropidae), and new records for the southwestern Atlantic Ocean. Rev Biol Trop 69(Suppl. 1):14–34

Gaitán E, Souto V (2020) Comunidades de macro-invertebrados bentónicos en el área del efectivo norte de merluza común (*Merluccius hubbsi*). Comparación entre los años 2012 y 2016. Frente Marítimo. Publicación de la Comisión Técnica Mixta del Frente Marítimo, p 53–77

Gaitán E, Giberto D, Escolar M, Schejter L, Bremec C (2013) Fauna bentónica asociada a recursos demersales en el Área del Tratado del Río de la Plata. Resultados de la Campaña de evaluación EH-06/12. Inf Invest INIDEP 11:12

Gaitán E, Giberto D, Escolar M, Bremec C (2014) Fauna bentónica asociada a los fondos de pesca en la plataforma patagónica entre 41° y 48°s. Resultados de la campaña de evaluación de merluza EH-04/13. Inf Invest INIDEP 35:19

García de la Rosa SB (1998) Estudios de las interrelaciones tróficas de dos elasmobranquios de la plataforma continental del Mar Argentino, en relación con las variaciones espacio-temporales y ambientales. *Squalus acanthias* (Squalidae) y *Raja flavirostris* (Rajidae). PhD Thesis, Universidad Nacional de Mar del Plata, Argentina

Genzano GN, Giberto D, Schejter L, Bremec C, Meretta P (2009) Hydroid assemblages from the Southwestern Atlantic Ocean (34–42°). Mar Ecol, 30:33–43

Giberto DA (2003) Benthic Diversity of the Rio de la Plata Estuary and Adjacent Marine Waters. Technical Report. PNUD Project/GefRLA/99/G31. Mar del Plata, Argentina, p 48

Giberto DA, Romero MV, Escolar M, Machinandiarena L, Bremec CS (2015) Diversidad de las comunidades bentónicas en las regiones de reclutamiento de la merluza común *Merluccius hubbsi* Marini, 1933. Revista Investigación y Desarrollo Pesquero 27:5–25

Giberto DA, Roux A, Bremec CS (2017a) Ecosistemas norpatagónicos: Golfo San Jorge y Península Valdés. Capítulo 4. In: Bremec CS, Giberto DA (eds) Comunidades bentónicas en regiones de interés pesquero de la Argentina. Publicación Especial INIDEP. Mar del Plata, Argentina, pp 47–56

Giberto DA, Belleggia M, Bremec CS (2017b) El bentos como alimento de peces comerciales. Capítulo 7. In: Bremec CS, Giberto DA (eds) Comunidades bentónicas en regiones de interés pesquero de la Argentina. Publicación Especial INIDEP, Mar del Plata, pp 93–108

Gotelli NJ, Colwell RK (2011) Estimating species richness. In: Magurran AE, Mcgill BJ (eds) Biological diversity: Frontiers in measuring and assessment, pp 39–54

Guerrero RA, Piola AR (1997) Masas de agua en la plataforma continental. In: Boschi EE (ed) El Mar Argentino y sus Recursos Pesqueros, Tomo 1:107–118

Haimovici M, Brunetti NE, Rodhouse PG, Csirke J, Leta RH (1998) *Illex argentinus*. FAO Fisheries Technical Paper, p 27–58

Halpern BS, Frazier M, Potapenko J, Casey KS, Koenig K, Longo C, Lowndes JS, Rockwood RC, Selig ER, Selkoe KA et al (2015) Spatial and temporal changes in cumulative human impacts on the world's ocean. Nat Commun 6(1):7615

Hinz H, Prieto V, Kaiser MJ (2009) Trawl disturbance on benthic communities: chronic effects and experimental predictions. Ecol Appl 19:761–773

Hurtado-García J, Manjón-Cabeza ME (2022) Species composition of sea stars (Echinodermata: Asteroidea) in the Patagonian Argentinian deep sea, including seven new records: connectivity with sub-Antarctic and Antarctic fauna. Polar Biol 45:1211–1228

Jenkins SR, Mullen C, Brand AR (2004) Predator and scavenger aggregation to discarded by-catch from dredge fisheries: importance of damage level. J Sea Res 51:69–76

Koen Alonso M, Crespo EA, Garcia NA, Mariotti P, Berón Vera B, Mora N (2001) Food habits of *Dipturus chilensis* (Pisces: Rajidae) off Patagonia, Argentina. ICES J Mar Sci 58:288–297

Lasta ML, Bremec CS (1998) *Zygochlamys patagonica* in the Argentine sea: a new scallop fishery. J Shellfish Res 17:103–111

Lasta M, Bremec C (1999) Vieira patagónica (*Zygochlamys patagonica* King and Broderip, 1832): una nueva pesquería en la plataforma continental Argentina. Rev Invest Desarr Pesq 12:5–18

Lastras G, Acosta J, Muñoz A, Canals M (2011) Submarine canyon formation and evolution in the Argentine continental margin between 44°30′S and 48°S. Geomorphology 128:116–136

Lauretta D, Penchaszadeh PE (2017) Gigantic oocytes in the deep sea black coral *Dendrobathypathes grandis* (Antipatharia) from the Mar del Plata submarine canyon area (southwestern Atlantic). Deep Sea Research Part I 128:109–114

Lauretta D, Vidos C, Martinez M, Penchaszadeh PE (2020) Brooding in the deep-sea sea anemone *Actinostola crassicornis* (Hertwig, 1882) (Cnidaria: Anthozoa: Actiniaria) from the southwestern Atlantic Ocean. Polar Biol 43:1353–1361

Link JS, Almeida FP (2002) Opportunistic feeding of longhorn sculpin (*Myxocephalus octodecemspinosus*): are scallop fishery discards an important food subsidy for scavengers on Georges Bank? Fish Bull 100:381–385

Lonardi AG, Ewing M (1971) Sediment transport and distribution in the Argentine basin. 4. Bathymetry of the continental margin, Argentine basin and other related provinces. Canyons and sources of sediments. Phys Chem Earth 4:81–121

López Gappa JJ (2000) Species richness of marine Bryozoa in the continental shelf and slope off Argentina (south-west Atlantic). Divers Distrib 6:15–27

López Gappa JJ, Landoni N (2005) Biodiversity of Porifera in the Southwest Atlantic between 35° and 56° S. Revista del Museo Argentino de Ciencias Naturales, ns 7(2):191–219

López Gappa JJ, Landoni NA (2009) Space utilization patterns of bryozoans on the Patagonian scallop *Psychrochlamys patagonica*. Sci Mar 73(1):161–171

Lucifora LO, Valero JL, Bremec CS, Lasta ML (2000) Feeding habits and prey selection by the skate *Dipturus chilensis* (Elasmobranchii: Rajidae) from the south-western Atlantic. J Mar Biol Assoc UK 80:953–954

Mabragaña E, Giberto DA (2007) Feeding ecology and abundance of two sympatric skates, the shortfin sand skate *Psammobatis normani* McEachran, and the smallthorn sand skate *P. rudis* Günther (Chondrichthyes, Rajidae), in the southwest Atlantic. ICES J Mar Sci 64:1017–1027

Mabragaña E, Giberto DA, Bremec CS (2005) Feeding ecology of *Bathyraja macloviana* (Rajiformes:Arhynchobatidae): a polychaete-feeding skate from the south-west Atlantic. Sci Mar 69:405–413

Madirolas A, Isla FI, Tripode M, Alvarez Colombo G, Cabreira A (2005) First results from the multibeam surveys carried out over the Argentine Continental shelf. Book of Abstracts FEMME Simrad Multibeam Users Conference, Dublin, Ireland, 26–29 April, 2005 (Extended abstract, p 2). Dublin, Ireland: Irish National Seabed Survey

Maggioni T, Taverna A, Reyna PB, Alurralde G, Rimondino C, Tatián M (2018) Deep-sea ascidians (Chordata, Tunicata) from the SW Atlantic: species richness with descriptions of two new species. Zootaxa 4526(1):001–028

Marrari M, Piola AR, Valla D (2017) Variability and 20-year trends in satellite-derived surface chlorophyll concentrations in large marine ecosystems around South and Western Central America. Front Mar Sci 4:372. https://doi.org/10.3389/fmars.2017.00372

Martinez M, Penchaszadeh P (2017) A new species of brooding Psolidae (Echinodermata: Holothuroidea) from deep-sea off Argentina, Southwestern Atlantic Ocean. Deep Sea Research, Part II 146:13–17

Martinez MI, Solís-Marín FA, Penchaszadeh PE (2014) *Benthodytes violeta*, a new species of a deep-sea holothuroid (Elasipodida: Psychropotidae) from Mar del Plata canyon (South-Western Atlantic Ocean). Zootaxa 3760(1):89

Martinez MI, Solís-Marín FA, Penchaszadeh PE (2019) First report of *Paelopatides* (Synallactida, Synallactidae) for the SW Atlantic, with description of a new species from the deep-sea off Argentina. Zool Anz 278:21–27

Martinez MI, Alba-Posse EJ, Lauretta D, Penchaszadeh PE (2020) Reproductive features in the sea cucumber *Pentactella perrieri* (Ekman, 1927) (Holothuroidea: Cucumariidae): a brooding hermaphrodite species from the southwestern Atlantic Ocean. Polar Biol 43:1383–1389

Mauna C, Acha ME, Lasta ML, Iribarne O (2011) The influence of a large SW Atlantic frontal system on epibenthic community composition, trophic guilds, and diversity. J Sea Res 66:39–46

Moore JM, Nishi E, Rouse GW (2017) Phylogenetic analyses of Chaetopteridae (Annelida). Zool Scr 46(5):596–610

Muñoz A, Acosta J, Cristobo J, Druet M, Uchupi E, Group A (2013) Geomorphology and shallow structure of a segment of the Atlantic Patagonian margin. Earth Sci Rev 121:73–95

Nishi E, Hickman CP Jr, Bailey-Brock JH (2009) *Chaetopterus* and *Mesochaetopterus* (Polychaeta: Chaetopteridae) from the Galapagos Islands, with descriptions of four new species. Proc Acad Natl Sci Phila 158(1):239–259

Ocampo EH, Farías NE, Luppi TA (2014) New record of the deep-sea crab *Ethusina abyssicola* from the Mar del Plata Canyon, Argentina. N Z J Zool 41:218–221

Olguin N, Ocampo EH, Farias NE (2015) New record of *Paralomis spinosissima* Birstein & Vinogradov (Decapoda: Anomura: Lithodidae) from Mar del Plata, Argentina. Zootaxa 3957:2

Parker G, Paterlini MC, Violante R (1997) El fondo marino. In Boschi EE (ed) Antecedentes históricos de las exploraciones en el mar y las características ambientales. El Mar Argentino y sus Recursos Pesqueros, Tomo 1:65–87

Pastorino G (2016) First report of the family Laubierinidae Waren & Bouchet, 1990 (Gastropoda: Tonnoidea) in the southwestern Atlantic. Molluscan Res 36:108–111

Pereira E, Doti BL (2017) *Edotia abyssalis* n. sp. from the Southwest Atlantic Ocean, first record of the genus (Isopoda, Valvifera, Idoteidae) in the deep sea. Zoologischer Anseiger 268:19–31. https://doi.org/10.1016/j.jcz.2017.04.007

Pereira E, Roccatagliata D, Doti BL (2019) *Xiphoarcturus* – a new genus and two new species of the family Antarcturidae (Isopoda: Valvifera) from the Mar del Plata submarine canyon and its phylogenetic relationships. Arthropod Syst Phylo 77:303–323. https://doi.org/10.26049/ASP77-2-2019-07

Pereira E, Roccatagliata D, Doti BL (2020) On the antarcturid genus *Fissarcturus* (Isopoda: Valvifera): description of *Fissarcturus argentinensis* n. sp., first description of the male of *Fissarcturus patagonicus* (Ohlin, 1901), and biogeographic remarks on the genus. Zool Anz 288:168–189. https://doi.org/10.1016/j.jcz.2020.08.002

Pereira E, Doti BL, Roccatagliata D (2021) A new species of *Pseudione sensu lato* (Isopoda: Bopyridae) on a squat lobster host from the deep south-west Atlantic. Zootaxa 4996:363–373. https://doi.org/10.11646/zootaxa.4996.2.10

Portela J, Acosta J, Cristobo J, Muñoz A, Parra S, Ibarrola T, Del Río JL, Vilela R, Ríos P, Blanco R, Almón B, Tel E, Besada V, Viñas L, Polonio V, Barba M, Marín P (2012) Management strategies to limit the impact of bottom trawling on VMEs in the high seas of the SW Atlantic. In: Cruzado A (ed) Marine ecosystem. InTech, pp 199–228

Portela J, Cristobo J, Ríos P, Acosta J, Parra S, Del Río JL, Tel E, Polonio V, Muñoz A, Patrocinio T, Vilela R, Barba M, Marín P (2015) A first approach to assess the impact of bottom trawling over vulnerable marine ecosystems on the high seas of the Southwest Atlantic. In: Lo JH, Blanco JA, Roy S (Ed.), Biodiversity in ecosystems: linking structure and function. InTech, 272–298

Razik S, Govin A, Chiessi CM, von Dobeneck T (2015) Depositional provinces, dispersal, and origin of terrigenous sediments along the SE South American continental margin. Mar Geol 363:261–272

Risaro J, Williams G, Pereyra D, Lauretta D (2020) *Umbellula pomona* sp. nov., a new sea pen from Mar del Plata submarine canyon (Cnidaria: Octocorallia: Pennatulacea). European Journal of Taxonomy 720:121–143

Rivadeneira PR, Martinez MI, Penchaszadeh PE, Brogger MI (2020) Reproduction and description of a new genus and species of deep-sea asteriid sea star (Echinodermata; Asteroidea) from the southwestern Atlantic. Deep-Sea Res I Oceanogr Res Pap 163:103348

Roccatagliata D (2020) On the deep-sea lampropid *Platytyphlops sarahae* n. sp. from Argentina, with remarks on some morphological characters of Cumacea. Zool Anz 286:135–145. https://doi.org/10.1016/j.jcz.2020.03.009

Roccatagliata D, Alberico NA (2015) Two new cumaceans (Crustacea: Peracarida) from the southwest Atlantic with remarks on the problematic genus *Holostylis* Stebbing, 1912. Mar Biodivers 46:163–181. https://doi.org/10.1007/s12526-015-0349-1

Romanelli JP (2017) Ecología trófica de tres especies de granaderos, *Coelorinchus fasciatus*, *Macrourus carinatus* y *Macrourus holotrachys*, presentes en el Océano Atlántico Sudoccidental. MSc Thesis, Universidad Nacional de Mar del Plata, Argentina

Romero MV, Schejter L, Bremec CS (2017) Epibiosis y bioerosión en invertebrados bentónicos marinos. In: Bremec CS, Giberto DA (eds) Comunidades bentónicas en regiones de interés pesquero de la Argentina. Publicación Especial INIDEP, Mar del Plata, pp 109–129

Romero MV, Casadio S, Bremec CS, Giberto DA (2022) Sclerobiosis: a term for colonization of marine hard substrates. Ameghiniana 59(4):265–273. https://doi.org/10.5710/AMGH.21.06.2022.3486

Rouse GW (2001) Chaetopteridae. In: Rouse G, Plejiel F (eds) Polychaetes. Oxford University Press, New York, pp 256–260

Ruocco NL, Lucifora LO, Díaz de Astarloa JM, Bremec C (2009) Diet of the white-dotted skate, *Bathyraja albomaculata*, in waters of Argentina. J Appl Ichthyol 25:94–97

Sánchez F, Mabragaña E (2002) Características biológicas de algunas rayas de la región sud patagónica. Informe Técnico INIDEP 48:1–15

Sanchez N, Pastorino G (2020) The North Atlantic Conoidean gastropod *Theta lyronuclea* (Raphitomidae) in deep-waters of the southwestern Atlantic. Malacologia 60:33–40

Sánchez F, Marí NR, Bernardele JC (2009) Distribución, abundancia y alimentación de pintarroja *Schroederichthys bivius* Müller and Henle, 1838 en el Océano Atlántico sudoccidental. Rev Biol Mar Oceanog 44:453–466

Sánchez MA, Giberto D, Schejter L, Bremec C (2011) The Patagonian scallop fishing grounds in shelf break frontal areas: the non assessed benthic fraction. Lat Am J Aquat Res 39(1):167–171

Scenna LB, García de la Rosa SB, Díaz De Astarloa JM (2006) Trophic ecology of the Patagonian skate, *Bathyraja macloviana*, on the Argentine continental shelf. ICES J Mar Sci 63:867–874

Schejter L (2000) Alimentación de la vieira patagónica *Zygochlamys patagonica* (King and Broderip, 1832) en el banco Reclutas (39°S–55°W) durante un período anual. Degree Thesis, Universidad Nacional de Mar del Plata, p 44

Schejter L, Bremec CS (2006) Benthic richness in scallop beds *Zygochlamys patagonica* (King and Broderip, 1832) as primary settlement substrate. J Shellfish Res 25(1):305–306

Schejter L, Bremec C (2007a) Benthic richness in the Argentine continental shelf: the role of *Zygochlamys patagonica* (Mollusca: Bivalvia: Pectinidae) as settlement substrate. J Mar Biol Assoc UK 87:917–925

Schejter L, Bremec C (2007b) Did the epibenthic bycatch at the Patagonian scallop assemblage change after ten years of fishing? In: Proceedings of the 16th International Pectinid Workshop, Halifax, Canadá. J Shellfish Res 26:1341–1343

Schejter L, Bremec CS (2008) ¿En cuánto contribuyen los epibiontes de moluscos a la riqueza específica bentónica asociada al frente de talud del Mar Argentino? In: VII CLAMA, Congreso Latinoamericano de Malacología, Valdivia, Chile, p 149–150

Schejter L, Bremec C (2009) Epibiosis contest at *Zygochlamys patagonica* fishing grounds: which is the winner? In: 17th International Pectinid Workshop, Santiago de Compostela, España. Resúmenes, p 143–144

Schejter L, Bremec C (2013) Composition, richness and characterization of the benthic community in a non-fished area at the Patagonian Scallop fishing grounds, Argentina. In: 19th International Pectinid Workshop, Florianópolis, Brasil. Proceedings, p 124–125

Schejter L, Escolar M (2013) Volutid shells as settlement substrates and refuge in soft bottoms of the SW Atlantic Ocean. Pan-Am J Aquat Sci 8(2):104–111

Schejter L, Giberto DA (2022) Filling biodiversity knowledge gaps: Sponges (Porifera: Demospongiae) Recorded off San Jorge Gulf (Argentina), SW Atlantic Ocean. 2nd international electronic conference on diversity (IECD 2022)-new insights into the biodiversity of plants, animals and microbes session animals diversity. https://doi.org/10.3390/IECD2022-12407 (registering DOI). 15-31 March 2022

Schejter L, Mantelatto F (2015) The hermit crab *Sympagurus dimorphus* (Anomura: Parapaguridae) at the edge of its range in the SW Atlantic Ocean: population and morphometry features. J Nat Hist 49(33/34):2055–2066

Schejter L, Bremec C, Akselman R, Hernández D, Spivak ED (2002) Annual feeding cycle of the Patagonian scallop *Zygochlamys patagonica* (King and Broderip, 1832) in Reclutas bed (39°S-55°W), Argentine Sea. J Shellfish Res 21:553–559

Schejter L, Calcinai B, Cerrano C, Bertolino M, Pansini M, Giberto D, Bremec C (2006) Porifera from the Argentine Sea: diversity in Patagonian scallop beds. Ital J Zool 73:373–385

Schejter L, Bertolino M, Calcinai B, Cerrano C, Bremec C (2008a) Los moluscos como sustrato de asentamiento de esponjas en áreas del frente de talud del Mar Argentino. En: VII CLAMA, Valdivia, Chile. Resúmenes, p 150

Schejter L, Bremec C, Hernández D (2008b) Comparison between disturbed and undisturbed areas of the Patagonian scallop (*Zygochlamys patagonica*) fishing ground "Reclutas" in the Argentine Sea. J Sea Res 60:193–200

Schejter L, Bremec C, Waloszek D, Escolar M (2010) Recently settled stages and larval developmental mode of the bivalves *Zygochlamys patagonica* and *Hiatella meridionalis* in the Argentine Sea. J Shellfish Res 29(1):63–67

Schejter L, Bertolino M, Calcinai B, Cerrano C, Bremec C (2011a) Epibiotic sponges on the hairy triton *Fusitriton magellanicus* in the SW Atlantic Ocean, with the description of *Myxilla (Styloptilon) canepai* sp. nov. Aquatic Biol 14:9–20

Schejter L, Escolar M, Bremec C (2011b) Variability in epibiont colonization of shells of *Fusitriton magellanicus* (Gastropoda) on the Argentinean shelf. J Mar Biol Assoc UK 91:897–906

Schejter L, Escolar M, Remaggi C, Álvarez-Colombo G, Ibañez P, Bremec CS (2012a) By-catch composition of the Patagonian scallop fishery: the fishes. Lat Am J Aquat Res 40(4):1094–1099

Schejter L, Chiesa IL, Doti BL, Bremec C (2012b) Mycale (Aegogropila) magellanica (Porifera: Demospongiae) in the southwestern Atlantic Ocean: endobiotic fauna and new distributional information. Sci Mar 76:753–761

Schejter L, Escolar M, Marecos A, Bremec C (2013) Seventeen years assessing biodiversity at *Zygochlamys patagonica* fishing grounds in the shelf break system, Argentina. In: 19th International Pectinid Workshop, Florianópolis, Brasil. Resúmenes, p 46–47

Schejter L, Escolar M, Marecos A, Bremec C (2014) Asociaciones faunísticas en las unidades de manejo del recurso "vieira patagónica" en el frente de talud durante el período 1998–2009. Inf. Invest. INIDEP N°13/2014, p 29

Schejter L, Schwartz M, Bremec CS (2015) Registro del coral de piedra *Desmophyllum dianthus* (Esper, 1794) (Scleractinia, Caryophylliidae) en áreas del frente de talud del Mar Argentino. Revista de Investigación y Desarrollo Pesquero 26:89–95

Schejter L, Escolar M, Giberto D (2016) Comunidad de invertebrados bentónicos en áreas de pesca y de reserva en bancos de vieira patagónica: Estado general en el año 2015 y comparación con datos del 2013. Inf. Invest. INIDEP N°38/2016, p 14

Schejter L, Bremec CS, Escolar M, Giberto DA (2017) Plataforma externa y talud continental. Capítulo 5. In: Bremec CS, Giberto DA (eds) Comunidades bentónicas en regiones de interés pesquero de la Argentina. Publicación Especial INIDEP, Mar del Plata, pp 57–75

Schejter L, Acuña FH, Garese A, Cordeiro RT, Perez CD (2018) Sea Pens (Cnidaria: Pennatulacea) from Argentine waters: new distributional records and first report of associated anemones. Pan-American Journal of Aquatic Sciences 3:292–301

Schejter L, Mauna C, Pérez CD (2021) New record and range extension of the primnoid octocoral *Verticillata castellviae* in the Southwest Atlantic Ocean. Mar Fish Sci 34:275–281

Schwartz M, Escolar M, Marecos A, Campodónico S (2016) Supervivencia de invertebrados bentónicos capturados incidentalmente en la pesquería de vieira patagónica. Inf Téc INIDEP 95:24

Shimabukuro V (2009) Hábitos alimentarios y dentición de *Bathyraja albomaculata* (Norman, 1937) (Chondrichthyes: Rajidae). MSc Thesis, Universidad Nacional de Mar del Plata, Argentina

Siegwald J, Oskars T, Pastorino G, Malaquias MAE (2020) A new *Scaphander* species from the deep sea of Argentina. Bull Mar Sci 96(1):111–126

Souto V (2009) Estructura y producción de la comunidad de la vieira *Zygochlamys patagonica* en el banco "Reclutas" (39° S) entre los años 1995 y 2006. Degree Thesis. Facultad de Ciencias Exactas y Naturales, Universidad Nacional de Mar del Plata, p 47

Steinmann L, Baques M, Wenau S, Schwenk T, Spiess V, Piola AR, Bozzano G, Violante R, Kasten S (2020) Discovery of a giant cold-water coral mound province along the northern Argentine margin and its link to the regional Contourite depositional system and oceanographic setting. Mar Geol 427:106223

Stöckli R, Vermote E, Saleous N, Simmon R, Herring D (2007) The blue marble next generation-a true color earth dataset including seasonal dynamics from MODIS, vol 87. Published by the NASA Earth Observatory, p 49

Suby A, Elías R, Fernández M, Bremec C, Giberto D (2019a) Biodiversidad y productividad macrobentónica infaunal del golfo San Jorge y litoral de Chubut, Argentina (43°–47°S). XVIII Congreso Latinoamericano de Ciencias del Mar-COLACMAR, 4–8 Noviembre, Mar del Plata, Argentina. Libro de Resúmenes, p 759

Suby A, Elías R, Romero MV, Giberto D (2019b) Biodiversidad y productividad macrobentónica del frente de Península Valdés, Argentina (42°–44°S). XVIII Congreso Latinoamericano de Ciencias del Mar-COLACMAR, 4–8 Noviembre, Mar del Plata, Argentina. Libro de Resúmenes, p 760

Teso V, Martínez M, Lauretta D, Pastorino G, Urteaga D, Averbuj A, Brogger M, Arrighetti F, Rivadeneira P, Flores J, Pertossi R, Sánchez N, Pacheco L, di Luca J, Sánchez Antelo C, Risaro J, Ciocco R, Penchaszadeh PE (2020) Growing up in the deep-sea protected development in deep-sea invertebrates: a case study in the southwestern Atlantic Ocean. Environment coastal and off-shore magazine, Deep-Sea Special Edition, p 48–51

Villa A, Colonello J, Belleggia M (2018) Ecología trófica de *Schroederichthys bivius* en el Golfo San Jorge y aguas adyacentes. Jornadas Nacionales de Ciencias del Mar. Buenos Aires, Argentina. 30 July-3 August, 2018

Walossek D (1991) *Chlamys patagonica* (King and Broderip, 1832), a long "neglected" species from the shelf off the Patagonia coast. In: Shumway SE, Sandifer P (eds) An international compendium of scallop biology and culture. The World Aquaculture Society, pp 256–263

Wilckens H, Miramontes E, Schwenk T, Artana C, Zhang W, Piola AR, Kasten S (2021) The erosive power of the Malvinas current: influence of bottom currents on morpho-sedimentary features along the northern Argentine margin (SW Atlantic Ocean). Mar Geol 439:106539

Chapter 7
Fisheries in the Patagonian Shelf-Break Front

Daniela Alemany, Anabela Zavatteri, Nicolás Prandoni, and Analía Giussi

Abstract The Patagonian Shelf-Break Front (PSBF) holds significant global importance as a crucial fishing zone, distinguished by its remarkable species diversity and abundance. Fisheries within the PSBF encompass not only Argentine fleets operating within and beyond the Argentine Exclusive Economic Zone (AEEZ) but also fleets from multiple distant countries actively engaged in fishing within this highly productive region of the Southwest Atlantic Ocean. The primary concentration of fishing activities predominantly occurs within a critical region referred to as the "Blue Hole," situated between the 200-meter isobath and the boundary of the AEEZ. This highly productive area attracts intensive distant-country fishing operations, notably targeting the Argentine shortfin squid by the jigging fleet and demersal fish by the trawler fleet. A substantial portion of the fishing resources linked with the PSBF comprises straddling species, freely traversing between the AEEZ and international waters without specific fishing regulations governing activities in the high seas. Presently, there is an absence of an international or multilateral agreement to effectively manage this area. The Argentine shortfin squid stands as the most significant fishing species in terms of catches along the PSBF. The considerable fishing endeavors of multiple countries in the region are primarily focused on exploiting this particular resource. In high seas areas beyond national jurisdiction, the concentration of fisheries presents substantial impediments to comprehending both human activities and the natural variability of species. These challenges pri-

Supplementary Information The online version contains supplementary material available at https://doi.org/10.1007/978-3-031-71190-9_7.

D. Alemany (✉)
Instituto de Investigaciones Marinas y Costeras (IIMyC, UNMdP-CONICET), Facultad de Ciencias Exactas y Naturales, Universidad Nacional de Mar del Plata (UNMdP), Consejo Nacional de Investigaciones Científicas y Técnicas (CONICET), Mar del Plata, Argentina

Instituto Nacional de Investigación y Desarrollo Pesquero (INIDEP), Mar del Plata, Argentina
e-mail: dalemany@inidep.edu.ar

A. Zavatteri · N. Prandoni · A. Giussi
Instituto Nacional de Investigación y Desarrollo Pesquero (INIDEP), Mar del Plata, Argentina

E. M. Acha et al. (eds.), *The Patagonian Shelfbreak Front*, Aquatic Ecology Series 13, https://doi.org/10.1007/978-3-031-71190-9_7

marily stem from inherent temporal and spatial constraints. Notably, the PSBF and the Blue Hole stand out as vulnerable zones due to heightened fishing activities carried out by distant countries' fleets. This situation may potentially exacerbate existing concerns, such as those related to climate change. The impact of these fisheries on biodiversity and the intricate local food web remains poorly understood, highlighting the critical necessity for in-depth investigations within the PSBF. Such an inquiry is of paramount significance as it is poised to furnish essential data crucial for the integrated and sustainable management of this region alongside its concurrent fisheries operations.

Acronyms and Abbreviations

AEEZ	Argentine Exclusive Economic Zone
CITES	Convention on International Trade in Endangered Species of Wild Fauna and Flora
CMS	Convention on the Conservation of Migratory Species of Wild Animals
EAF	Ecosystem Approach to Fisheries
FAO	Food and Agriculture Organization of the United Nations
FT	Freezer-trawlers
IPOA Sharks	International Action Plan for the Conservation and Management of Sharks
ITQ	individual transferable quotas
IT	Ice-trawlers
J	Jiggers
MSC	Marine Stewardship Council
PAN Sharks	National Action Plan for the Conservation and Management of Chondrichthyans
PSBF	Patagonian Shelf-break Front
SPS	South Patagonian Stock
TAC	total allowable catches
UK	United Kingdom
USSR	Union of Soviet Socialist Republics

7.1 Introduction

The Southwest Atlantic is a worldwide recognized fishing ground, both for its diversity in fishing resources and for its high biomasses. This is due in part to the presence of several frontal systems, important oceanographic features that are not only used by a great diversity of marine organisms but are also used by human activities such as fisheries. In this sense, the Patagonian Shelf-Break Front (PSBF) represents an important fishing area in the Southwest Atlantic Ocean, as the distribution of fishing fleets and fishing effort are positively associated with it (Waluda et al. 2008; Alemany et al. 2014, 2016).

As mentioned in previous chapters, the PSBF is a dynamic and complex oceanographic area, bathymetrically defined by the continental slope, and located under Argentine sovereignty and also international waters. In that sense, fisheries from Argentina and from several other distant countries operate along it. Hundreds of vessels of different flags (mostly from China, Taiwan, South Korea, and Spain) move thousands of miles to fish along the outer edges of the Argentine Exclusive Economic Zone (AEEZ), at international waters of the PSBF, where high abundance of species of commercial value occur. Thus, these fisheries are composed by distant countries' fleets that operate in the high seas, far from their landing home ports (FAO 2020).

Fisheries that develop in the PSBF area are among the most important in the world. Its relevance dates back to the 1980s given the interest of white meat from fish and squids, highly demanded for their quality and abundance. Towards the end of the decade, the incorporation of large trawling vessels to the fishing activity allowed the capture of fish populations that were in a virgin state or close to it, which made the yields extremely relevant. Furthermore, the introduction of jigger vessels in the area encouraged the development of an important fishery exploitation of squid, a very abundant resource in the PSBF.

Regarding the Argentine fleet, that fishes in the AEEZ and in the PSBF, information on its distribution, target species, and catches are available from national databases. On the other hand, the information of international vessels fishing in the adjacent area of the AEEZ is derived from FAO statistics, being often difficult to acquire and compile detailed relevant data.

7.2 Main Fisheries Resources

Historically, the main species caught in the AEEZ and the PSBF are cephalopods as the Argentine shortfin squid (*Illex argentinus*) and the Patagonian squid (*Doryteuthis gahi*), demersal fish species including the Argentine hake (*Merluccius hubbsi*), the long tail hake (*Macruronus magellanicus*), the southern hake (*Merluccius australis*), the Patagonian toothfish (*Dissostichus eleginoides*), the southern blue whiting (*Micromesistius australis*), the red cod (*Salilota australis*), the rock cod (*Patagonotothen ramsayi*), grenadiers (*Macrourus* spp.), and the pink cusk eel (*Genypterus blacodes*); among bivalves the Patagonian scallop (*Zygochlamys patagonica*). Aforementioned species are monitored annually and most of them are in a sustainable state (Giussi et al. 2022).

The majority of the marine resources have a broad distribution in waters of the Argentine continental shelf where important processes of their life cycles take place (Sánchez and Bezzi 2004), but these species also take advantage of the PSBF to feed, reproduce or migrate (see Chap. 5). Regarding the marine species of commercial interest, three main fisheries operate along the PSBF, the Argentine shortfin squid fishery, the demersal fish fishery, and the Patagonian scallop fishery.

7.2.1 Management

Several regulations and management measures, implemented by the Argentine fisheries enforcement authority in the AEEZ, have led to the sustainable development of fisheries in the area. In recent years, since biodiversity has gained importance in the field of fisheries, single species management based on catch quotas has been replaced by the Ecosystem Approach to Fisheries (EAF) and the Argentine Federal Fisheries Council has adopted it since 2006. The EAF represents a holistic management approach since it recognizes the need to guarantee healthy and productive ecosystems, while maintaining the quality and value of the fishing activity (sustainable exploitation; Seitune et al. 2022).

Argentina has a complex and extensive legal framework for the management and exploitation of marine ecosystems, with a scope for the sustainable use of fishing resources and conservation. The wide range of regulations or management measures (https://www.magyp.gob.ar/sitio/areas/pesca_maritima/) to mitigate the impact of fisheries industry include total allowable catches (TAC), multi-specific individual transferable quotas (ITQ), a system for opening and closing fishing seasons, permanent or temporary fishing closed areas (to protect spawning, breeding and mating grounds for most of the target species), conservation related fisheries National Action Plans (to prevent, discourage, and eliminate Illegal, Unreported and Unregulated Fishing; for the conservation and management of chondrichthyans; to reduce birds, marine mammals and turtles interaction with fisheries), among others (e.g., Giussi et al. 2022; Seitune et al. 2022). However, all these measures are applicable within the AEEZ but do not apply to international waters.

Given the geographical location of the PSBF, one of the greatest fisheries challenges is the integrated management outside the AEEZ, at High Seas, where there are no regulations, control, or monitoring of straddling marine resources (e.g., the Argentine shortfin squid *Illex argentinus*; Chen et al. 2007). Although legal tools are in place for management, enforcement is somehow difficult, considering the vast extension of the area and the sovereignty conflict over the Malvinas archipelago.

In addition to the wide range of management measures mentioned above, some Argentine fisheries operating at the PSBF have been certified to meet international best practice for sustainable fishing. In that sense, the long tail hake and the Patagonian scallop are two examples of certified fisheries, while the Argentine shortfin squid, the Patagonian toothfish and the Argentine hake fisheries are undergoing Fishery Improvement Proyects (FIP). Consequently, they have been the subject of more in-depth review processes of their biological and fishing aspects.

7.3 Distant Countries' Fishing Fleets from 1970 to 2019

In this section, the information of fishing vessels that come from distant countries and operate along the PSBF, adjacent to the AEEZ, is derived from FAO statistics for the period 1970–2019.

The history of the development of fishing in the PSBF is related to the nature of the available resources and the process of discovery of these fishing grounds, which began to be exploited in the mid-70 s. The evolution of catches in the area has been highly variable during the last 50 years showing four maximum catch peaks that exceed 800,000 t per year reported by the distant countries' fleet in 1988, 1999, 2007, and 2015 (Fig. 7.1). In the 1970s and 1980s, Asian and European trawling fleets mainly from the former USSR, Japan, and Poland, began to visit the area and reported catches of less than 100,000 t of finfish until 1979. From that moment and for the following 6 years, Poland became the main actor in the region, fishing up to 250,000 t in 1983.

Towards the mid-80 s, Asian countries started to fish squids in the area by introducing "jiggers," vessels with a fishing method that had never been performed in the Southwest Atlantic area before. This highlighted the relevance of the PSBF as a world squid fishing ground. Japan, Taiwan, and Korea were the main actors until 2000, with catches per country ranging from 100,000 to 200,000 t per year. In the late 1990s, Japan stopped operating in the area while China began to have a presence at the PSBF, increasing its jigger fleet every year. At that time, the jigger fishery started to develop rapidly and trawling continued to be carried out mostly by Spanish-flagged vessels and the United Kingdom fleet from 2000 onwards (Fig. 7.1). The number of distant countries with relevant participation in the fishery at the PSBF showed a reduced trend in the last two decades of the period.

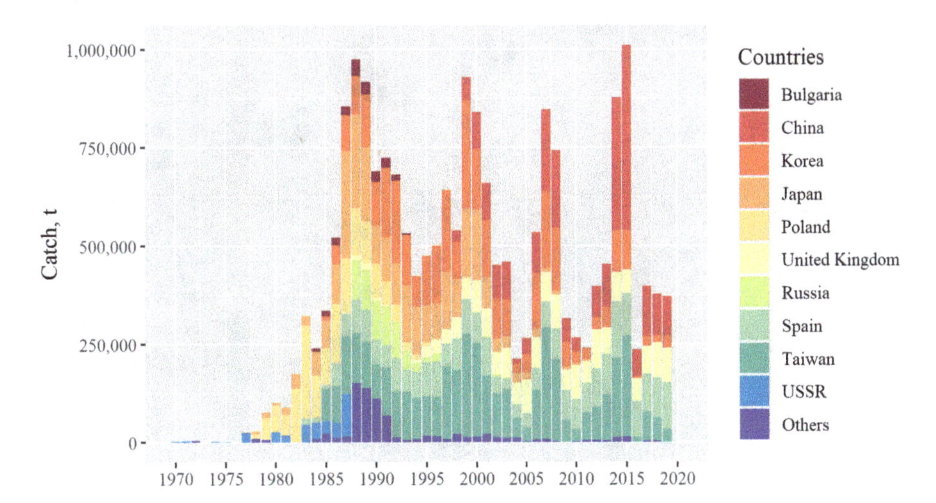

Fig. 7.1 Total catches declared by distant countries in the PSBF (FAO statistics outside the AEEZ) between 1970 and 2019

Historically, vessels of different flags have operated variably. Eastern European countries registered the highest finfish and cephalopods catches until the early 1990s. Particularly, the fishing fleets targeting finfish modified the stock structure of the southern blue whiting and the long tail hake, showing the former species high levels of overexploitation. However, some management measures (e.g., restricted area and TAC, among others) took into account the stock structure and the reduction of fishing effort which contributed to a slow recovery of certain assessed fish stocks. Landing declarations of the rest of the bony fish species have evidenced a high variability. Since in those years there were no biological studies of the species and no estimates of population size, catches increased without any regulation over a period of years depending on the availability of fish. Fish stocks should probably have had lower levels of harvest to remain sustainable.

When the evolution of the fisheries is analyzed from the perspective of resources (Fig. 7.2), the first available statistics from FAO showed that between 1978 and 1981, reported catches were less than 100,000 t, 90% of which correspond to the southern blue whiting fished by Polish and USSR trawlers. From 1980s to early 1990s, this species was the primary target of the distant countries' fleets, showing the largest catches among finfishes. The southern blue whiting has a wide distribution, including the Pacific and Atlantic Oceans, and two spawning grounds were identified in both (Wöhler et al. 2004). Probably, both populations share the same area in the southern part of the Argentine shelf, nearly 54 °S. Given the high fishing pressure to which it was subjected, its abundance declined below the reference point of biological sustainability, with catches falling below 10,000 t. In 2010, a restricted area was implemented in its spawning ground resulting in an evident recovery of the stock abundance and population structure (Zavatteri and Giussi 2020).

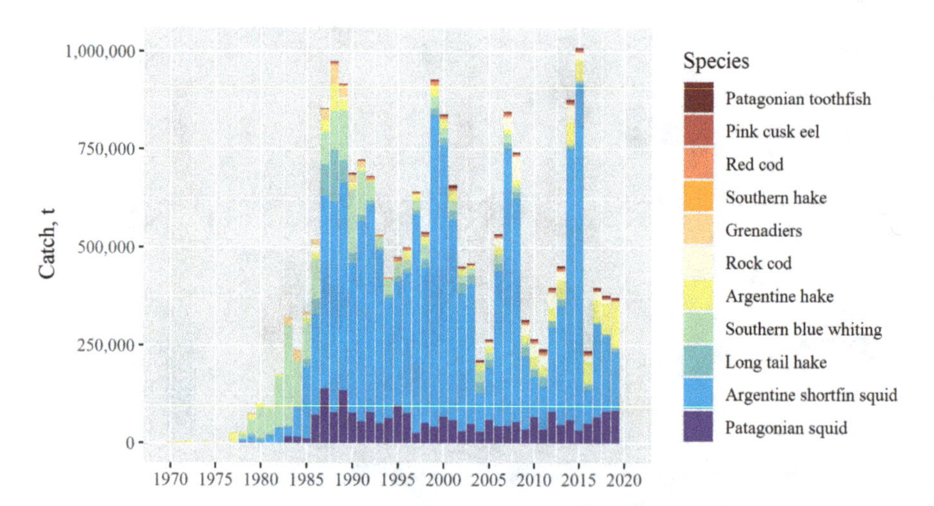

Fig. 7.2 Total catches per species declared by the distant countries fleet from 1970 to 2019 in the PSBF

From the 1990s, catches of other finfish species, the long tail hake and the Argentine hake, became relevant, oscillating at 100,000 t per year. The abundance of the first one increased rapidly during the year 2000 due to relevant recruitments, but then the biomass remained almost stable at a similar level to the beginning of the fishery. In terms of the Argentine hake, from 2010 its catches increased in similar relation to its abundance (Santos and Villarino 2022), given several management rules implemented at the end of the 1990s in its spawning and nursery grounds at the Argentine shelf that contributed to rebuild the stock structure of the species (Irusta et al. 2016).

The Patagonian toothfish fishery began around 1990, when catches reached a maximum of 2000 t per year. After a period without commercial interest, in 2000 several distant countries' fleets declared extraordinary landings, which probably affected the subsequent values declared for the species. This fishery is composed of a trawling fleet that operates mainly in the AEEZ, while some longliners with smaller catches directly fish for this species in international waters.

Other finfish species commercially exploited in the PSBF, mainly by the Spanish fleet, were the pink cusk eel, the red cod, and rock cod. Particularly, the pink cusk eel population was overexploited, and nowadays it is mainly detected in schools on the Argentine continental shelf, on which rigorous management and protection measures have been established. The high fishing effort applied for several years to this species changed the structure of the population resulting in smaller catches.

In terms of the red cod, maximum catches were reported between 1998 and 2001 when vessels from Argentina and other distant countries increased their fishing effort. After that, landings of this species showed a decreasing trend with a short period of higher catches due to some Argentine regulations. The minimum declaration of red cod was registered in 2019 (Gorini and Giussi 2020). In the PSBF, high catches of southern hake occurred in the late 1980s. During 1984 and 1988, landings grew up abruptly reaching the maximum for the species of 13,000 t in 1988. Subsequent catches fell down, followed by a period of stability with an average catch of 4500 t until 2010; the minimum catch was registered in 2019, indicating a dangerous situation for the resource (Gorini and Giussi 2020). About the rock cod, during 2005 and 2018, and given the development of an important market in countries of eastern Europe, this species occupied the second place of finfish catches.

Regarding cephalopods, large-scale commercial fisheries in the Southwest Atlantic started in the 1980s, exploiting two abundant squid species, the Argentine shortfin squid *Illex argentinus* and the Patagonian squid *Doryteuthis gahi* (Csirke 1987). The Argentine shortfin squid is a typical straddling stock, and it is also the most abundant squid species in the Southwest Atlantic, inhabiting the shelf and shelf-break between 22 °S and 55 °S (Jereb and Roper 2010). The South Patagonian Stock (SPS) constitutes about 95% of the total stock being the basis of the fishery that operates in the PSBF, since the Argentine and the high seas fisheries exploit it. In the late 1970s, as mentioned previously, a new fishing method (jiggers) allowed the increase of total annual catches of the Argentine shortfin squid which raised to values between 250,000 and 500,000 t per year at the end of the 1980s and 1990s, reaching in 1999 and 2015 maximum values around 1,000,000 t (Brunetti et al.

1999; Arkhipkin et al. 2015; Fig. 7.2). Annual captures of this squid species have fluctuated in the last decades due to fishing effort and changes in abundance related to oceanographic conditions that affect its recruitment and survival (Brunetti et al. 1998; Rodhouse et al. 2014; Torres Alberto et al. 2021).

The Patagonian squid, with a wide distributional range over the continental shelf and the PSBF, supports the other industrial cephalopod fisheries in the region, being also the second most important commercial loliginid squid fishery worldwide (Pineda et al. 1998; Jereb and Roper 2010; Arkhipkin et al. 2013). Even though the species inhabits a large area of the Atlantic, the largest concentration commercially exploited since the beginning of the 1980s is located along the PSBF, in relation to waters of the Malvinas Current (Arkhipkin et al. 2004; Arkhipkin et al. 2013, 2015). To the east and south of Malvinas Islands, fishing has been carried out exclusively by trawlers operating with bottom nets with small mesh liners, organized by issuing annual licenses to vessels that can operate during two fishing seasons (February–April and August–October; Arkhipkin et al. 2008, 2013, 2015). The annual abundance of the species is closely related to the strength of Malvinas Current; in those years when the current is more intense, part of the population moves to the north forming aggregations on the PSBF between 45 °S and 47 °S (Arkhipkin et al. 2004), where they are also fished by distant countries' fleets operating in the area.

It is worth noting that both squid species, the Argentine shortfin squid and the Patagonian squid, are protected by several management and conservation measures such as temporal fishing bans during periods of pre-spawning migrations and spawning.

In terms of species caught by distant countries in the PSBF (Fig. 7.3), the Argentine shortfin squid is the main target of the Chinese, Korean, Japanese, Taiwanese, and Russian fleets. On the other hand, Bulgaria, Poland, the Soviet

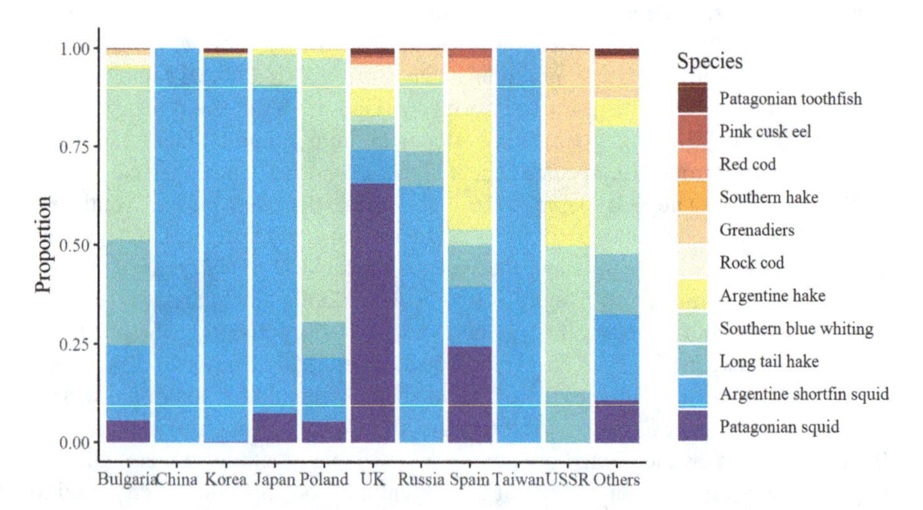

Fig. 7.3 Proportion of species declared by distant countries' fleets in the PSBF (FAO statistics outside the AEEZ) between 1970 and 2019

Union, and Spain focus their fishing effort on finfish, mainly targeting the long tail hake, the southern blue whiting, and the Argentine hake. The UK fleet mainly fish Patagonian squid and to a lesser extent the Argentine shortfin squid, the long tail hake, the Argentine hake, and the rock cod.

7.4 Argentine and Distant Countries' Fleets from 2000 to 2019

In this section, information of fisheries from the Argentine fleet (available at https://www.magyp.gob.ar/sitio/areas/pesca_maritima/desembarques/) is incorporated in the following figures, selecting FAO statistical squares within the AEEZ overlapping with the PSBF and also from distant countries' fleets operating in the adjacent area of the AEEZ. The involvement of the Argentine fleet was not significant at the beginning of the fishery in the Southwest Atlantic, but during the 1980s several agreements were celebrated with other countries in order to develop several fisheries (Giussi et al. 2016).

During 2000 and 2019, mainly 6 countries, Argentina, China, Korea, Spain, United Kingdom, and Taiwan operated in the PSBF (Fig. 7.4). Catches averaged around 600,000 t, with the highest values recorded in 2000, 2001, and 2015, which exceeded 1000,000 t. In 2000 and 2001, Argentina was the fishing country reporting the largest catches at the PSBF (ca. 350,000 to 500,000 t) while Taiwan and China were the distant countries that reported the highest proportion of the total catches, the former in 2007, 2008, 2014, and 2015 while the latter in the last 2 years.

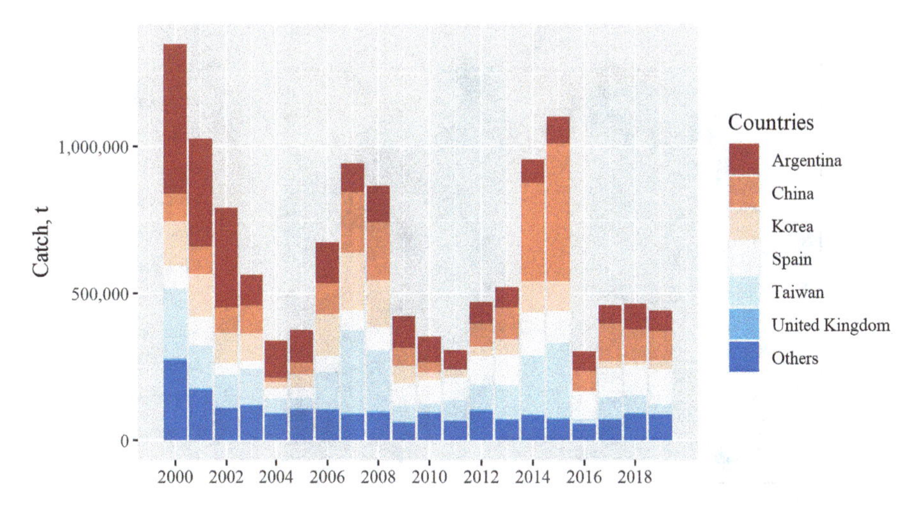

Fig. 7.4 Total catches declared by Argentina and distant countries' fleets fishing in the PSBF between 2000 and 2019

In the last decades, annual captures from the Argentine national fleet oscillated between 2000 and 377,000 t in the period 1987–2017 (Giussi et al. 2022). The distant countries' fleets of China, Korea, Taiwan, and Spain (with more than 300 vessels) operate along the PSBF, intensely fishing the Argentine shortfin squid since 1990 (Arkhipkin et al. 2023; Ivanovic et al. 2023).

When analyzing the proportion of species declared by each fishing country at the PSBF between 2000 and 2019, the reported target species vary widely (Figs. 7.5 and 7.6). Among finfish, the Argentinian fleet that operates at the PSBF mainly reported the Southern blue whiting and long tail hake, followed by the Southern hake, the Argentine hake, grenadiers, and the Patagonian toothfish. Spain reports more participation in the Argentine hake, the pink cusk eel, and the Rock cod fisheries. Regarding the species declared by Korea and the UK, the proportion of finfish is low, being the Patagonian toothfish, the Argentine hake, the rock cod, and the pink cusk eel the most fished by the former and the long tail hake, the Argentine hake, the Southern blue whiting, and the Southern hake by the latter.

Regarding cephalopod species, Chinese and Taiwanese vessels fish exclusively the Argentine shortfin squid at the PSBF, followed by Korea, while the UK fleet and vessels from other countries (grouped as "Others") mainly participate in the Patagonian squid fishery.

The Patagonian scallop fishery deserves a special mention as the Argentine fleet is the only one fishing this important commercial resource in the PSBF (Figs. 7.5 and 7.6). Moreover, it was the first Argentine fishery and the first scallop fishery in the world to achieve the Marine Stewardship Council (MSC) certification. The Patagonian scallop is a sedentary resource that distributes in association with the PSBF (Bogazzi et al. 2005) forming beds along it, which are commercially exploited.

Between 1950 and 1994, only Argentina reported catches of the Patagonian scallop, from 5 t in 1970 to 14,000 t in 1981 (FAO 2021). During this time, catches were

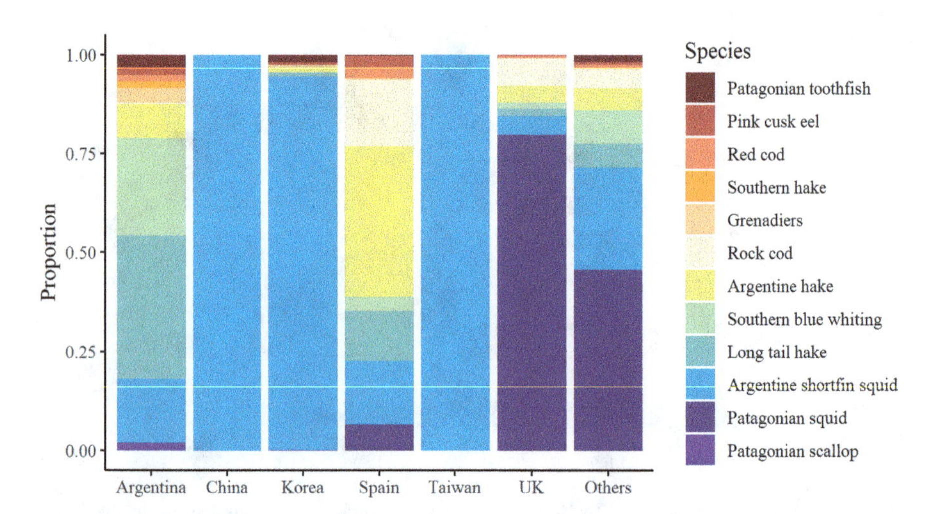

Fig. 7.5 Proportion of species declared by country from 2000 to 2019 in the PSBF

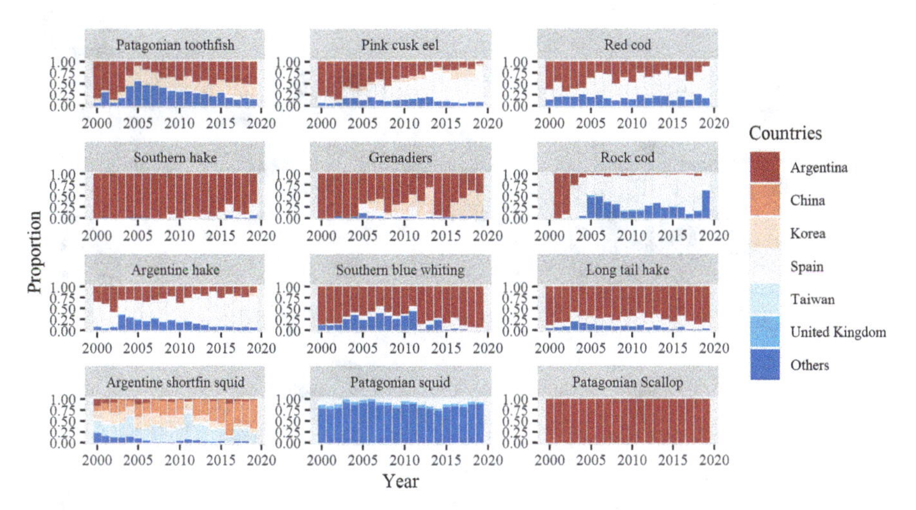

Fig. 7.6 Proportion of catch per species discriminated by fishing countries between 2000 and 2019 in the PSBF

obtained as bycatch of the bottom trawl fishery directed to finfish and the Patagonian scallop fishery began to develop in 1996. In that sense, total catches ranged between 22,000 and 80,000 t from 1996 to 2019, representing 3000–11,000 t of muscle production (Campodónico et al. 2019; FAO 2021). As in other pectinid fisheries, catch volumes are strongly related to the recruitment success (Soria et al. 2016; Campodónico et al. 2019).

Since 1995, Argentina has conducted annual studies to assess the Patagonian Scallop recruitment and the biomass available for commercial fishing grounds (Campodónico et al. 2019). Two massive events of recruitment were detected up to the present, corresponding to reproductive seasons 1994–1995 (Valero 2002) and 2000–2001 (Lasta et al. 2001). The last event (2000–2001) resulted in catches registered between 2005 and 2008, which were the highest of the whole period (50,000–80,000 t total catch; 6000–11,000 t of muscle). Since 2009, the fishery has experienced a decline of catches due to the lack of massive recruitments. As a result, the biomass available for fishing was related to recruitments restricted to each fishing ground exclusively. This declining trend resulted in total catches below 30,000 t, registered for the first time in 2018 (27,000 t) and 2019 (22,000 t).

7.4.1 Fishing Fleets at the Patagonian Shelf-Break Front

Given the distance to ports in Argentina and distant countries and time at sea, the industrial fisheries that operate in the PSBF are generally composed of medium and large-sized factory vessels. Depending on the fishing gear used, three types of vessels fish in the PSBF area: jiggers, longliners, and trawlers (Fig. 7.7a). During the

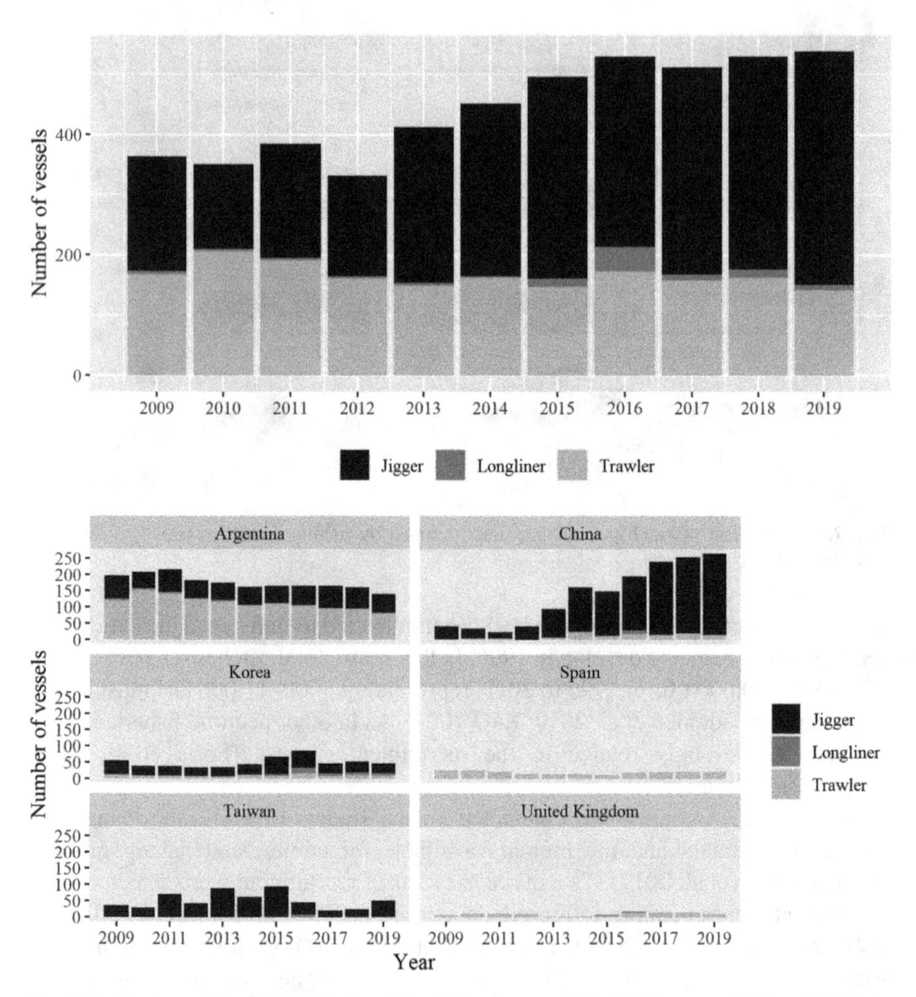

Fig. 7.7 Number of vessels discriminated by fishing gear in the PSBF between 2009 and 2019. A, top panel: by vessel type; B, bottom panel: by country

past decade (2009–2019), the number of vessels increased from ca. 360 to 530 at the PSBF, and particularly the jigging fleet doubled its vessels from 200 to 400 in 2019. From 2013, jigging vessels are the most important fleet operating at the PSBF. On the other hand, the variability in the number of longliners and trawlers was low, but during 2016 the longliner fleet doubled its vessels. Trawling fisheries operate at the PSBF with an average of 180 fishing vessels.

In terms of fleet type by fishing country, the Argentine fleet operates mainly with trawlers when targeting fish, but when targeting cephalopods, it uses jiggers. In that sense, when targeting the Argentine shortfin squid in the AEEZ, the fishing season runs from February 1 to August 31 (first semester), when the species is migrating on the continental shelf, while a closed season is applied between September and

January to protect juveniles (Brunetti et al. 2000). Chinese and Taiwanese fleets are mainly composed of jigger vessels according to its main reported species, the Argentine shortfin squid, while the Korean fleet is also mainly composed of jiggers and, to a lesser extent, by trawlers and longliners (Fig. 7.7b). In contrast, the Spanish fleet only operates with trawler vessels (less than 50 fishing vessels), like the UK (ca. 20 fishing vessels), although the latter also reported longliners in 2015 y 2016.

7.5 Spatial Distribution of Argentine and Distant Countries' Fishing Fleets

Fishing activities of the Argentine fleet in the PSBF are mainly carried out by jiggers (J), ice trawlers (IT) and freezer trawlers (FT) (Fig. 7.8; according to criteria of the Fisheries Management Area of the Argentinean National Undersecretary of Fisheries, Martínez Puljak et al. (2010). These fleets operate in almost the entire Argentine continental shelf but, in certain months of the year, they distribute and concentrate their fishing effort along the PSBF (Alemany et al. 2014).

The jigging fleet, targeting the Argentine shortfin squid, shows the greatest coupling with the PSBF between April and June (Alemany et al. 2014); this fleet showed a marked seasonality, given the fishing season and the migration cycle of its target species. The Argentine shortfin squid makes extensive and seasonal migrations along the PSBF and outer shelf, mainly south of 44 °S in autumn, and north of this latitude in winter (Brunetti et al. 2000). In that sense, the highest squid concentrations are associated with the PSBF (Brunetti et al. 1998; Bazzino et al. 2005), and also its fishery. The ice-trawler fleet that fishes the Argentine hake operates throughout the year in almost the entire continental shelf, however from April to June, the highest fishing effort locates near the PSBF (Alemany et al. 2014, 2016). Regarding the fishing activity of the freezer-trawler fleet, it was mainly concentrated in two main areas of the PSBF, at its northern part (ca. 39 °S–42 °S) targeting the Patagonian scallop along the front (Bogazzi et al. 2005), and at its southern part (ca. 48 °S–54 °S), targeting mainly the long tail hake from April to June and from August to December.

Regarding distant countries' fleets operating at que PSBF, high fishing effort is identified along the outer limit of the AEEZ. Both the jigging fleet and the trawling fleet concentrated at two main areas, the most important one between ca. 45 °S and 47 °S (known as the "Blue Hole") and the other one at 42 °S (Fig. 7.9). The cumulative (2016–2019) apparent fishing effort at these zones ranged between 1000 and 5000 hours per km^2 for each type of fishery. Given the trawling operation, its effort is mainly distributed at depths less than 300 m, while jiggers operate also at deeper waters (up to 4000 m).

Among the distant countries operating at the PSBF and between 2016 and 2019, China contributed 82% of the fishing effort of the jigging fleet, followed by Korea (10%) and Taiwan (8%). In terms of trawling, China (48%), Korea (25%) and Spain (20%) accumulate more than 90% of the apparent fishing effort in the area.

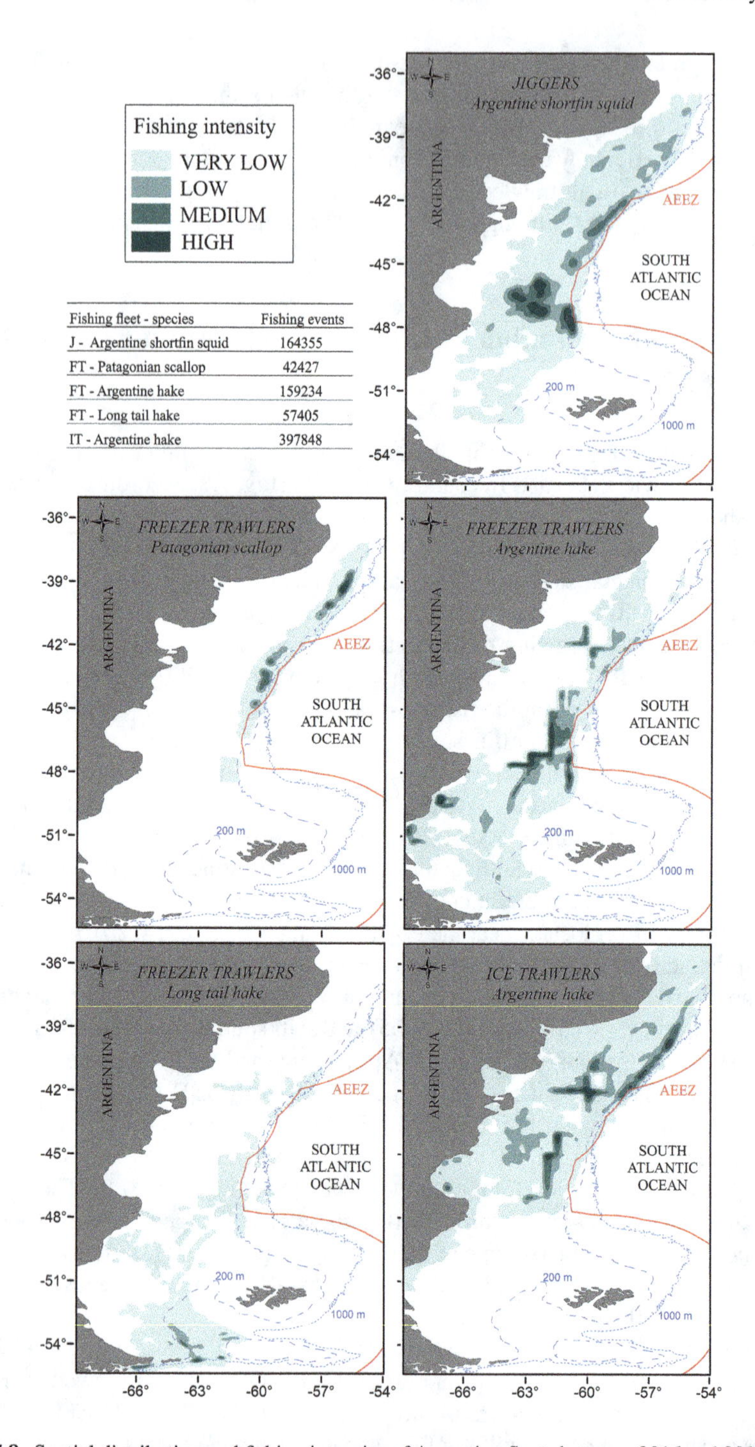

Fig. 7.8 Spatial distribution and fishing intensity of Argentine fleets between 2016 and 2019 estimated from VMS records (see Alemany et al. 2016 for methodological details). (**a**) Jiggers; (**b**) Freezer-trawlers targeting the Patagonian scallop; (**c**) Freezer-trawlers targeting the Argentine hake; (**d**) Freezer-trawlers targeting the long tail hake; (**e**) Ice-trawlers targeting the Argentine hake. Red line showing the Argentine Exclusive Economic Zone

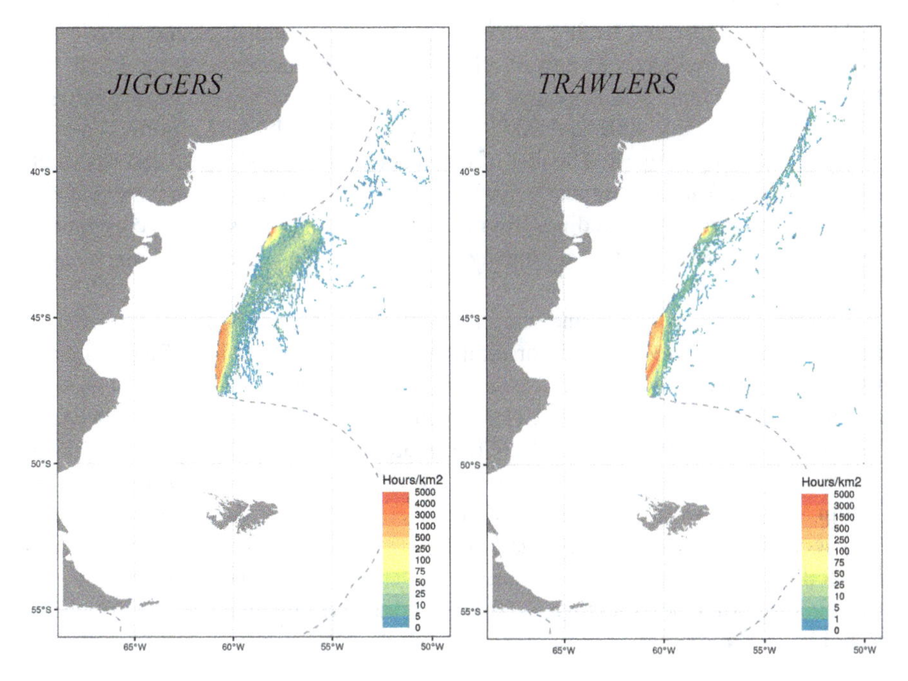

Fig. 7.9 Spatial distribution and fishing intensity of distant countries' fleets (2016–2019). Apparent fishing effort (hours; resolution 5 × 5 km) in the PSBF (calculated with the combination of Automatic Identification System AIS data on vessel location and machine learning techniques). Source: Global Fishing Watch. Analysis and cartography: S. Fermepin and V. Falabella (WCS). (**a**) Jiggers targeting the Argentine shortfin squid *I. argentinus*; (**b**) Trawlers targeting demersal fish species. Grey dash line showing the Argentine Exclusive Economic Zone

7.6 Sharks and Rays

In the area a huge number of species of sharks and rays are caught by distant countries' fleet; however, its composition is difficult to identify since they may not be disaggregated in declaration forms, as well as they could be erroneously classified due to the use of diverse common names adopted by fishing vessels from different countries and, thus, languages.

Chondrichthyans landings reports began in the 1950s with very low catches; however, it showed a gradual increase to over 75,000 t in 2010. Spain and Korea have relevant participation in this fishery. The Spanish fleet has declared catches of sharks and rays since the mid-1980s, and the Korean fleet focused on catching rays. Among the species that have been identified are those belonging to the Rajidae family, declared as "Rays, stingrays or manta nap." Other countries with less participation were Portugal, the UK, and Taiwan. Regarding shark species, the most relevant is the blue shark (*Prionace glauca*), followed by *Isurus oxyrinchus* and the spurdog *Squalus acanthias*. The blue shark is declared mainly by Spain, Portugal, and Taiwan.

In Argentina directed fishing activities targeting rays and sharks are prohibited; their capture is incidental as part of other fisheries. In that sense, at the PSBF and for the period 2000–2019, Argentine total commercial landings of chondrichthyans were ca. 900 t, being 2001 and 2002 the years with the highest reported catches, 170 t and 95 t respectively. The rest of the years of the period, mean catches oscillated between 25 and 40 t annually. Eight species of cartilaginous fish were reported in Argentine commercial landings, four shark species and four skate species; among them *Mustelus schmitti*, *Zearaja brevicaudata*, *Squatina* spp., and *Squalus acanthias* were the most abundant.

In terms of management and conservation issues, Argentina is part of different treaties that are related, directly or indirectly, to the conservation of cartilaginous fish. Among these international instruments, it is worth mentioning two to which the Argentine Republic has adhered: CITES, which was approved by Law 22.344 (April 1981) and the Convention on the Conservation of Migratory Species of Wild Animals (CMS), approved by Argentina Law 23.918 (January 1992). Moreover, within the framework of the International Action Plan for the Conservation and Management of Sharks (IPOA Sharks) developed in 1998, the Federal Fisheries Council approved in 2009 the National Action Plan for the Conservation and Management of Chondrichthyans (sharks, rays, and chimeras) in the Argentine Republic (PAN Sharks).

7.7 Conclusions and Final Remarks

The Patagonian Shelf-Break Front is an important fishing ground worldwide, characterized by a great diversity and abundance of species. Those of commercial interest, such as the Argentine shortfin squid (*Illex argentinus*, Ivanovic et al. 2016), the Argentine hake (*Merluccius hubbsi*, Irusta et al. 2016), the southern species (long tail hake *Macruronus magellanicus*, Giussi et al. 2016; Patagonian toothfish *Dissostichus eleginoides*, Martínez and Wöhler 2016; Southern blue whiting *Micromesistius australis*, Wöhler et al. 2004), the Patagonian scallop (*Zygochlamys patagonica*, Campodónico et al. 2019), and other charismatic species of special interest for conservation (e.g., birds and marine mammals, Falabella et al. 2009) make use of the PSBF at different stages of their life cycles. In that sense, this front not only attracts and concentrates species, but is also an area of intense industrial fishing (Alemany et al. 2016). Fisheries along the PSBF involve not only Argentine fleets operating inside and outside the AEEZ but also several distant countries' fleets fishing in this productive area of the Southwest Atlantic Ocean.

The highest concentration of fishing activity along the PSBF occurs mostly in a key area, the "Blue Hole," which is located between the 200 m isobath and the limit of the AEEZ (200 nautical miles from the coast). Thus, part of the Argentine continental shelf is in international waters or the high seas (see Falabella et al. 2013) as well as its resources of commercial interest. Even though Argentina has jurisdiction over the seabed up to 350 nautical miles (due to the extension of the continental shelf granted to the country by the United Nations Organization in 2016), most fishing activity in the area occurs in the water column. Particularly, in the Blue Hole, there is intense distant countries' fishing activity (e.g., Spanish fleet, Vilela et al. 2018), mainly focused on the capture of the Argentine shortfin squid by the jigging fleet (Cozzolino and Lasta 2016), and demersal fish by the trawler fleet (FAO 2020). In that sense, much of the fishing resources associated with the PSBF are straddling species, which swim freely between the AEEZ and international waters without fishing regulation on the high seas. To date, there is no international or multilateral agreement to manage this area (Arkhipkin et al. 2023).

It is indisputable that the Argentine shortfin squid is the most important fishing species in terms of catches along the PSBF, and the greatest fishing effort exerted by several fishing countries in the area is directed at this resource. Though this squid carries out almost its entire life cycle on the Argentine continental shelf, in certain months of the year, it is found along the PSBF, also inhabiting international waters on its migration towards higher depths. It is at this moment that distant countries' fleets capture it. Satellite images showing international vessels illuminating the limits of the AEEZ are well-recognized (Cozzolino and Lasta 2016), and the number of fishing vessels is one of the largest in the world ocean (Seto et al. 2023). Therefore, when fisheries are concentrated on the high seas in areas beyond national jurisdiction, scientific understanding of human activities and species natural variability is highly challenged and particularly limited, both in space and time.

Finally, the effects of trawling on the seabed, as well as illegal, unreported, and unregulated fishing, particularly in the high seas, constitutes threats to the environment (Hiddink et al. 2006; Halpern et al. 2008). In that sense, both the PSBF and the Blue Hole are vulnerable areas, since the marine organisms that make use of them are affected by the growth of fishing activity of the distant countries' fleets, and presumably by other factors such as climate change. The impact of fisheries on biodiversity and on the regional food web is unknown, so the study of the PSBF is of particular interest in order to provide information for the integrated and sustainable management of this region as well as the fisheries that operate there.

Acknowledgements The authors would like to thank Marcela Ivanovic, Silvana Campodónico, Eddie Aristizabal, and Gustavo Martínez Puljak for their valuable contribution in the preparation of this chapter. This is INIDEP contribution N° 2311.

References

Alemany D, Acha EM, Iribarne OO (2014) Marine fronts are important fishing areas for demersal species at the Argentine Sea (Southwest Atlantic Ocean). J Sea Res 87:56–67

Alemany D, Acha EM, Iribarne OO (2016) Distribution and intensity of bottom trawl fisheries in the Patagonian shelf large marine ecosystem and its relationship with marine fronts. Fish Oceanogr 25:183–192

Alemany D, Prandoni N, Ivanovic M, Acha EM (2022) Diversidad de peces y calamares en el área denominada Agujero Azul y zonas adyacentes. Inf Ases Transf 65. Instituto Nacional de Investigación y Desarrollo Pesquero (INIDEP), Mar del Plata, Argentina

Arkhipkin AI, Grzebielec R, Sirota AM et al (2004) The influence of seasonal environmental changes on ontogenetic migrations of the squid *Loligo gahi* on the Falkland shelf. Fish Oceanogr 13:1–9

Arkhipkin A, Middleton D, Barton J (2008) Management and conservation of a short-lived fishery resource: *Loligo gahi* around The Falkland Islands. Am Fish Soc Symp 49:1243–1252

Arkhipkin AI, Hatfield EMC, Rodhouse PGK (2013) *Doryteuthis gahi*, Patagonian long-finned squid. In: Rosa R, Odor R, Pierce G (eds) Advances in squid biology, ecology and fisheries. Nova Science Publishers, Inc, pp 123–157

Arkhipkin AI, Rodhouse PGK, Pierce GJ et al (2015) World squid fisheries. Rev Fish Sci Aquacult 23:92–252

Arkhipkin A, Nigmatullin CM, Parkyn DC, Winter A, Csirke J (2023) High seas fisheries: the Achilles' heel of major straddling squid resources. Rev Fish Biol Fish 33:453–474

Bazzino G, Quiñones RA, Norbis W (2005) Environmental associations of shortfin squid Illex argentinus (Cephalopoda: Ommastrephidae) in the Northern Patagonian Shelf. Fish Res 76(3):401–416

Bogazzi E, Baldoni A, Rivas A, Martos P, Reta R, Orensanz JM, Lasta M, Dell'Arciprete P, Werner F (2005) Spatial correspondence between areas of concentration of Patagonian scallop (*Zygochlamys patagonica*) and frontal systems in the southwestern Atlantic. Fish Oceanogr 14:359–376

Brunetti NE, Elena B, Rossi G, Ivanovic ML, Aubone A, Guerrero R, Benavides H (1998) Summer distribution, abundance, and population structure of *Illex argentinus* on the Argentine shelf in relation to environmental features. Afr J Mar Sci 20:175–186

Brunetti NE, Ivanovic ML, Sakai M (1999) Calamares de importancia comercial en la Argentina. Biología, distribución, pesquerías, muestreo biológico. Publicaciones Especiales Instituto Nacional de Investigación y Desarrollo Pesquero (INIDEP), Mar del Plata, Argentina, p 45. http://hdl.handle.net/1834/2336

Brunetti NE, Ivanovic ML, Aubone A, Rossi GR (2000) Calamar (*Illex argentinus*). In: Bezzi SI, Akselman R, Boschi EE (eds) Síntesis del estado de las pesquerías marítimas argentinas y de la Cuenca del Plata. INIDEP, Mar del Plata, pp 103–116

Campodónico S, Escolar M, García J, Aubone A (2019) Síntesis histórica y estado actual de la pesquería de vieira patagónica *Zygochlamys patagonica* (King 1832) en la Argentina. Biología, evaluación de biomasa y manejo. MAFIS 32(2):125–148

Chen CS, Chiu TS, Haung WB (2007) The spatial and temporal distribution patterns of the Argentine short-finned squid, *Illex argentinus*, abundances in the Southwest Atlantic and the effects of environmental influences. Zool Stud 46:111–122

Cozzolino E, Lasta CA (2016) Use of VIIRS DNB satellite images to detect jigger ships involved in the *Illex argentinus* fishery. Remote Sens Appl: Soc Environ 4:167–178

Csirke J (1987) Los recursos pesqueros patagónicos y las pesquerías de altura en el Atlántico sudoccidental. FAO Doc Téc Pesca N° 286, p 78

Falabella V, Campagna C, Croxall J (2009) Atlas del mar patagónico. Especies y espacios. Wildlife Conservation Society and BirdLife International, Buenos Aires. Available at: https://atlas-marpatagonico.org/

Falabella V, Campagna C, Krapovickas S (2013) Faros del mar patagónico. Áreas relevantes para la conservación de la biodiversidad marina. Resumen ejecutivo 2013. Buenos Aires, p 55

FAO (2020) Worldwide review of bottom fisheries in the high seas in 2016. FAO fisheries and aquaculture technical paper vol 657. Rome. 342 pp

FAO (2021) FAO Yearbook. Fishery and Aquaculture Statistics 2019/FAO annuaire. Statistiques des pêches et de l'aquaculture 2019/FAO anuario. Estadísticas de pesca y acuicultura 2019, p 110

Giussi AR, Zavatteri A, Di Marco EJ, Gorini F, Bernardele J, Marí N (2016) Biology and fisheries of long tail hake from Atlantic Ocean (*Macruronus magellanicus*). Rev Invest Des Pesq 28:55–82

Giussi AR, Prodoscimi L, Carozza CR, Navarro GS (2022) Estado de los recursos pesqueros bajo administración exclusiva de la República Argentina. Aportes para el informe SOFIA 2022. Versión corregida. Inf Ases y Transf INIDEP 12:12

Gorini FL, Giussi AR (2020) Actualización de la estadística pesquera de peces demersales australes en el atlántico sudoccidental (Período 2007–2019). Inf Tec of INIDEP 28:63

Halpern BS, Walbridge S, Selkoe KA, Kappel CV, Micheli F, D'Agrosa C, Bruno JF, Casey KS, Ebert C, Fox HE et al (2008) A global map of human impact on marine ecosystems. Science 319:948–952

Hiddink JG, Jennings S, Kaiser MJ, Queiros AM, Duplisea DE, Piet GJ (2006) Cumulative impacts of seabed trawl disturbance on benthic biomass, production, and species richness in different habitats. Can J Fish Aquat Sci 63:721–736

Irusta G, Macchi GJ, Louge E, Rodrigues KA, D'Atri L et al (2016) Biology and fishery of the Argentine hake (*Merluccius hubbsi*). Rev Invest Des Pesq INIDEP 28:9–36

Ivanovic M, Elena B, Rossi G, Buono M (2016) Distribución, estructura poblacional y patrones migratorios del calamar (*Illex argentinus*, Ommastrephidae). Inf Ases Transf INIDEP 69:13

Ivanovic M, Aubone A, Rossi G, Prandoni N, Buono M, McInnes M, Elena B, Tapia T, Pappi A, Allega I, Cozzolino E (2023) Calamar argentino. Pesquería 2022. Inf Tec INIDEP 2:27

Jereb P, Roper CFE (2010) Cephalopods of the world, an annotated and illustrated catalogue of cephalopod species known to date. Volume 2. Myopsid and Oegopsid squids. FAO Species Catalogue for Fishery Purposes 4, p 605

Lasta M, Hernández D, Bogazzi E, Campodónico S (2001) Vieira patagónica, Unidad Norte de Manejo-CTMFM, evaluación de biomasa año 2001. Inf Tec INIDEP 9(01):25

Martínez PA, Wöhler OC (2016) Hacia la recuperación de la pesquería de la merluza negra (*Dissostichus eleginoides*) en el Mar Argentino: un ejemplo de trabajo conjunto entre el sector de la administración, la investigación y la industria. Frente Marítimo 24:115–124

Pineda SE, Brunetti NE, Scarlatto N (1998) Calamares loligínidos (Cephalopoda, Loliginidae). In: Boschi EE (ed) Los moluscos de interés pesquero. Cultivos y estrategias reproductivas de bivalvos y equinodermos. INIDEP El Mar Argentino y sus Recursos Pesqueros 2: 13–36

Rodhouse P, Pierce G, Nichols OC, Sauer W, Arkhipkin A, Laptikhovsky V, Sadayasu K (2014) Environmental effects on cephalopod population dynamics: implications for management of fisheries. In: Advances in marine biology, vol 67. Academic Press, pp 99–233

Sánchez RP, Bezzi SI (2004) Los peces marinos de interés pesquero. Caracterización biológica y evaluación del estado de explotación. INIDEP. El Mar Argentino y sus recursos pesqueros 4, p 359

Santos B, Villarino MF (2022) Evaluación del estado de explotación del efectivo sur de 41° S de merluza (*M. hubbsi*) y estimación de la captura biológicamente aceptable para 2021. Inf Tec INIDEP 41:47

Seitune D, Battistuzzi J, Degracia Torres P, Vera MJ, Vera Morales E (2022) Políticas públicas para el abordaje del cambio climático y las pesquerías en el Océano Atlántico Sudoccidental. In: Buratti CC, Chidichimo MP, Cortés F et al (eds) Estado del conocimiento de los efectos del cambio climático en el Océano Atlántico Sudoccidental sobre los recursos pesqueros y sus implicancias para el manejo sostenible. Ministerio de Agricultura, Ganadería y Pesca, Buenos Aires, pp 179–219

Seto KL, Miller NA, Kroodsma D, Hanich Q, Miyahara M, Saito R, Boerder K, Tsuda M, Oozeki Y, Urrutia SO (2023) Fishing through the cracks: the unregulated nature of global squid fisheries. Sci Adv 9(10):eadd8125

Soria G, Orensanz JM, Morsán EM, Parma AM, Amoroso RO (2016) Scallops biology, fisheries, and management in Argentina. In: Shummway SE, Parsons GJ (eds) Scallops: biology, ecology, aquaculture, and fisheries. Elsevier Science, Oxford, pp 1019–1046

Torres Alberto ML, Bodnariuk N, Ivanovic M, Saraceno M, Acha EM (2021) Dynamics of the confluence of Malvinas and Brazil currents, and a southern Patagonian spawning ground, explain recruitment fluctuations of the main stock of *Illex argentinus*. Fish Oceanogr 30:127–141

Valero J (2002) Analysis of temporal and spatial variation in growth and natural mortality estimation with an Integrated Dynamic Model in the Patagonian scallop (*Zygochlamys patagonica*). MSc thesis. Seattle: School of Aquatic and Fisheries Sciences, University of Washington, p 154

Vilela R, Conesa D, del Rio JL, López-Quílez A, Portela J, Bellido JM (2018) Integrating fishing spatial patterns and strategies to improve high seas fisheries management. Mar Pol 94:132–142

Waluda CM, Griffiths HJ, Rodhouse PG (2008) Remotely sensed spatial dynamics of the Illex argentinus fishery, Southwest Atlantic. Fish Res 91:196–202

Wöhler OC, Cassia MC, Hansen AJ (2004) Caracterización biológica y evaluación del estado de explotación de la polaca (*Micromesistius australis*). In: Boschi EE (ed) El Mar Argentino y sus recursos pesqueros, vol 4. Los peces marinos de interés pesquero. Caracterización biológica y evaluación del estado de explotación. INIDEP, Mar del Plata, Argentina, pp 283–305

Zavatteri A, Giussi AR (2020) Evaluación de la abundancia de polaca (*Micromesistius australis*) en el Atlántico Sudoccidental. Período 1987–2019. Recomendación de la CBA para el año 2021. Inf Tec INIDEP 35:22

Chapter 8
Seabirds in the Argentine Continental Shelf and Shelf-Break

Marco Favero, Juan Pablo Seco Pon, Jesica Paz, Maximiliano Hernandez, and Sofía Copello

Abstract The Patagonian shelf-break represents the transition zone from the relatively shallow waters of the continental shelf to the deep plains exceeding 2000 m depth. Along this pronounced slope, shelf-edge and wind-driven upwelling fronts support high and persistent levels of primary and secondary productivity, attracting an important abundance and diversity of marine megafauna, including seabirds and marine mammals. Among seabirds, albatrosses, large petrels, and shearwaters are highly migratory seabirds capable of ranging vast distances and abundant in this region of the planet. Seabird species may use these fronts as primary foraging areas if their prey are at a disadvantage while exposed to thermal, haline, or nutritional stresses. Recent surveys conducted in the northern Patagonian shelf and shelf-break areas show a spatial correlation between seabird abundance and species richness with confluence zones near the slope, and the differential use of some species and composition of seabird assemblages in waters either north or south of the Malvinas-Brazil confluence. The Patagonian shelf and shelf-break, and its biodiversity, is exposed to a variety of threats originated or enhanced by anthropogenic activities. In the past four decades, many seabird populations and other megafauna have experienced declines product of the widespread human perturbation and losses of original marine and breeding habitats. Compared to other bird species, the conservation status of seabirds is rapidly declining, with more than one third of the 326 extant seabird species listed as threatened with extinction. Pelagic species, chiefly Procellariforms (albatrosses, petrels and shearwaters), are particularly threatened, and their populations have declined faster than coastal species. The exploration and exploitation of non-renewable resources, activities that are showing an expansion over the Patagonian shelf and the shelf-break, have the potential to alter the at-sea distribution of pelagic seabirds at micro and meso-scale. Regarding the exploitation

Supplementary Information The online version contains supplementary material available at https://doi.org/10.1007/978-3-031-71190-9_8.

M. Favero (✉) · J. P. Seco Pon · J. Paz · M. Hernandez · S. Copello
Grupo Ecología y Conservación de Aves Marinas y Costeras, Instituto de Investigaciones Marinas y Costeras (IIMyC, UNMDP – CONICET), Mar del Plata, Argentina

E. M. Acha et al. (eds.), *The Patagonian Shelfbreak Front*, Aquatic Ecology Series 13, https://doi.org/10.1007/978-3-031-71190-9_8

of renewable resources, the distribution of fishing effort is highly overlapped with foraging seabirds, showing impacts that range from food supplementation through scavenging behind vessels, to resource competition and incidental mortality. Albatrosses and petrels are susceptible to threats operating throughout their wide distribution ranges extending across national boundaries into international waters. Recent evidence has also demonstrated that the negative effects of seabird bycatch and other stressors affecting species on land and at sea can be exacerbated by the effects of climate change and its concomitant changes in atmospheric circulation, water masses, and prey distribution. These environmental changes have even the potential of generating significant shifts in the distribution of human activities and the overlap with seabirds. Addressing this issue likely represents one of the major environmental conservation challenges that will require well-informed management practices and the implementation of meaningful policy responses.

8.1 Introduction

The waters of the Southwest Atlantic, in particular on the continental shelf and shelf-break, are widely recognized as one of the most productive temperate marine ecosystems in the world, holding a massive biomass of marine megafauna, including seabirds and marine mammals (Croxall & Wood 2002; Gil et al. 2019). Ocean biophysical processes influence the distribution and behavior of marine organisms, effect that has been generally attributed to enhanced prey concentration and/or availability in waters affected by frontal systems, eddies, and other oceanographic processes (Acha et al. 2004; Ballance et al. 2006; Cox et al. 2008; Pelletier et al. 2012; Boyd et al. 2016). Seabird at-sea distribution is determined by physical and biological factors operating at variable scales and levels of ecological organization, but often corresponding with environmental variability operating at macro and mega scales (Serratosa et al. 2020).

The continental shelf of the Southwest Atlantic is used by more than 70 resident, seasonal resident or migratory seabird species, both pelagic (e.g., albatrosses, petrels, and shearwaters) and coastal (e.g., cormorants, skuas, gulls, terns). Other species, traditionally not listed as seabirds (e.g., shorebirds, oystercatchers, grebes), could be added to the list as important users of coastal environments. While the coastal environments are largely dominated by gulls (Laridae), terns (Sternidae), and cormorants (Phalacrocoracidae), among others, that rarely stray far from the shorelines, other seabirds like albatrosses (Diomedeidae), petrels, shearwaters, fulmars and prions (Procellariidae), storm-petrels (Hydrobatidae), and diving-petrels (Pelecanoididae) are characterized by their extensive use of oceanic waters and extreme flying and migratory capabilities. Penguins (Spheniscidae) may also perform extensive foraging trips, varying among species and the time of season. In addition to this feature, seabirds show a range of physiological capabilities to explore the water column with penguins as the more extreme divers (Ponganis et al. 1999).

Among seabirds, albatrosses and petrels are those typically considered pelagic. Albatrosses are very large seabirds, with *Diomedea* species exceeding three meters (11 ft) in wingspan, the largest flying bird in the planet. These seabirds spend a substantial part of their lives at sea. Immature albatrosses may spend a decade or more at sea until they are ready to breed. Once reaching the adult stage, they will get back to land almost exclusively to breed and molt. In general, these seabirds show a concerning conservation status that can be only understood by referring to certain characteristics of their life histories, including the low productivity of their populations as a result of a low reproductive frequency (in some species every other year or twice every 3 years), the laying of a single egg per reproductive season, and a delayed maturity with some species maturing at 10–15 years of age. Most albatrosses are long-lived species, with records exceeding 65 years of life (Brooke 2004); in 2021 the US Fish and Wildlife Service posted the hatch of a chick from *Wisdom*, a 70-year-old female Laysan albatross (*Phoebastria immutabilis*) breeding on Sand Island, Midway Atoll. These extreme life histories make albatrosses very sensitive to increased mortality resulting from the exposure to at-sea and/or land-based threats. This is clearly expressed in the marked worsening of their conservation status observed in the last few decades in comparison with other groups of birds (Croxall et al. 2012; Phillips et al. 2016). This deterioration triggered in 1999 the negotiations for the establishment of the Agreement on the Conservation of Albatrosses and Petrels, a binding instrument of the Convention on the Conservation of Migratory Species of Wild Animals. The Agreement entered into force in 2004 as a multilateral instrument seeking to improve the conservation status of albatrosses and petrels through the coordination of international efforts (ACAP 2020).

The concentration of prey along the Southwest Atlantic continental shelf and shelf-break is exploited by a number of seabirds such as albatrosses and petrels (Fig. 8.1), which mostly breed in the archipelagos of the extreme south of South America (Diego Ramírez archipelago, Isla de los Estados), Islas Malvinas, Islas Georgias del Sur, and Islas Sandwich del Sur. Some species such as the Black-browed albatross (*Thalassarche melanophris*), the Wandering albatross (*Diomedea exulans*), the Southern Giant Petrel (*Macronectes giganteus*) or the White-chinned Petrel (*Procellaria aequinoctiallis*), while largely breeding in these archipelagos, primarily feed on the Patagonian shelf and along the slope from 60° to 35° south, while nonbreeders and juveniles even use waters farther north in the South Atlantic. In general terms, natural prey for albatrosses includes midwater fish, mollusks and crustaceans captured at or near the water surface, usually diving not deeper than two to three meters, although recent studies have reported Black-browed albatrosses reaching 19 m depth (Guilford et al. 2022). Other Procellariforms, smaller in size, like shearwaters (*Ardenna* spp.) and diving petrels (*Pelecanoides* spp.) can explore deeper in the water column, reaching some 30 meters depth (Tickell 2000). Prey can be obtained by these large predators thousands of kilometers from the colony along the continental shelves and other frontal areas, while other small and medium-sized petrels may feed near or in association with neritic fronts. Others like the Southern and Northern Giant petrels (*Macronectes giganteus* and *M. halli*, respectively) may alternate the use of pelagic foraging areas with coastal predation and scavenging,

Fig. 8.1 Albatross (Diomedeidae) species present in the continental shelf and high seas of the SW Atlantic Ocean: (**a**) Black-browed albatross (*Thalassarche melanophris*), (**b**) Black-browed albatross dominating seabird assemblages attending trawl fisheries, (**c**) Light-mantled albatross (*Phoebetria palpebrata*), (**d**) Wandering albatross (*Diomedea exulans*), (**e**) Grey-headed albatross (*Thalassarche chrysostoma*), and (**f**) Southern Royal albatross (*Diomedea epomophora*). Photo credits M. Favero (**a**, **c**–**f**) and JP Seco Pon (**b**)

both at sea and on land, in association with seabird and marine mammal breeding colonies.

It is very well known that, besides their natural prey, many seabird species learn, during their long lives and time spent at sea, to take advantage of highly predictable resources generated by human activities like the conspicuous fishing operations producing offal (byproduct of the processing of the catch) and discards (portion of the capture with low commercial value released back to the sea) (Favero et al. 2011; Collet et al. 2015; Collet & Weimerskirch 2020). Although this food subsidy could be quickly read as beneficial for albatrosses and petrels, in fact it generates a fatal attraction to fishing operations that increases the risk of incidental mortality, by far recognized as the most important impact globally threatening many species with extinction. While attending fishing vessels seeking for bait, offal and discards, albatrosses, petrels, and other seabirds get hooked, entangled, or collide with vessels or the fishing gear. Incidental mortality rates could appear low on a vessel scale but are surely non sustainable for seabird populations at a fleet, regional or global scale (Croxall et al. 2012; Favero & Seco Pon 2014; Phillips et al. 2016).

8.2 Seabird Movement and Distribution

The study of the at-sea distribution of megafauna has changed dramatically over the last decades with the advent of new techniques and technologies to acquire and process data, both from the environment and the megafauna. In the early 1900s, data on species distributions used to be collected from ad libitum samplings, the capture of individuals, and stranding records during routine or occasional voyages and even in whaling and fishing expeditions. By the mid-1900s, survey methods progressively standardized, allowing the quantitative analysis and comparison of data in a systematic way, and the documentation of data on distribution and oceanographic features linked at more precise temporal and spatial scales. Towards the end of the century, such analyses were enhanced by the development of remote sensing technologies introduced in the field of oceanography and ecology, allowing the acquisition of very refined data covering large spatial and temporal scales. In addition, the evolution of new and sophisticated analytical methods converted early investigations based on simple correlations to multivariate analysis and modeling, now allowing sophisticated studies on distribution and habitat use (Balance et al. 2006).

Albatrosses and large petrels (among others) are highly migratory seabirds that even during the breeding season may travel thousands of kilometers in a single foraging trip. Earlier references on albatrosses covering very long distances at sea come from the XVIII and early XIX Centuries, including Samuel Taylor Coleridge's Rime of the Ancient Mariner (Coleridge 1798; Barwell 2014). In 1887 a carcass of a Wandering albatross was found on an Australian beach with a message around its neck telling of sailors stranded on Crozet Islands, some 5000 kilometers away in the SW Indian Ocean. That message was written just 45 days before the bird was found stranded ashore, constituting one of the first tangible records of the impressive journeys that Wandering albatrosses perform (De Roy et al. 2008). Mark-recapture techniques allowed in the mid/late 1900s the first estimations of migratory and foraging ranges in albatrosses, thought to range up to 1500–2000 km from the nest during foraging trips. The foraging range of pelagic seabirds remained largely speculative until the late 1900s with the first publication of a satellite tracked Wandering albatrosses (*Diomedea exulans*) breeding in (again) Crozet Island. This study showed the amazing flying capabilities of albatrosses, covering between 3600 and 15,000 km in a single foraging trip during an incubation shift, flying at speeds reaching 80 km per hour and over distances of up to 900 km per day (Jouventin & Weimerskirch 1990). With time and the development of smaller and more sophisticated telemetry devices, not only gathering geographical position but also behavior and environmental data, the information available at present comprise a large number of seabird species from a range of populations (BirdLife International 2024).

In general terms, the distribution patterns in albatrosses and petrels can be characterized as follows: during the time off breeding seabirds disperse widely from breeding grounds; some of them distributed in the Southern Hemisphere can even circumnavigate Antarctica twice, staying prolonged periods of time foraging in rich waters like those affected by the Malvinas or Benguela currents in the West and East Atlantic, respectively. Immature individuals show similar distributions than mature nonbreeders until they reach maturity and get back to breeding grounds. During courtship and incubation, breeders can also perform very long foraging trips for several days, but these journeys become temporally and spatially more restricted during the early chick rearing. Later in the season, as chicks grow up and are left alone in their nests, foraging trips of adult birds regain in extension. These changes in the use of foraging habitats throughout the annual cycle could be exemplified with the Black-browed albatross, a species dominating the seabird assemblages in the South Atlantic. Throughout the breeding season, these albatrosses prefer neritic (0–500 m) waters, using shelf-break and upper shelf-slope (500–1000 m) and oceanic (>1000 m) waters to a lesser extent. Black-browed albatrosses prefer areas with steeper bathymetric relief and, during incubation, warmer sea surface temperatures, although some individuals from Georgias del Sur Islands also show preference over oceanic waters of intense mesoscale turbulence like the Brazil-Malvinas confluence zone (Wakefield et al. 2021). In general terms, the shelf-break can be seen as an important boundary for the distribution of this species (Fig. 8.2a). In winter, Black-browed albatrosses from Malvinas largely remain in the Patagonian shelf and waters off Uruguay and Brazil, an area that is also heavily used by the White-chinned petrel population from Islas Georgias del Sur (Copello et al. 2013; Paz et al. 2021a, Fig. 8.2b).

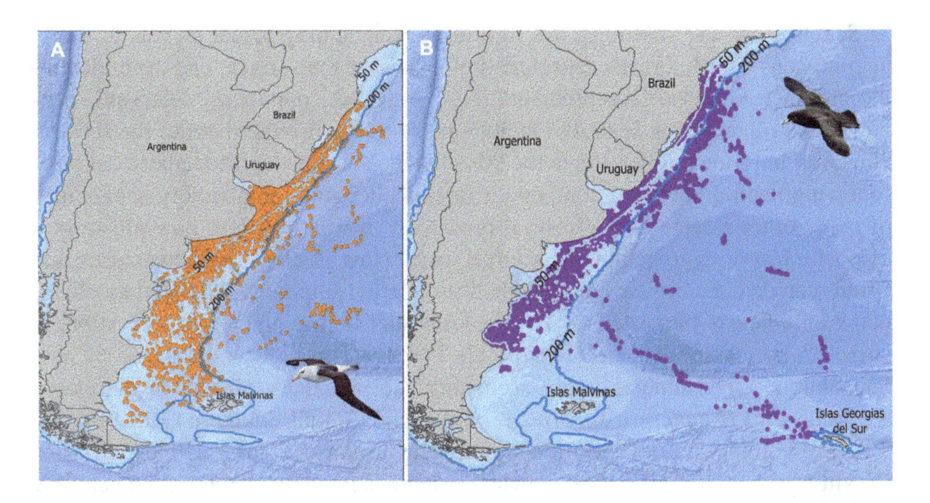

Fig. 8.2 Nonbreeding distribution of Black-browed albatrosses (*Thalassarche melanophris*) (**a**) and White-chinned petrels (*Procellaria aequinoctialis*) (**b**). Although not fully confirmed, literature and movements of individuals indicate that Black-browed albatrosses are from colonies in Islas Malvinas and White-chinned petrels from colonies in Islas Georgias del Sur, respectively. (Adapted from Copello et al. 2013; Paz et al. 2021a, b)

8.3 Finding Food in an Apparent Featureless Ocean

As said, pelagic seabirds and other offshore marine megafauna are highly mobile and capable of ranging vast distances, although foraging effort at sea is usually concentrated in space and time. Discrete habitat features have been identified over shelf-sea environments as important areas offering enhanced foraging opportunities, including fronts and offshore banks (Bost et al. 2009; Benjamins et al. 2015). The patchy distribution of predators in the oceans is expected to match the organization of prey, but this has proved challenging to demonstrate at fine scales (Cox et al. 2008; Torres et al. 2008). At a large scale, the oceanic areas with features favoring the foraging of seabirds (and other megafauna) typically occur in a persistent and predictable manner, and predators can learn the locations at which accessing prey is more likely (Cox et al. 2008). Behavioral changes in seabirds are typically described in association with increases in primary and secondary productivity or specific changes in the circulation of water masses operating at different spatial and temporal scales. In addition, different aspects of prey availability (e.g., prey abundance, vertical distribution, predictability) may affect species specific foraging capabilities and strategies, including diving capabilities, prey type, and changes in energetic requirements along the annual cycle (Langton et al. 2011). Targeted search patterns have been shown to coincide with the occurrence of a number of specific habitats that are repetitively visited, and these behaviors have been shown to develop as individuals mature (Grecian et al. 2018).

Seabirds can use multiple cues to find food, including visual signals to detect aggregations of fish, crustaceans, and mollusks near the surface, and even identify the presence of subsurface predators (e.g., large fish, pinnipeds, cetaceans, penguins, and other diving seabirds) facilitating preys by driving them to the surface. However, until the publication in the 90s of evidence showing Procellariform birds following natural chemical compounds (and challenging the olfactory capabilities of many other marine and terrestrial predators), it was not clear how albatrosses could find prey aggregations at a large scale. Now is clear that albatrosses and petrels can detect aromatic compounds like dimethyl sulphides generated by phytoplankton when is harvested by zooplankton, and that many species are strongly attracted by its presence (Nevitt et al. 1995; Nevitt 2008). In Antarctic and Subantarctic environments, other aromatics like pyrazines, present in macerated Antarctic krill (*Euphausia superba*, a primary prey in all Southern Ocean trophic webs), also trigger species specific and consistent responses in several Procellariforms (Nevitt et al. 2004). This new evidence has demonstrated that the foraging behavior of albatrosses is scale dependent. Over macro-mega scales of thousands of kilometers, albatross dispersion is influenced by large-scale ocean productivity and water mass distributions, finding large areas using navigational cues that are not yet well understood, but quite likely strongly linked to learning from experience and copying from more experienced conspecifics. At this large coarse scale, the olfactive landscape superimposed to the marine landscape can reflect oceanographic and bathymetric features like the shelf-break and the presence of oceanic fronts of different scales. Once seabirds reach these general areas, at smaller scales of tens to hundreds square kilometers, they can recognize a hotspot by a change on the olfactive landscape, triggering what is defined as an area-restricted search involving the following of odor plumes in combination with visual monitoring of prey and activity of other seabirds and megafauna (Nevitt 2000; Hyrenbach et al. 2002).

8.4 Drivers of Seabird Biodiversity in the Continental Shelf and Shelf-Break

The shelf-break represents the transition zone from the relatively shallow waters of the continental shelf to the deep plains exceeding 2000 m depth (Simpson & Sharples 2012). Along this pronounced slope, shelf-edge fronts and wind-driven upwelling fronts support high and persistent levels of primary and secondary productivity, attracting an important abundance and diversity of marine megafauna (Cox et al. 2008; Acha et al. 2015). Shelf-break fronts occur at the interface between on-shelf and open-ocean waters and are marked by strong gradients in salinity, and occasionally temperature. High levels of primary productivity are typically sustained, sometimes perennially, attracting planktivorous grazers alongside large numbers of pelagic fish (Genin 2004; Greer et al. 2015). The physical features of frontal systems not only offer unique opportunities for a range of organisms but may also impose physiological challenges to potential prey. Some predators may use these fronts as primary foraging areas if their prey are at a disadvantage while exposed to thermal, haline, or nutritional stresses (Ballance et al. 2006). Depending upon the lateral extent and topography of the adjacent continental shelf, these features may be far from land, and sometimes inaccessible to foragers like coastal seabirds (e.g., gulls, terns, cormorants). In the Pacific coasts of South America, the shelf-break can be very close to land (even less than 10 nm), but is much further away on the Atlantic side, exceeding 200 nm and making it very difficult for coastal seabirds to reach. However, albatrosses, petrels, among other Procellariforms, can perform far-ranging foraging trips even when constrained by breeding duties, reaching these important areas throughout the complete annual cycle (Robertson & Gales 1998).

Seabirds can be relatively more abundant in major frontal systems, with composition of assemblages varying according to regions and water masses. The usual variables shown in the literature as correlated with seabird abundance and richness, also important in structuring the occurrence and biomass of prey, are sea-surface temperature, sea surface salinity, chlorophyll-a concentration, distance to the coast, and bathymetry (see Serratosa et al. 2020; Paiva et al. 2010; Evans et al. 2021). However, at mesoscale, water temperature seems to act as the primary determinant of biogeographic patterns and ecosystem processes (Sala et al. 2017; Wakefield et al. 2021). Depth distribution plays a key role in prey accessibility, particularly for those seabirds like albatrosses and large petrels with limited diving capabilities, hence feeding at or near the surface. Surface convergent zones at shelf-edge fronts, upwelling fronts, and tidal-mixing fronts are areas that have been directly linked to shallow prey aggregations and frequently used by these seabirds and other top predators (Cox et al. 2008; Stevick et al. 2008). Recent surveys conducted in northern Patagonian shelf and shelf-break show the spatial correlation between seabird abundance and species richness with the Malvinas—Brazil confluence zone near the slope, and the differential use of some species and composition of seabird assemblages in waters either north or south of such convergence (Hernandez et al. 2024, Fig. 8.3).

Distribution patterns in seabirds can be species specific and vary depending on the sex, age, breeding condition, and even individual behavior (Weimerskirch et al. 1997; Patrick & Weimerskirch 2014; Orben et al. 2018; Paz et al. 2021a; Paz 2022). Beyond generalizations that could be done relating oceanographic features and the structure of seabird assemblages, different species may show particular preferences in the selection of foraging habitats. Some albatrosses have been reported associated with frontal systems and in particular the Antarctic Polar Front Zone, an area with persistent biophysical conditions characterized by particular thermal ranges, elevated primary productivity, and frequent manifestation of mesoscale thermal fronts (Scales et al. 2016), while other pelagic species prefer highly productive areas favored by the runoff of large rivers (Scheffer et al. 2012; Zamon et al. 2014; Thompson et al. 2021). Foraging areas may also vary according to the breeding

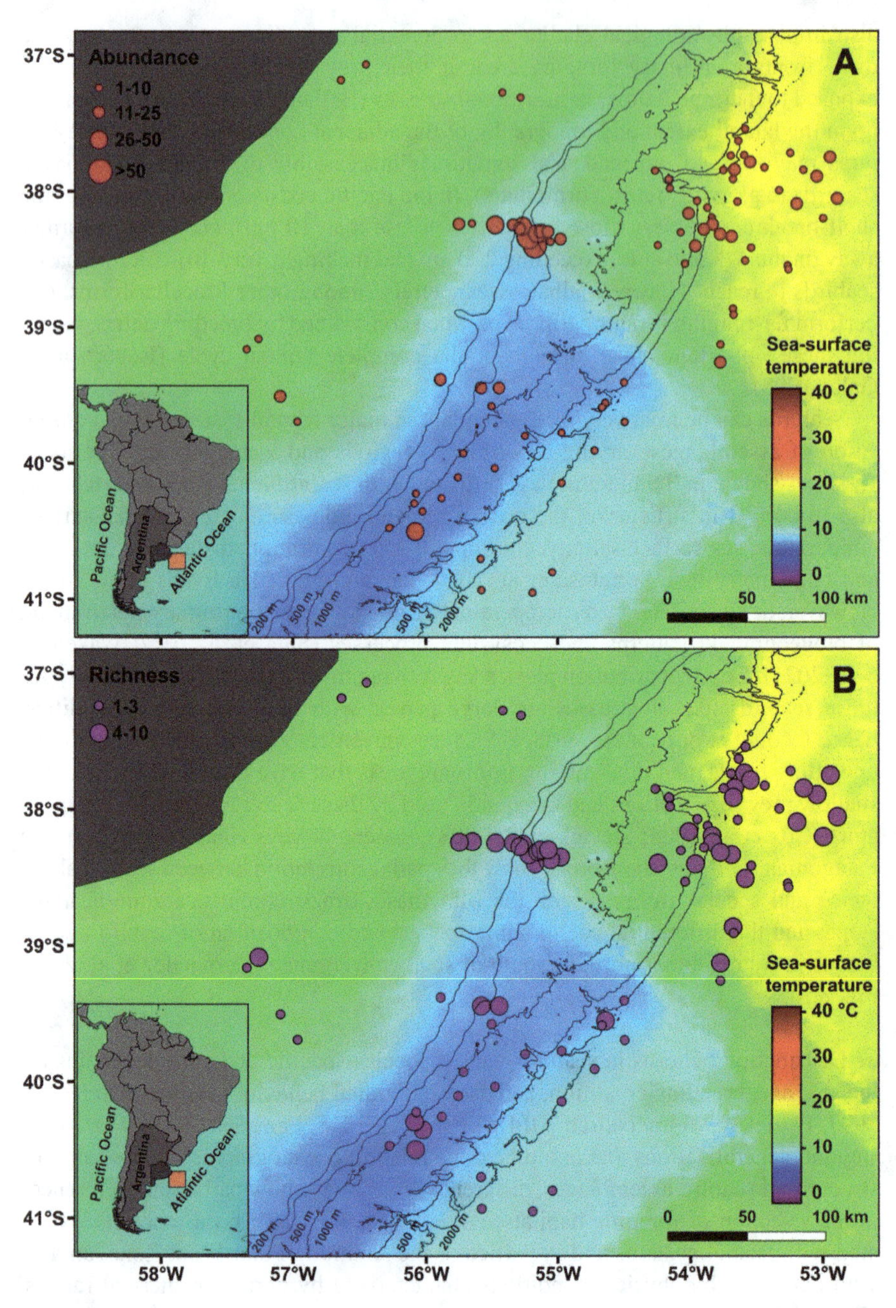

Fig. 8.3 Spatial distribution of seabird abundance (number of individuals per count, top panel) and species richness (number of species identified per count, low panel) during at sea observations in northern Argentine waters, against the average sea surface temperature landscape in May–June. Dominant seabirds in assemblages were Black-browed albatrosses (*Thalassarche melanophris*), White-chinned petrels (*Procellaria aequinoctialis*), Atlantic petrels (*Pterodroma incerta*), and Prions (*Pachyptila* spp.) altogether accounting for ca. 80% of total individuals. (Adapted from Hernandez et al. 2024)

condition of albatrosses, for example showing partial spatial segregation between successful and failed breeders (Ponchon et al. 2021).

In Patagonia, the shelf-break works as the major distribution boundary for a number of pelagic seabirds. Southern Giant petrels forage in environments characterized by high productivity, warm sea surface temperature, and shallow waters. Juvenile giant petrels mainly use productive neritic waters with occurrence of thermal fronts, while adults use environments without thermal fronts, remaining on the continental shelf (Quintana et al. 2010; Copello et al. 2011). Female giant petrels tend to spend more time in the shelf break, exploiting deeper waters than males preferring coastal environments (Blanco & Quintana 2014; Blanco et al. 2015, 2017, Fig. 8.4). Considering species breeding in offshore islands, the Black-browed albatrosses (*Thalassarche melanophris*) from Islas Malvinas are dominant in seabird assemblages along the Patagonian shelf, showing strong preference for shelf waters off Argentina, Uruguay, and Brazil in the Atlantic, as well as Chile and further north in the Pacific. Such distribution gets more restricted to waters around colonies during the breeding season (during early stages of chick rearing) but remains wide in nonbreeding adult and immature individuals (Copello et al. 2013; Paz et al. 2021a).

Adult and immature Black-browed albatrosses also show significant differences in the oceanographic conditions associated with foraging areas, with adults foraging in deeper and colder waters compared to immature individuals (Paz 2022, Fig. 8.5). Differences in habitat preferences can be significant even at individual scale; for example, the nonbreeding distribution of Black browed albatrosses in the northern Patagonian shelf and shelf-break shows individuals preferring the use of the mouth of Rio de la Plata, while others can be strongly associated to the shelf-break (Copello et al. 2013; Paz et al. 2021a, Fig. 8.6a). Smaller migrant species like the Great shearwater (*Ardenna gravis*) may show individuals associated to the 50 m isobath, the 200 m isobath, and even beyond the shelf-break (Paz et al. 2021b, Fig. 8.6b).

Despite the above referred differences, at a coarse scale, it results clear that for species like the Black-browed albatross, there is a clear preference for the use of waters on the continental shelf, with the shelf-break acting as a barrier. Opposite scenarios can be observed in large albatrosses like the Wandering albatross (*Diomedea exulans*); adults breeding in Georgias del Sur Islands can either use shelf waters around breeding colonies and deep offshore waters reaching the shelf-break, but rarely get through the shelf-break to visit the Patagonian shelf (Nicholls et al. 2002; Carneiro et al. 2022). Smaller and very abundant White-chinned petrels (*Procellaria aequinoctialis*) breeding in Islas Georgias del Sur, frequently visit the waters of the continental shelf and upwelling areas along the shelf-break, Banco Namuncurá/Burdwood Bank and east to Isla de los Estados and Diego Ramirez (Rexer-Huber 2017). However, in winter, individuals show preference for waters in the Patagonian shelf, spreading over the continental slope and deep waters while moving north to southern Brazil (Fig. 8.2b). The Patagonian shelf and shelf-break is also widely used, particularly during the austral winter, by typically Antarctic or Subantarctic species, e.g., the Southern fulmar (*Fulmarus glacialoides*), the Antarctic petrel (*Thalassoica antarctica*) and the Cape petrel (*Daption capense*), as

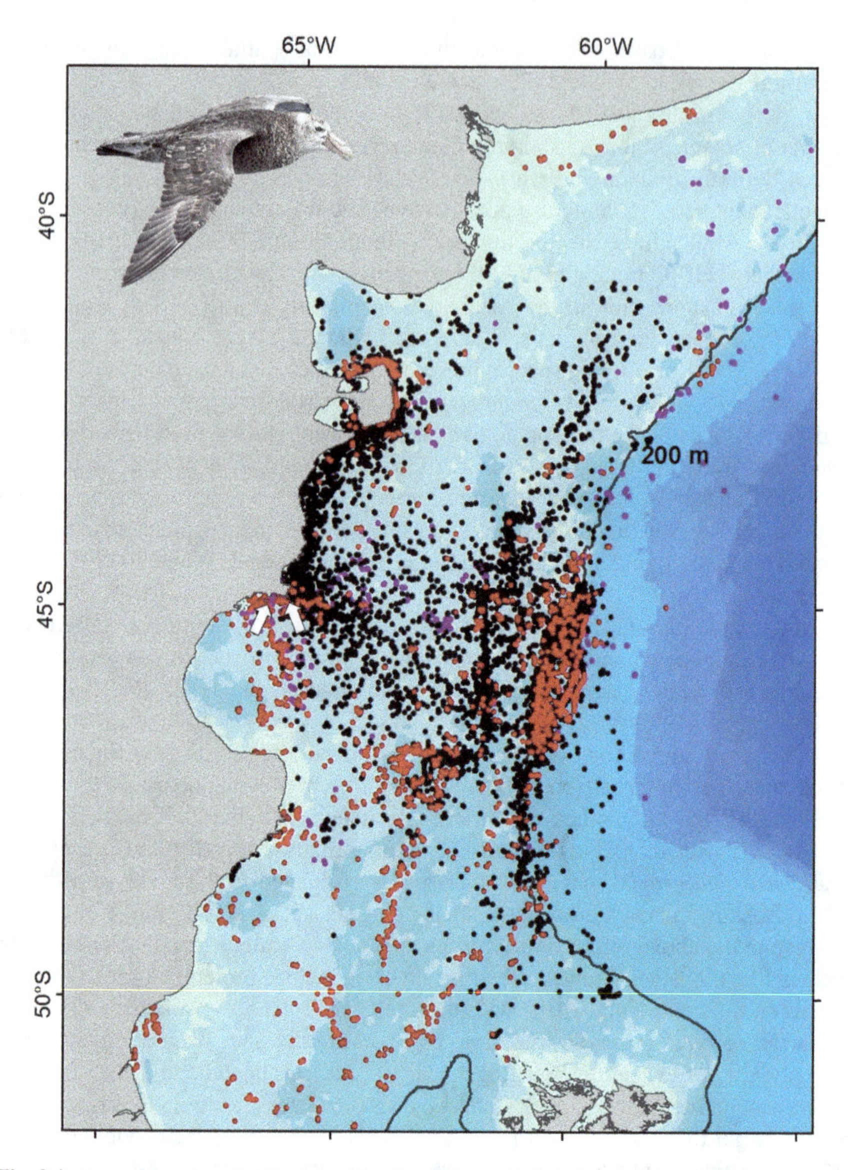

Fig. 8.4 Annual foraging distribution of Southern giant petrels (*Macronectes giganteus*) from Isla Arce and Isla Gran Robredo (location shown by two arrows). Different colors indicate the use of the Patagonian shelf and shelf break by adults (breeding black dots, wintering red dots) and juveniles (purple dots). (Adapted from Quintana et al. 2010; Blanco & Quintana 2014; Blanco et al. 2017; Blanco et al. 2022)

well as species breeding in waters off New Zealand like the Northern Royal albatross (*Diomedea sanfordi*), Southern Royal albatross (*D. epomophora*) and the White-capped albatross (*Thalassarche steadi*), the Southern Indian Ocean like the Sooty albatross (*Phoebetria fusca*), the Central-East Atlantic like the Spectacled

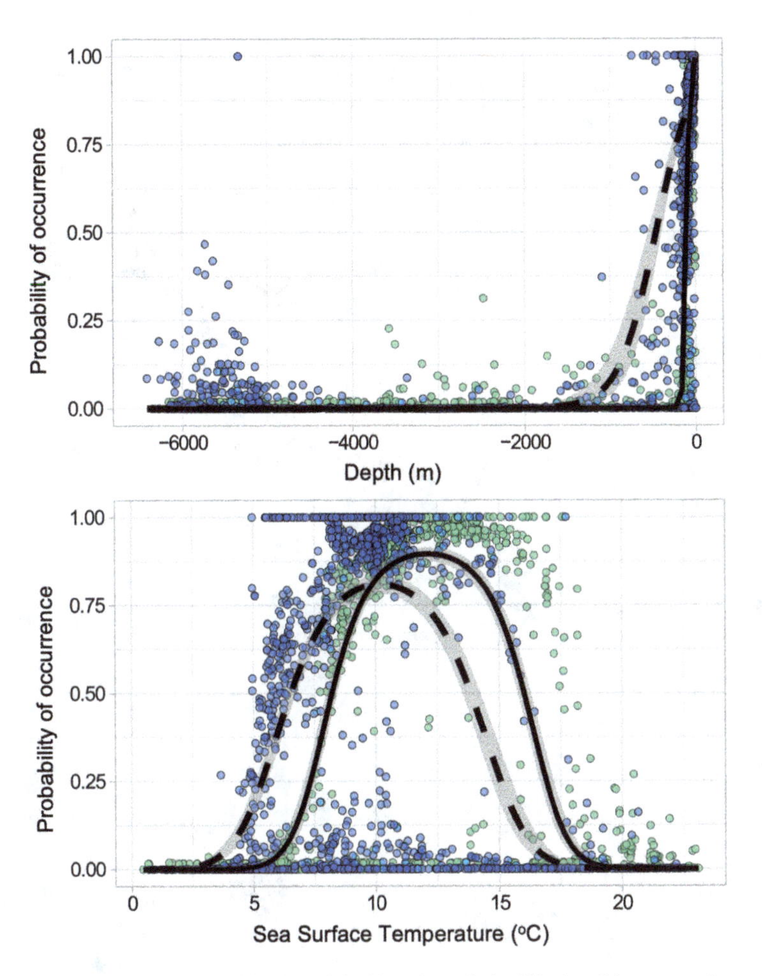

Fig. 8.5 Foraging habitat suitability for adult (blue dots, dashed line) and immature (green dots, continuous line) Black-browed albatrosses (*Thalassarche melanophris*) as results of models using sea depth and sea surface temperature. (Adapted from Paz 2022)

petrel (*Procellaria conspicillata*) and the Great shearwater (*Ardenna gravis*), and even species breeding in the northern hemisphere like the Cory's shearwater (*Calonectris borealis*) that migrate to the South Atlantic during the time off breeding.

8.5 Conservation Issues

The Patagonian shelf and shelf-break is exposed to a variety of threats originated or enhanced by anthropogenic activities. In the past four decades, many seabird populations and other marine megafauna have experienced declines product of the

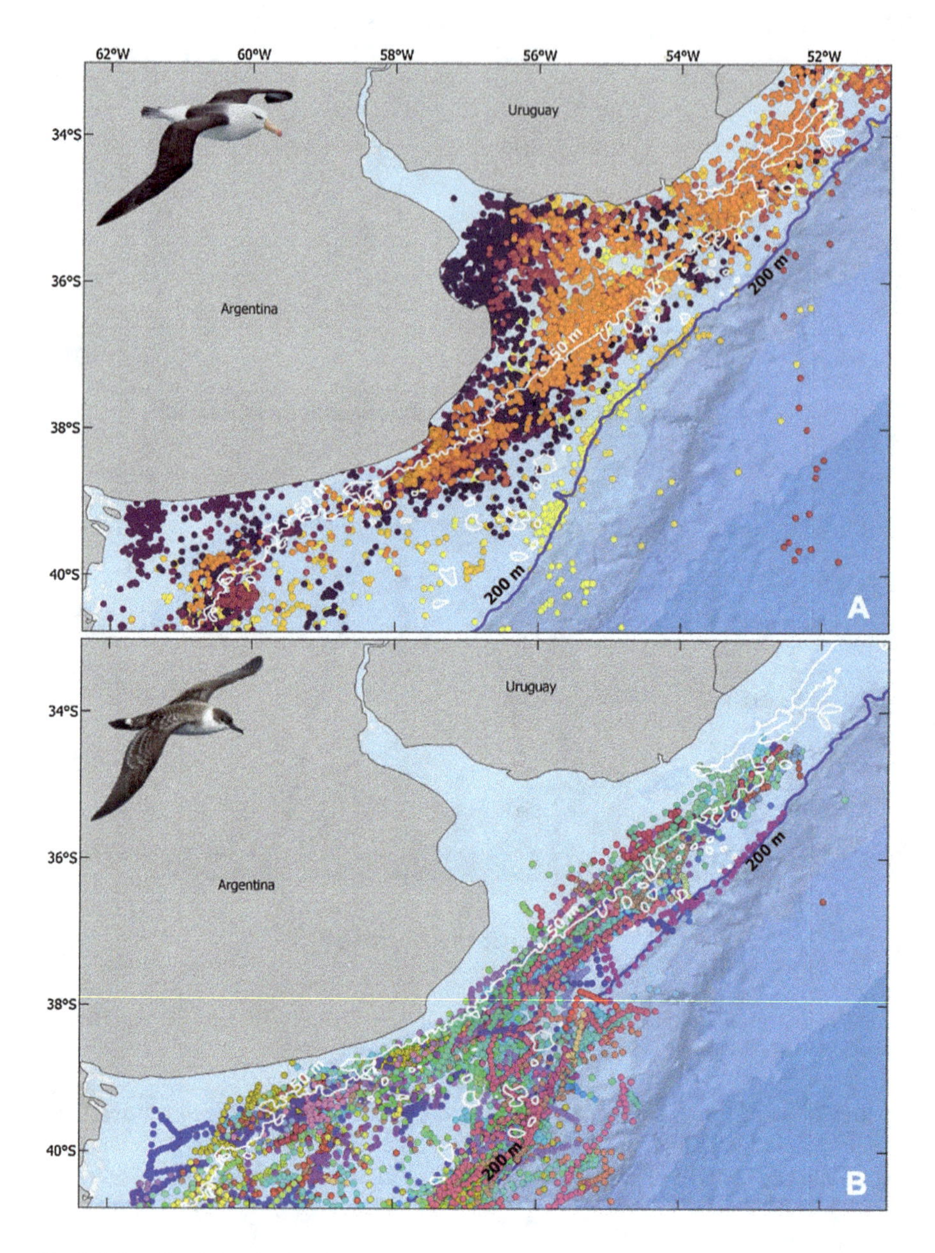

Fig. 8.6 Details of the distribution of (**a**) Black-browed albatrosses (*Thalassarche melanophris*) and (**b**) Great shearwaters (*Ardenna gravis*) in neighboring waters of Rio de la Plata including northern Argentina, Uruguay, and Northern Brazil (50 m and 200 m isobaths are shown). Dots of different colons indicate different tracked albatrosses. (Adapted from Copello et al. 2013, 2016; Paz et al. 2021a, b)

widespread human perturbation and losses of original marine habitats (Butchart et al. 2004; Mariano-Jelicich et al. 2022). Among these threats, introduced species, aquaculture, habitat degradation and pollution, unsustainable extraction of renewable resources (including the discard of non-commercial species and sizes), illegal fishing, incidental mortality of megafauna associated with fishing, and exploration and exploitation of non-renewable resources are frequently referred in the literature (Favero et al. 2011; Sequeira et al. 2019; Seco Pon et al. 2019; Phillips & Waluda 2020; Blanco et al. 2022, among others). Some of these threats also affect—or have the potential to affect—the distribution of seabirds at sea as a product of avoidance (e.g., to noise pollution) or attraction (e.g., to fishing operations) behavior.

Compared to other bird species, the conservation status of seabirds is rapidly declining, with more than one third of the 326 extant species listed as threatened with extinction by the International Union for the Conservation of Nature (Croxall et al. 2012; Phillips et al. 2016; Dias et al. 2019). More than half of the seabird species for which the population trend is known are declining, 27% are stable and only 16% of those are increasing. Although some gull species are showing increased population numbers, likely due to their capacity to exploit alternative food sources of anthropic origin (e.g., fishery discards, refuse dumps, slaughterhouses) (Yorio et al. 2016), pelagic species are more threatened, and their populations have declined faster than coastal species.

This difference is particularly pronounced in albatrosses and large petrels (ACAP 2022, Table 8.1). Although Black-browed albatross populations from Islas Malvinas have shown recoveries, largely attributed to better fishing practices and decreased fishing effort in some fleets (Catry et al. 2011; Favero et al. 2013), those from Georgias del Sur are decreasing at a rate of 5% per year. Some Grey-headed albatross colonies in Islas Georgias del Sur have shown declines between 2 and 4% per year over the last 20–30 years, while Wandering albatross populations from the same archipelago also showed declines of 4–5% since the late 90s (Poncet et al. 2017; Rackete et al. 2021).

The exploration and exploitation of non-renewable resources, activities that are showing an expansion over the Patagonian shelf and the shelf-break, have the potential to alter the at-sea distribution of pelagic seabirds at micro and meso-scale. In addition to marine natural sources of noise (e.g., wave action, earthquakes, rainfall), the sounds generated by human activities include vessel traffic, exploration and production of oil and gas, wind farms, and military operations and research. Oceanic noise generated by human activities has been widely recognized as a source of pollution and a global environmental issue (Nowacek et al. 2015; Seco Pon et al. 2019). Among them, seismic surveys produce some of the most intense sounds in the oceans, often operating over large areas and long periods (Clark and Gagnon 2006). This has been reported to significantly affect the distribution and abundance of seabirds using a range of foraging strategies (Pichegru et al. 2017; Seco Pon et al. 2019).

Almost all of the world's fisheries overlap spatially and temporally with foraging seabirds, with impacts that range from food supplementation (through scavenging behind vessels) to resource competition and incidental mortality. Recent studies analyzing carbon and nitrogen stable isotopic composition and DNA metabarcoding

Table 8.1 Population breeding size, conservation status and trends, and relative abundance of albatross and petrel species listed by the Agreement on the Conservation of Albatrosses and Petrels with distribution in waters of the SW Atlantic (ACAP 2022)

Species	Population[a] (breeding pairs)	Breeding region	IUCN[b] (2021)	Trend[c]	AB[d]
Tristan albatross *Diomedea dabbenena*	1500 (H)	Southeast Atlantic	**CR**	**DE**	+
Northern royal albatross *Diomedea sanfordi*	4000 (L)	S Pacific—New Zealand	**EN**	**DE**	++
Atlantic yellow-nosed albatross *Thalassarche chlororhynchos*	34,000 (L)	Southeast Atlantic	**EN**	**ST**	+
Grey-headed albatross *Thalassarche chrysostoma*	81,000 (M)	Circumpolar Subantarctic	**EN**	**DE**	++
Sooty albatross *Phoebetria fusca*	12,000 (LL)	S Indian Ocean—Subantarctic	**EN**	**DE**	+
Southern royal albatross *Diomedea epomophora*	8000 (L)	Subantarctic—New Zealand	**VU**	**ST**	++
Wandering albatross *Diomedea exulans*	9000 (H)	Circumpolar Subantarctic	**VU**	**DE**	++
White-chinned petrel *Procellaria aequinoctialis*	1,120,000 (LL)	Circumpolar Subantarctic	**VU**	**DE**	+++
Spectacled petrel *Procellaria conspicillata*	42,000 (H)	South (central) Atlantic	**VU**	**IN**	+
Light-mantled albatross *Phoebetria palpebrata*	16,000 (?)	Circumpolar Subantarctic	**NT**	**?**	++
Grey petrel *Procellaria cinerea*	87,000 (LL)	Circumpolar (Indian—Pacific)	**NT**	**DE**	++
White-capped albatross *Thalassarche steadi*	63,000 (?)	S Pacific—New Zealand	**NT**	**?**	+
Southern Giant petrel *Macronectes giganteus*	46,000 (M)	Circumpolar Antarctic and Subantarctic	**LC**	**IN**	+++

(continued)

Table 8.1 (continued)

Species	Population[a] (breeding pairs)	Breeding region	IUCN[b] (2021)	Trend[c]	AB[d]
Northern Giant petrel *Macronectes halli*	12,000 [(M)]	Circumpolar Subantarctic and S Pacific—New Zealand	**LC**	**IN**	++
Black-browed albatross *Thalassarche melanophris*	690,000 [(H)]	Circumpolar Subantarctic, S Pacific and SW Atlantic	**LC**	**IN**	+++

[a]Total breeding pairs (rounded to the thousands) as assessed in 2022 by ACAP. Accuracy is shown in parentheses as follows: (H) high, (M) medium, (L) low, (LL) very low
[b]Species global conservation status assessed by the International Union for the Conservation of Nature: (CR) critically endangered, (EN) endangered), (VU) vulnerable, (NT) near threatened, and (LC) least concern
[c]Population trend as assessed in 2022 by ACAP: IN increasing, ST stable, DE decreasing, and **?** unknown
[d]Relative abundance in the SW Atlantic: (+) frequent, (++) abundant, and (+++) very abundant

have confirmed the ongoing interactions with fisheries through the consumption of fishery discards and offal (Mariano-Jelicich et al. 2017; McInnes et al. 2017; Mariano-Jelicich et al. 2023). This seabirds-fisheries overlap, and the consequent use of resources can be understood as a proxy of risk of incidental mortality (i.e., bycatch), which may vary according to season, species, sex, age, populations, and fishing fleets (Jimenez et al. 2014; Clay et al. 2019; Corbeau et al. 2021). Albatrosses and petrels are susceptible to threats operating throughout their wide migratory and foraging ranges extending across national boundaries into international waters. Declines among slow-breeding albatrosses and petrels have been largely attributed to incidental mortality associated with fishing (commercial, artisanal and even small-scale), currently and largely recognized as the main at-sea threat (González-Zevallos & Yorio 2006; Suazo et al. 2013; Favero & Seco Pon 2014; Seco Pon et al. 2015; Phillips et al. 2016). Incidental mortality is not the only issue for seabirds attending vessels. Other effects may include changes in their foraging behavior and patterns of distribution since seabirds learn to find and follow fishing vessels (Oro et al. 2013), and also plastic pollution associated with debris generated by vessels (Copello & Quintana 2003; Phillips & Waluda 2020). Highly productive waters in the Patagonian shelf and shelf-break are heavily used by a range of fishing fleets, chiefly bottom and mid-water trawlers targeting the Argentine hake (*Merluccius hubbsi*), Hoki (*Macruronus magellanicus*), Southern blue whiting (*Micromesistius australis*) and the Argentine red shrimp (*Pleoticus muelleri*), among others, but also the benthic otter trawl net vessels targeting the Patagonian scallop (*Zygochlamys patagonica*) and a massive jigging fleet targeting the Argentine Shortfin Squid (*Illex argentinus*) along the continental shelf-break. This important fishing effort and the food subsidy generated in terms of offal and discards attract vast numbers of seabirds risking their lives while trying to take advantage of this predictable resource (Croxall & Wood 2002; Favero & Silva Rodriguez 2005).

Seabird incidental mortality in fisheries can be mitigated with the effective implementation of best fishing practices that reduce attraction to boats or minimize entanglement and/or collisions with fishing gear. The literature shows that finding practical solutions to the bycatch problem is feasible with political will and the collaboration of multiple stakeholders. One of the instruments developed at national level during the last decades has been the adoption of National Plans of Action based on the guidelines provided by the Food and Agriculture Organization (FAO 1999, 2009; Good et al. 2020). Argentina has adopted its own National Plan of Action—seabirds in 2010 (CFP 2010) and developed associated legislation and regulations aimed at protecting seabirds and other marine biodiversity. In terms of collaboration with neighboring countries, the recent approval of the Regional Plan of Action—Seabirds in 2022 by the Argentine—Uruguay *Comisión Técnica Mixta del Frente Marítimo* (www.ctmfm.org) represents a relevant progress in addressing seabird conservation in fisheries at an international scale (Domingo et al. 2022). Despite the progress achieved in this matter, more efforts and conservation actions are needed to enhance the implementation of these instruments and the uptake of current regulations (Favero & Seco Pon 2014), as well as the effective conservation of marine ecosystems worldwide (Gil et al. 2019; Good et al. 2020).

On top of the above referred issues, recent evidence shows that the negative effects of seabird bycatch, and other stressors affecting species on land and at sea, can be exacerbated by the effects of climate change and its concomitant changes in atmospheric circulation, water masses, and prey distribution (Weimerskirch et al. 2012; Lascelles et al. 2016; Pardo et al. 2017). The forecasts for the austral oceans predict warmer and stronger weather systems, increased frontal systems, and oceanic eddies, leading to changes in trophic webs, thus the foraging areas used by top predators that may or may not fully adapt depending on the flexibility expressed by each species (Constable et al. 2014). These environmental changes have even the potential of generating significant shifts in the distribution of fishing effort and the increase of overlap with seabirds and the risk of bycatch (Kroodsma et al. 2018; Krüger et al. 2018). Addressing this issue likely represents one of the major environmental conservation challenges that will require well-informed management practices and the implementation of meaningful policy responses (Cox et al. 2008).

Over the last few decades, the management of marine environments evolved from single species-based protocols and strategies to more comprehensive methods at ecosystem scale that incorporate the precautionary approach to conservation, and even the use of top predators as indicators (Einoder 2009). These management strategies have primarily focused on the protection of fixed areas, leading for example to the designation of marine protected areas aimed at reducing the overlap with spatially explicit threats, like overfishing, bycatch or the exploration and extraction of non-renewables, that may cause mortality and/or alter the access to habitats (Pichegru et al. 2010; Hooker et al. 2011; Brooks et al. 2020). Area closures are frequently understood as an extreme management measure for fisheries, and this can be seen as beneficial for seabirds by completely removing the incidental mortality within those areas. Nevertheless, these measures can increase the fishing effort in immediate neighboring waters (an effect known as fishing the line), exacerbating the negative effect of bycatch and other threats right outside the closed area (Copello et al. 2016). Although static protected areas have been a keystone of conservation, they have in recent years demonstrated to be insufficient (Oestreich 2020), with dynamic ocean management emerging as a more adaptive approach and conservation strategy to accommodate dynamic habitats and highly migratory species, spatially and temporally adjusted in response to the variable nature of the ocean and users (Lewison et al. 2015; Maxwell et al. 2015; Oestreich 2020). Implementing dynamic ocean management requires monitoring approaches that allow the creation of flexible boundaries and measures implemented, which has been challenging so far (Hazen et al. 2018).

Over the last few decades, Argentina has created more than 60 coastal and marine protected areas at provincial, interjurisdictional, or national level. Most of these reserves are labeled as coastal or coastal-marine, and only three of them are located offshore (the Yaganes Marine Protected Area and the Namuncurá (Burdwood) Bank I and II Marine Protected Areas), constituting Argentina's National System of Marine Protected Areas. A recent review highlighted that most protected areas in Argentina show deficient administration and management (Gil et al. 2019). This limited performance in the implementation of regulations also applies to the management of industrial fisheries affecting seabirds, including deficient data gathering and uncertainties regarding the uptake of current conservation measures. The relatively recent introduction of many local fisheries into certification schemes may facilitate the implementation of an ecosystem approach to fisheries management, although a recent review highlighted the need of strengthening the processes (Seco Pon et al. 2018). In addition, Argentina, as well as other coastal countries in the region, experiences an important fishing effort in waters beyond national jurisdictions, just across the continental shelf-break. Although some of these fisheries may be regulated by Regional Fisheries Management Organizations administrating fisheries in international waters, other important fishing fleets in international waters are not regulated by such multilateral instruments, not even counting issues associated with illegal, unregulated, and unreported fisheries (Favero et al. 2022). International collaboration remains a key factor to address transboundary threats to biodiversity and improve the management of human activities at sea, actually or potentially affecting the marine wildlife and their environments.

Acknowledgements This project was mainly funded by the National Agency for the Promotion of Science and Technology (Agencia Nacional de Promoción Científica y Técnica, Argentina PICT 2017-1761), the National Research Council (Consejo Nacional de Investigaciones Científicas y Técnicas, Argentina PIP CONICET 101031CO), and the Instituto de Investigaciones Marinas y Costeras (UNMDP-CONICET). The authors thank P Yorio, MP Silva Rodriguez, and book editors for their helpful feedback provided during the review process, F Quintana and G Blanco for kindly providing data on Southern Giant petrel distribution, and R Ronconi for the data showing the distribution of Great shearwaters.

References

ACAP (2020) Agreement on the Conservation of Albatrosses and Petrels. Text of the Agreement. https://www.acap.aq/documents/instruments/206-agreement-on-the-conservation-of-albatrosses-and-petrels/file

ACAP (2022) Agreement on the Conservation of Albatrosses and Petrels. Report on Progress with the Implementation of the Agreement 2018–2021. MoP7 Doc 10 Rev 1. Seventh Meeting of the Parties. 9–13 May 2022. https://acap.aq/documents/meeting-of-the-parties/mop7/mop7-meeting-documents/4011-mop7-doc-10-implementation-report-2018-2021/file

Acha EM, Mianzan HW, Guerrero RA, Favero M, Bava J (2004) Coastal marine fronts at the southern cone of South America. Physical and ecological processes. J Mar Syst 44:83–105

Acha EM, Piola A, Iribarne O, Mianzan H (2015) Ecological processes at marine fronts. Springer International Publishing, London. https://doi.org/10.1007/978-3-319-15479-4

Ballance LT, Pitman RL, Fiedler PC (2006) Oceanographic influences on seabirds and cetaceans of the eastern tropical Pacific: a review. Prog Oceanogr 69:360–390

Barwell G (2014) Albatross. Reaktion Books Ltd, London, p 208

Benjamins S, Dale A, Hastie GD, Waggitt JJ, Lea MA, Scott BE, Wilson B (2015) Confusion reigns? A review of marine megafauna interactions with tidal-stream environments. Oceanogr Mar Biol 53:1–54

BirdLife International (2024) Seabird Tracking Database. http://www.seabirdtracking.org/. Downloaded May 2024

Blanco GS, Quintana F (2014) Differential use of the Argentine shelf by wintering adults and juveniles southern giant petrels, *Macronectes giganteus*, from Patagonia. Estuar Coast Shelf Sci 149:151–159

Blanco GS, Pisoni JP, Quintana F (2015) Characterization of the seascape used by juvenile and wintering adult southern Giant petrels from Patagonia Argentina. Estuar Coast Shelf Sci 153:135–144

Blanco GS, Sánchez-Carnero N, Pisoni JP, Quintana F (2017) Seascape modeling of southern giant petrels from Patagonia during different life-cycles. Mar Biol 164:1–14

Blanco GS, Tonini MH, Gallo L, Dell'Omo G, Quintana F (2022) Tracking the exposure of a pelagic seabird to marine plastic pollution. Mar Pollut Bull 180:113767

Bost CA, Cotte C, Bailleul F, Cherel Y, Charrassin JB, Guinet C, Ainley DG, Weimerskirch H (2009) The importance of oceanographic fronts to marine birds and mammals of the southern oceans. J Mar Syst 78:363–376

Boyd IL, Wanless S, Camphuysen KCJ (2016) Top predators in marine ecosystems: their roles in monitoring and management. Cambridge University Press, Cambridge

Brooke ML (2004) Albatrosses and petrels across the world. Oxford University Press, Oxford, p 499

Brooks CM, Chown SL, Douglass LL, Raymond BP, Shaw JD, Sylvester ZT, Torrens CL (2020) Progress towards a representative network of Southern Ocean protected areas. PLoS One 15:e0231361

Butchart SHM, Stattersfield AJ, Bennun LA, Shutes SM, Akcakaya HR, Baillie JEM, Stuart SN, Hilton-Taylor C, Mace GM (2004) Measuring global trends in the status of biodiversity: red list indices for birds. PLoS Biol 2:e383

Carneiro AP, Dias MP, Oppel S, Pearmain EJ, Clark BL, Wood AG, Clavelle T, Phillips RA (2022) Integrating immersion with GPS data improves behavioural classification for wandering albatrosses and shows scavenging behind fishing vessels mirrors natural foraging. Anim Conserv 25:627–637. https://doi.org/10.1111/acv.12768

Catry P, Forcada J, Almeida A (2011) Demographic parameters of black-browed albatrosses *Thalassarche melanophris* from The Falkland Islands. Polar Biol 34:1221–1229

CFP (2010) Plan de Acción Nacional para Reducir la Interacción de Aves marinas con Pesquerías en la República Argentina. Consejo Federal Pesquero / Federal Fisheries Council, p 133. https://cfp.gob.ar/wp-content/uploads/2017/09/PANAVES.pdf

Clark, CW, Gagnon GC (2006) Considering the temporal and spatial scales of noise exposures from seismic surveys on baleen whales. International Whaling Commission Scientific Committee document SC/58/E9, Cambridge, UK

Clay TA, Small C, Tuck GN, Pardo D, Carneiro APB, Wood AG, Croxall JP, Crossin GT, Phillips RA (2019) A comprehensive large-scale assessment of fisheries bycatch risk to threatened seabird populations. J Appl Ecol 56:1882–1893

Coleridge ST (1798) The rhyme of the ancient mariner. In: Lyrical ballads, with a few other poems, London

Collet J, Weimerskirch H (2020) Albatrosses can memorize locations of predictable fishing boats but favour natural foraging. Proc R Soc B 287:20200958

Collet J, Patrick SC, Weimerskirch H (2015) Albatrosses redirect flight towards vessels at the limit of their visual range. Mar Ecol Prog Ser 526:199–205

Constable AJ, Melbourne-Thomas J, Corney SP, Arrigo KR, Barbraud C, Barnes DKA, Bindoff NL, Boyd PW, Brandt A, Costa DP, Davidson AT, Ducklow HW, Emmerson L, Fukuchi M, Gutt J, Hindell MA, Hofmann EE, Hosie GW, Jacob S, Johnston MN, Kawaguchi S, Kokubun N, Koubbi P, Lea MA, Makhado A, Massom RA, Meiners K, Meredith MP, Murphy EJ, Nicol S, Reid K, Richerson K, Riddle MJ, Rintoul SR, Smith WO, Southwell C, Stark JS, Sumne M, Swadling KM, Takahashi KT, Trathan PN, Welsford DC, Weimerskirch H, Westwood KJ, Wienecke BC, Wolf-Gladrow D, Wright SW, Xavier JC, Ziegler P (2014) Climate change and Southern Ocean ecosystems I: how changes in physical habitats directly affect marine biota. Glob Chang Biol 20:3004–3025

Copello S, Quintana F (2003) Marine debris ingestion by southern Giant petrels and its potential relationships with fisheries in the southern Atlantic Ocean. Mar Pollut Bull 46:1513–1515

Copello S, Dogliotti AI, Gagliardini DA, Quintana F (2011) Oceanographic and biological landscapes used by the southern Giant petrel during the breeding season at the Patagonian Shelf. Mar Biol 158:1247–1257

Copello S, Seco Pon JP, Favero M (2013) Use of marine space by Black-browed albatrosses during the non-breeding season in the Southwest Atlantic Ocean. Estuar Coast Shelf Sci 123:34–38

Copello S, Blanco G, Seco Pon JP, Quintana F, Favero M (2016) Exporting the problem: issues with fishing closures in seabird conservation. Mar Policy 74:120–127

Corbeau A, Collet J, Orgeret F, Pistorius P, Weimerskirch H (2021) Fine-scale interactions between boats and large albatrosses indicate variable susceptibility to bycatch risk according to species and populations. Anim Conserv 24:689–699

Cox SL, Embling CB, Hosegood PJ, Votier SC, Ingram SN (2008) Oceanographic drivers of marine mammal and seabird habitat-use across shelf seas: a guide to key features and recommendations for future research and conservation management. Estuar Coast Shelf Sci. https://doi.org/10.1016/j.ecss.2018.06.022

Croxall JP, Wood AG (2002) The importance of the Patagonian Shelf for top predator species breeding at South Georgia. Aquat Conserv: Mar Freshw Ecosyst 12:101–118

Croxall JP, Butchart SHM, Lascelles B, Stattersfield AJ, Sullivan BJ, Symes A, Taylor P (2012) Seabird conservation status, threats and priority actions: a global assessment. Bird Conserv Int 22:1–34

De Roy T, Jones M, Fitter J (2008) Albatross, their world, their ways. David Bateman Ltd., Auckland

Dias MP, Martin R, Pearmain EJ, Burfield IJ, Small C, Phillips RA, Yates O, Lascelles B, Garcia Borboroglu P, Croxall JP (2019) Threats to seabirds: a global assessment. Biol Conserv 237:525–537

Domingo A, Favero M, Navarro G, Sánchez R, Tombesi ML (2022) Plan de Acción Regional para Reducir la Interacción de Aves Marinas con las Pesquerías en el área del Tratado del Río de la Plata y su Frente Marítimo. Serie Publicaciones Especiales de la Comisión Técnica Mixta del Frente Marítimo, No. 2, p 105

Einoder LD (2009) A review of the use of seabirds as indicators in fisheries and ecosystem management. Fish Res 95:6–13

Evans R, Lea MA, Hindell MA (2021) Predicting the distribution of foraging seabirds during a period of heightened environmental variability. Ecol Appl 31:e02343

FAO (1999) International plan of action for reducing incidental catch of seabirds in longline fisheries. Food and Agriculture Organization of the United Nations, Rome
FAO (2009) Best practices to reduce incidental catch of seabirds in capture fisheries. FAO Technical Guidelines for Responsible Fisheries: No. 1, Suppl. 2, FAO, Rome
Favero M, Seco Pon JP (2014) Challenges in seabird bycatch mitigation. Anim Conserv 17:532–533
Favero M, Silva Rodriguez MP (2005) Estado actual y conservación de aves pelágicas que utilizan la plataforma continental Argentina como área de alimentación. Hornero 20:95–110
Favero M, Gandini P, Blanco G, García G, Copello S, Seco Pon JP, Frere E, Quintana F, Yorio P, Rabuffetti F, Cañete G (2011) Seabird mortality associated to freshies in the Patagonian shelf: effect of discards in the occurrence of interactions with fishing gear. Anim Conserv 14:131–139
Favero M, Blanco G, Copello S, Seco Pon JP, Patterlini C, Mariano-Jelicich R, García GO, Berón MP (2013) Seabird by-catch in the Argentinean demersal longline fishery, 2001–2010. Endanger Species Res 19:187–199
Favero M, Seco Pon JP, Copello S (2022) Conservación de biodiversidad en aguas más allá de las jurisdicciones nacionales: la interacción entre albatros y pesquerías comerciales como estudio de caso. En 'Conservación y uso sostenible de la diversidad biológica marina. Aportes interdisciplinarios para un régimen internacional en zonas situadas fuera de la jurisdicción nacional'. Editorial Universidad del Rosario
Genin A (2004) Bio-physical coupling in the formation of zooplankton and fish aggregations over abrupt topographies. J Mar Syst 50:3–20
Gil MN, Giarratano E, Barros VR, Bortolus A, Codignotto JO, Delfino Schenke R, Góngora ME, Lovrich G, Monti AJ, Pascual MS, Rivas AL, Tagliorette A (2019) Southern Argentina: the Patagonian continental shelf. In: World seas: an environmental evaluation. Elsevier Academic Press Inc
González-Zevallos D, Yorio P (2006) Seabird use of discards and incidental captures at the Argentine hake trawl fishery in the Golfo San Jorge, Argentina. Mar Ecol Prog Ser 316:175–183
Good SD, Baker GB, Gummery M, Votiera SC, Phillips RA (2020) National plans of action (NPOAs) for reducing seabird bycatch: developing best practice for assessing and managing fisheries impacts. Biol Conserv 247:108592
Grecian WJ, Lane JV, Michelot T, Wade HM, Hamer KC (2018) Understanding the ontogeny of foraging behaviour: insights from combining marine predator bio-logging with satellite derived oceanography in hidden Markov models. J R Soc Interface 15:20180084
Greer AT, Cowen RK, Guigand CM, Hare JA (2015) Fine-scale planktonic habitat partitioning at a shelf-slope front revealed by a high-resolution imaging system. J Mar Syst 142:111–125
Guilford T, Padget O, Maurice L, Catry P (2022) Unexpectedly deep diving in an albatross. Curr Biol 32:R26–R28
Hazen EL, Scales KL, Maxwell SM, Briscoe DK, Welch H, Bograd SJ, Bailey H, Benson SR, Eguchi T, Dewar H, Kohin S, Costa DP, Crowder LB, Lewison RL (2018) A dynamic ocean management tool to reduce bycatch and support sustainable fisheries. Sci Adv 4. https://doi.org/10.1126/sciadv.aar3001
Hernandez MM, Favero M, Seco Pon JP (2024) Effect of environmental variability on seabird assemblages across the Brazil–Malvinas confluence during the austral winter. Mar Biol 171:51
Hooker SK, Canadas A, Hyrenbach KD, Corrigan C, Polovina JJ, Reeves RR (2011) Making marine protected area networks effective for marine top predators. Endanger Species Res 13:203–218
Hyrenbach KD, Fernández P, Anderson DJ (2002) Oceanographic habitats of two sympatric North Pacific albatrosses during the breeding season. Mar Ecol Prog Ser 233:283–301
Jiménez S, Phillips RA, Brazeiro A, Defeo O, Domingo A (2014) Bycatch of great albatrosses in pelagic longline fisheries in the Southwest Atlantic: contributing factors and implications for management. Biol Conserv 171:9–20
Jouventin P, Weimerskirch H (1990) Satellite tracking of wandering albatrosses. Nature 343:746–748

Kroodsma DA, Mayorga J, Hochberg T, Miller NA, Boerder K, Ferretti F, Wilson A, Bergman B, White TD, Block BB, Woods P, Sullivan B, Costello C, Worm B (2018) Tracking the global footprint of fisheries. Science 359:904–908

Krüger L, Ramos JA, Xavier JC, Grémillet D, González-Solís J, Petry MV, Phillips RA, Wanless RM, Paiva VH (2018) Projected distributions of Southern Ocean albatrosses, petrels and fisheries as a consequence of climatic change. Ecography 41:195–208

Langton R, Davies IM, Scott B (2011) Seabird conservation and tidal stream and wave power generation: information needs for predicting and managing potential impacts. Mar Policy 35:623–630

Lascelles BG, Taylor PR, Miller MGR, Dias MP, Oppel S, Torres L, Hedd A, Le Corre M, Phillips RA, Shaffer SA, Weimerskirch H, Small C (2016) Applying global criteria to tracking data to define important areas for marine conservation. Divers Distrib 22:422–431

Lewison RL, Crowder LB, Wallace BP, Moore JE, Cox T, Zydelis R, Mcdonald S, Dimatteo A, Dunn DC, Kot CY, Bjorkland R, Kelez S, Soykan C, Stewart KR, Sims M, Boustany A, Read AJ, Halpin P, Nichols WJ, Safina C (2015) Global patterns of marine mammal, seabird, and sea turtle bycatch reveal taxa-specific and cumulative megafauna hotspots. Proc Natl Acad Sci 111:5271–5276

Mariano-Jelicich R, Copello S, Seco Pon JP, Favero M (2017) Long-term changes in Black-browed albatrosses diet as a result of fisheries expansion: an isotopic approach. Mar Biol 164:148

Mariano-Jelicich R, Berón MP, Copello S, Dellabianca NA, García G, Labrada-Martagón V, Paso Viola MN, Paz J, Riccialdelli L, San Martin A, Seco Pon JP, Torres MA, Favero M (2022) Functional diversity of marine organisms. Marine megafauna: marine turtles, seabirds and marine mammals. In: Pan J, Guinder V, Pratolongo P (eds) Marine biology: a functional approach to the oceans and their organisms. Science Publishers, pp 297–324

Mariano-Jelicich R, Seco Pon JP, Copello S, Favero M (2023) Distribution and diet of cape petrels (*Daption capense*) attending fishing vessels off the Patagonian continental shelf during the non-breeding season in austral winter: insights from on-board censuses and stable isotope analysis. Polar Biol. https://doi.org/10.1007/s00300-023-03144-6

Maxwell SM, Hazen EL, Lewison RL, Dunn DC, Bailey H, Bograd SJ, Briscoe DK, Fossette S, Hobday JA, Bennett M, Benson S, Caldwell MR, Costa DP, Dewar H, Eguchi T, Hazen L, Kohin S, Sippel T, Crowder LB (2015) Dynamic Ocean management: defining and conceptualising real-time management of the ocean. Mar Policy 58:42–50

McInnes JC, Jarman SN, Lea MA, Raymon B, Deagle BE, Phillips RA, Catry P, Stanworth A, Weimerskirch H, Kusch A, Gras M, Cherel Y, Maschette D, Alderman R (2017) DNA Metabarcoding as a marine conservation and management tool: a circumpolar examination of fishery discards in the diet of threatened albatrosses. Fron Mar Sci 4:00277

Nevitt GA (2000) Olfactory foraging by Antarctic procellariiform seabirds: life at high Reynolds numbers. Biol Bull 198:245–253

Nevitt GA (2008) Sensory ecology on the high seas: the odor world of the procellariiform seabirds. J Exp Biol 211:1706–1713

Nevitt GA, Veit RR, Kareiva P (1995) Dimethyl sulphide as a foraging cue for Antarctic procellariiform seabird. Nature 376:680–682

Nevitt GA, Reid K, Trathan P (2004) Testing olfactory foraging strategies in an Antarctic seabird assemblage. J Exp Biol 207:3537–3544

Nicholls DG, Robertson CJR, Prince PA, Murray MD, Walker KJ, Elliott GP (2002) Foraging niches of three Diomedea albatrosses. Mar Ecol Prog Ser 231:269–277

Nowacek DP, Clark CW, Mann D, Miller PJO, Rosenbaum HC, Golden JS, Jasny M, Kraska J, Southall BL (2015) Marine seismic surveys and ocean noise: time for coordinated and prudent planning. Front Ecol Environ 13:378–386

Oestreich WK, Chapman MS, Crowder LB (2020) A comparative analysis of dynamic management in marine and terrestrial systems. Front Ecol Environ 18:496–504. https://doi.org/10.1002/fee.2243

Orben RA, O'Connor AJ, Suryan RM, Ozaki K, Sato F, Deguchi T (2018) Ontogenetic changes in at-sea distributions of immature short-tailed albatrosses *Phoebastria albatrus*. Endanger Species Res 35:23–37

Oro D, Genovart M, Tavecchia G, Fowler MS, Martínez-Abraín A (2013) Ecological and evolutionary implications of food subsidies from humans. Ecol Lett 16:1501–1514

Paiva VH, Geraldes P, Ramírez I, Meirinho A, Garthe S, Ramos JA (2010) Oceanographic characteristics of areas used by Cory's shearwaters during short and long foraging trips in the North Atlantic. Mar Biol 157:1385–1399

Pardo D, Forcada J, Wood AG, Tuck GN, Ireland L, Pradel R, Croxall JP, Phillips RA (2017) Additive effects of climate and fisheries drive ongoing declines in multiple albatross species. Proceedings of the National Academy of Sciences. Proc Natl Acad Sci 114. https://doi.org/10.1073/pnas.1618819114

Patrick SC, Weimerskirch H (2014) Personality, foraging and fitness consequences in a long-lived seabird. PLoS One 9:e87269

Paz JA (2022) Conservación del Albatros Ceja Negra en la Plataforma Continental Argentina: interacciones con la actividad pesquera, idoneidad de hábitat y priorización espacial. Doctoral Thesis Dissertation. Universidad Nacional de Mar del Plata

Paz JA, Seco Pon JP, Krüger L, Favero M, Copello S (2021a) Is there sexual segregation in habitat selection by Black-browed albatrosses wintering in the Southwest Atlantic? Emu. https://doi.org/10.1080/01584197.2020.1869910

Paz JA, Ronconi RA, Seco Pon JP, Copello S, Ryan PG, Favero M (2021b) Spatial overlap and effect of fishing effort on the foraging behavior of the Great Shearwater (*Ardenna gravis*) on the Argentine Continental Shelf. III World Seabird Conference. Hobart

Pelletier L, Kato A, Chiaradia A, Ropert-Coudert Y (2012) Can thermoclines be a cue to prey distribution for marine top predators? A case study with little penguins. PLoS One 7:e31768

Phillips RA, Waluda CM (2020) Albatrosses and petrels at South Georgia as sentinels of marine debris input from vessels in the Southwest Atlantic Ocean. Environ Int 136:105443

Phillips RA, Gales R, Baker GB, Double MC, Favero M, Quintana F, Tasker M, Weimerskirch H, Uhart M, Wolfaardt A (2016) The conservation status and priorities for albatrosses and large petrels. Biol Conserv 201:169–183

Pichegru L, Gremillet D, Crawford RJM, Ryan PG (2010) Marine no-take zone rapidly benefits endangered penguin. Biol Lett 6:498–501

Pichegru L, Nyengera R, McInnes AM, Pistorius P (2017) Avoidance of seismic survey activities by penguins. Sci Rep 7:16305

Poncet S, Wolfaardt AC, Black A, Browning S, Lawto K, Lee J, Passfield K, Strange G, Phillips RA (2017) Recent trends in numbers of wandering, Black-browed and Grey-headed albatrosses breeding at South Georgia. Polar Biol 40:1347–1358

Ponchon A, Gamble A, Tornos J, Delord K, Barbraud C, Travis JMJ, Weimerskirch H, Boulinier T (2021) Similar at-sea behaviour but different habitat use between failed and successful breeding albatrosses. Mar Ecol Prog Ser 678:183–196

Ponganis P, Starke LN, Horning M, Kooyman GL (1999) Development of diving capacity in emperor penguins. J Exp Biol 202:781–786

Quintana F, Dell'Arciprete OP, Copello S (2010) Foraging behavior and habitat use by the southern Giant petrel on the Patagonian Shelf. Mar Biol 157:515–525

Rackete C, Poncet S, Good SD, Phillips RA, Passfield K, Trathan P (2021) Variation among colonies in breeding success and population trajectories of wandering albatrosses *Diomedea exulans* at South Georgia. Polar Biol 44:221–227

Rexer-Huber K (2017) White-chinned petrel distribution, abundance and connectivity: New Zealand populations and their global context. Report to New Zealand Department of Conservation. Parker Conservation, Dunedin, p 13

Robertson G, Gales R (1998) Albatross biology and conservation. Surrey Beatty and Sons, Chipping Norton

Sala JE, Pisoni JP, Quintana F (2017) Three-dimensional temperature fields of the north Patagonian Sea recorded by Magellanic penguins as biological sampling platforms. Estuar Coast Shelf Sci 189:203–215

Scales KL, Miller PI, Ingram SN, Hazen EL, Bograd SJ, Phillips RA (2016) Identifying predictable foraging habitats for a wide-ranging marine predator using ensemble ecological niche models. Divers Distrib 22:212–224

Scheffer A, Bost CA, Trathan PN (2012) Frontal zones, temperature gradient and depth characterize the foraging habitat of king penguins at South Georgia. Mar Ecol Prog Ser 465:281–297

Seco Pon JP, Copello S, Tamini L, Mariano-Jelicich R, Paz J, Blanco G, Favero M (2015) Seabird conservation in fisheries: current state of knowledge and conservation needs for Argentine high-seas fleets. In: Seabirds and songbirds: habitat preferences, conservation and migratory behavior. Nova Science Publishers, New York, pp 45–89

Seco Pon JP, Paz JA, Mariano-Jelicich R, García GO, Copello S, Berón MP, Blanco G, Flaminio JL, Favero M (2018) Certification schemes in Argentina fisheries: opportunities and challenges for seabird conservation. In: Mikkola H (ed) Seabirds. InTechOpen, London, pp 25–46

Seco Pon JP, Bastida J, Giardino G, Favero M, Copello S (2019) Seabirds east of Tierra del Fuego, Argentina during a 3D seismic survey. Ornitol Neotrop 30:103–111

Sequeira A, Hays GC, Sims DW, Eguíluz VM, Rodríguez JP, Heupel MR, Harcourt R, Calich H, Queiroz N, Costa DP, Fernández-Gracia J, Ferreira LC, Goldsworthy SD, Hindell MA, Lea M-A, Meekan MG, Pagano AM, Shaffer SA, Reisser J, Thums M, Weise M, Duarte CM (2019) Overhauling Ocean spatial planning to improve marine megafauna conservation. Front Mar Sci 6:00639

Serratosa J, Hyrenbach KD, Miranda-Urbina D, Portflitt-Toro M, Luna N, N & G Luna-Jorquera. (2020) Environmental drivers of seabird at-sea distribution in the eastern South Pacific Ocean: assemblage composition across a longitudinal productivity gradient. Fron Mar Sci 6(2019):00838

Simpson JH, Sharples J (2012) Introduction to the physical and biological oceanography of shelf seas. Cambridge University Press, p 424

Stevick PT, Incze LS, Kraus SD, Rosen S, Wolff N, Baukus A (2008) Trophic relationships and oceanography on and around a small offshore bank. Mar Ecol Prog Ser 363:15–28

Suazo CG, Schlatter RP, Arriagada AM, Cabezas L, Ojeda J (2013) Fishermen's perceptions of interactions between seabirds and artisanal fisheries in the Chonos archipelago, Chilean Patagonia. Oryx 47:184–189

Thompson DR, Goetz KT, Sagar PM, Torres LG, Kroeger CE, Sztukowski LA, Orben RA, Hoskins AJ, Phillips RA (2021) The year-round distribution and habitat preferences of Campbell albatross (*Thalassarche impavida*). Aquat Conserv: Mar Freshw Ecos 31:2967–2978

Tickell WLN (2000) Albatrosses. Midas Printing, p 448

Torres LG, Read AJ, Halpin P (2008) Fine-scale habitat modelling of a top marine predator: do prey data improve predictive capacity? Ecol Appl 18:1702–1717

Wakefield ED, Miller DL, Bond SL, Le Bouard F, Carvalho PC, Catry P, Dilley BJ, Fifield DA, Gjerdrum C, González-Solís J, Hogan H, Laptikhovsky V, Merkel B, Miller JAO, Miller PI, Pinder SJ, Pipa T, Ryan PM, Thompson LA, Thompson PM, Matthiopoulos J (2021) The summer distribution, habitat associations and abundance of seabirds in the sub-polar frontal zone of the Northwest Atlantic. Prog Oceanogr 198:102657

Weimerskirch H, Cherel Y, Cuenot Chaillet F, Ridoux V (1997) Alternative foraging strategies and resource allocation by male and female wandering albatrosses. Ecology 78:2051–2063

Weimerskirch H, Louzao M, de Grissac S, Delord K (2012) Changes in wind pattern alter albatross distribution and life-history traits. Science 335:211–214

Yorio P, Olinto Branco J, Lenzi J, Luna-Jorquera G, Zavalaga C (2016) Distribution and trends in Kelp Gull (*Larus dominicanus*) coastal breeding populations in South America. Waterbirds 39:114–135

Zamon JE, Phillips EM, Guy TJ (2014) Marine bird aggregations associated with the tidally driven plume and plume fronts of the Columbia River. Deep-Sea Res Part II: Top Stud Oceanogr 107:85–95

Chapter 9
Patagonian Shelf-Break Front: The Ecosystem Services Hot-Spot of the South West Atlantic Ocean

Paulina Martinetto, Carolina Kahl, Daniela Alemany, and Florencia Botto

Abstract Marine fronts are typically associated with relatively sharp changes in temperature and/or salinity and are characterized by their high nutrient concentrations and phytoplankton biomass. Additionally, top-down and bottom-up processes that involve primary production propagate the structure to the entire ecosystem. As we have seen in the previous chapters, the Patagonian shelf-break front (PSBF) shows marine consumers and biogeochemical cycles coupled to its high primary production. For instance, different groups of vertebrates are coupled to primary production through trophic interactions, and the large amount of phytoplankton photosynthesizing constitutes the so-called biological pump capturing CO_2 from the atmosphere. The high biological abundance as well as the high rates of the different ecological processes occurring in the PSBF has been suggested to support a higher provision of marine ecosystem services (ES) in comparison with adjacent areas. In this sense, the PSBF can be seen as a hot-spot of ecosystem services. This chapter explores this hypothesis by revisiting the evidence, identifying gaps, and proposing further lines of research.

Supplementary Information The online version contains supplementary material available at https://doi.org/10.1007/978-3-031-71190-9_9.

P. Martinetto (✉) · F. Botto
Facultad de Ciencias Exactas y Naturales, Instituto de Investigaciones Marinas y Costeras (IIMyC, UNMdP-CONICET), Universidad Nacional de Mar del Plata (UNMdP), Consejo Nacional de Investigaciones Científicas y Técnicas (CONICET), Mar del Plata, Argentina

C. Kahl
Depto. Oceanografía, Servicio de Hidrografía Naval, Buenos Aires, Argentina

D. Alemany
Instituto de Investigaciones Marinas y Costeras (IIMyC, UNMdP-CONICET), Facultad de Ciencias Exactas y Naturales, Universidad Nacional de Mar del Plata (UNMdP), Consejo Nacional de Investigaciones Científicas y Técnicas (CONICET), Mar del Plata, Argentina

Instituto Nacional de Investigación y Desarrollo Pesquero (INIDEP), Mar del Plata, Argentina

E. M. Acha et al. (eds.), *The Patagonian Shelfbreak Front*, Aquatic Ecology Series 13, https://doi.org/10.1007/978-3-031-71190-9_9

Acronyms and Abbreviations

Cant	anthropogenic carbon
ES	ecosystem services
NbS	nature based solutions
PSBF	Patagonian shelf break front
SSH	sea surface height
SSS	sea surface salinity
SST	sea surface temperature

9.1 Introduction

There is no doubt of the dependency of nature for human wellbeing, and there is also no doubt on the rapid decline of natural ecosystems. In fact, the last IPCC report makes clear the interconnection among nature, humans, and climate and the necessity of preserving between 50% and 80% of natural ecosystems to ensure a healthy and habitable planet (IPCC 2022). The last IPBES report also highlights that "nature and its vital contributions to people, which together embody biodiversity and ecosystem functions and services, are deteriorating worldwide … many of nature's contributions to people are essential for human health and their decline thus threatens a good quality of life" (Díaz et al. 2019).

Covering two thirds of the planet, the ocean hosts vast biodiversity and modulates the global climate system by regulating cycles of heat, water, and elements including carbon. Humans have a strong and long-lasting relationship with the ocean. By providing key resources, the oceans have allowed humans to spread all over the planet. Today, over 50% of the world human population lives within 200 km from the coast, with nearly a 30% living closer than 100 km (Kummu et al. 2016). Thus, marine systems are central to many cultures whose livelihoods depend on the food, minerals, energy, and employment that they provide.

Marine stakeholders such as fishermen, nongovernmental organizations, and governmental and intergovernmental agencies involved in the management of natural resources pursue sustainability and conservation of marine environments often under different frameworks, and thus, their interests can often be perceived as opposed; however, to generate successful environmental policies it is essential to include the different points of view (Cáceres et al. 2015). Traditional approaches to ocean management and conservation often fail to integrate social and ecological dimensions focusing only on one side. The Ecosystem Services (ES) approach may help to integrate both dimensions given that it set the focus on the benefits that humans obtain from nature, including ecological as well as social and economic dimensions (MEA 2005; Fisher et al. 2009). However, its application in oceanic socio-ecosystems presents some difficulties (Cognetti and Maltagliati 2010; Townsend et al. 2018). In particular, the fact that humans do not inhabit the open ocean generates a temporal and spatial uncoupling between the provision of services by the ocean and their use by people on land, which can mask the origin of these services. Thus, marine ES studies in general address functional features (Armstrong et al. 2012; Thurber et al. 2014) and rarely include components within the social dimension (Sagebiel et al. 2016).

Marine fronts constitute discrete features that interrupt an apparently continuous and smoothly varying ocean. Such oceanic features are perceived and chosen by marine organisms as foraging, reproductive, recruitment, and migratory areas. In this sense, marine fronts structure life in the ocean. They are considered as hot spots of marine life as they often hold maximum biodiversity and high primary and secondary production (Worm et al. 2005; Mann and Lazier 2006). The combination of light, turbulence, temperature, and nutrients occurring in frontal areas is optimal to initiate and sustain phytoplankton blooms. In turn, these areas are important for other ecosystem processes and components that are linked to primary production through top-down and bottom-up processes (Falkowski et al. 1998; Benoit-Bid and NcManus 2012). For instance, marine mammals, turtles, birds, and fishes are coupled to primary production through trophic interactions (Mann and Lazier 2006), and the high abundance of phytoplankton uptakes large amounts of atmospheric CO_2 through the so-called biological pump. Even certain types of fisheries (e.g., bottom trawling and jigger fleets) are closely coupled with marine fronts (Alemany et al. 2014). A well-studied example is the correlation between the alternation in the regimes of sardine-anchovy and the salmon production with frontal areas of the California current (Woodson and Litvin 2015): fronts positively affect total ecosystem biomass, production, and fishery abundance and yield. Moreover, a global analysis demonstrates that at the scale of Large Marine Ecosystems primary production determines fisheries catches (Chassot et al. 2010). It is in this context that it has been recently proposed that marine fronts can be also seen as ES hot-spots (Martinetto et al. 2020; Belkin 2021; Pittman et al. 2021).

In a recent article, we revised the evidence that links the production of marine ES in the PSBF with primary production concluding that the PSBF indeed could be considered as a hotspot of ES (Martinetto et al. 2020). Here we updated the information and discuss its scope within the global change context.

9.2 The Ecosystem Functions and Services at the Patagonian Shelf Break Front

The coupling between the PSBF structure and the ecosystem functions and services can be illustrated as a cascade where in the upper part are located the biological, physical, and chemical features (environment) while in the lower part are the beneficiaries (socio-economic system) (Fig. 9.1). In turn, management decisions generated in the socio-economic system, specifically those related to the exploitation of

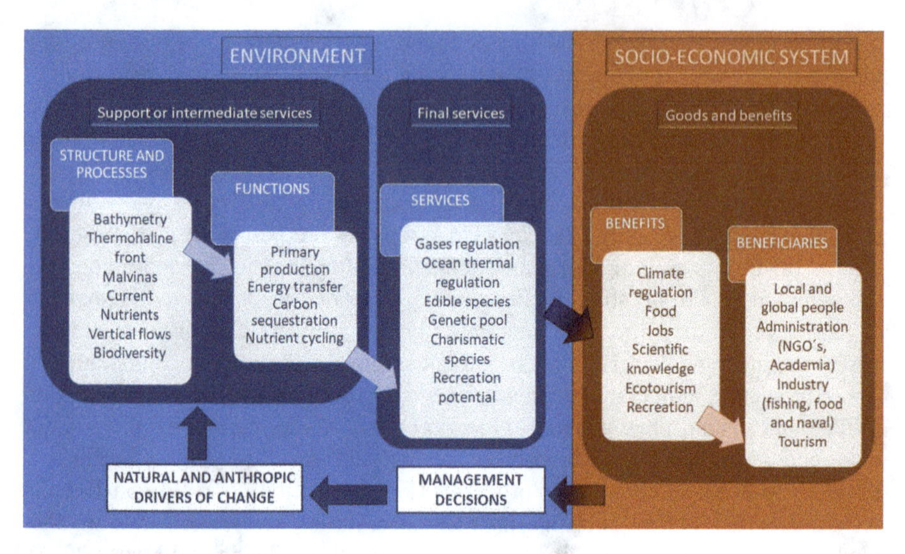

Fig. 9.1 Ecosystem service production functions cascade for the socio-ecological system of the Patagonian shelf-break front. (Adapted from Martinetto et al. 2020)

natural resources and to climate change adaptation and mitigation actions (or no actions), affect the environmental components and consequently the provision of ES.

The specific structure, processes, and functions that have place in the environment make possible the final services. In particular, the PSBF has a combination of features optimal for the development of primary production and the production of coupled services. Light and water masses organize life in the ocean. In the formation of the PSBF, bathymetry plays a fundamental role determining the direction and encounter of water masses (see Chaps. 1 and 2 for details). The Argentine Sea has a continental shelf that descends smoothly up to 200 meters depth to then precipitate sharply up to 4000–5000 m depth. This sharp decline is known as the shelf-break and behind that are the ocean basins. These bottom features make possible the PSBF, a thermohaline front resulting from the encounter of the warm and less salty waters in the continental shelf with the cold, saltier, and nutrient rich water coming from the Malvinas Current (Chap. 2). The rise of nutrients to the euphotic zone supports the steady development of phytoplankton, making the PSBF the most productive area in the Argentine Sea (Chap. 3). In turn, the bloom of primary producers and zooplankton at the PSBF (Chap. 4) attract a great diversity and abundance of fishes and squids (Chap. 5), which take advantage of the PSBF to feed, reproduce and/or migrate; some of these nekton species sustain important fisheries in the area. Thus, most of the marine ES are supported directly or indirectly by primary production.

Cultural services are perhaps the most difficult to perceive in the ocean. Given that in the marine realm, far from the coasts, cultural services are provided by the ocean as a whole and not by a particular region; its identification with the PSBF is almost negligible or, at least, weak. Perhaps the most evident cultural service is its role for science. The compilation in this book of the many studies done under the hypothesis that the PSBF is the main primary production hotspot or the "backbone" of primary production in the Argentine Sea and that energy is exported from the PSBF to the other regions in the continental shelf proves the importance that the PSBF has for marine science.

9.3 Climate Regulation at the Patagonian Shelf-Break Front

One of the most acknowledged ES provided by the ocean is the regulation of the greenhouse gas effects through the capture of large amounts of CO_2, mainly by primary producers (Stocker 2015). It is estimated that for the last decade, the ocean absorbed 2.8 Pg of atmospheric CO_2 per year (Friedlingstein et al. 2022), contributing to mitigate global warming. Marginal seas play an important role in the global carbon cycle by linking the different carbon reservoirs (Gattuso et al. 1998). In the global CO_2 budget, the estimations suggest that continental shelves uptake within 0.19–0.45 $PgCyr^{-1}$ (Cai et al. 2006; Chen and Borges 2009; Cai, 2011; Bauer et al. 2013; Chen et al. 2013; Dai et al. 2013; Laruelle et al. 2014), and records of pCO_2 show a tendency for enhanced shelf uptake of atmospheric CO_2 (Laruelle et al. 2018).

In particular, the Patagonian Sea in the Southwestern Atlantic Ocean is an area of high CO_2 uptake (Bianchi et al. 2009; Padin et al. 2010), where the biological pump is the dominant process determining the sea-air CO_2 balance (Kahl et al. 2017). As a result of the high anthropogenic CO_2 uptake (16 $TgCyear^{-1}$, Bianchi et al. 2009), it has been estimated that in the Argentine Sea the pH has decreased from the industrial era to the present by 0.1, equivalent to a rate of −0.002 $year^{-1}$ (Kahl 2018; Orselli et al. 2018), rate comparable with pH records for different regions of the globe (Bates et al. 2014) and with outputs of global climate models (Hartin et al. 2016). In fact, estimations of anthropogenic carbon (Cant) content indicate penetration of Cant into the water column throughout the entire PSBF, with values between 40 and 90 $\mu mol\ kg^{-1}$. That carbon content results in a decrease in pH from the pre-industrial period to 2006 between 0.11 and 0.17, thus observing a certain degree of acidification (Kahl 2018).

Updated estimations presented here indicate that, while the PSBF covers only ~11% of the Argentine Sea it absorbs ~23% of the total atmospheric CO_2 captured being the area of maximum CO_2 uptake at the Patagonian Shelf Large Marine Ecosystem. On average, the PSBF region acts as a CO_2 sink throughout the year (Fig. 9.2, Table 9.1) with an annual mean of −9.2 mmol $m^{-2}\ d^{-1}$ and intense uptake during summer (−12.6 mmol $m^{-2}\ d^{-1}$, Table 9.1) and autumn (−13.9 mmol $m^{-2}\ d^{-1}$, Table 9.1). Except for spring, all the mean values represent more intense uptake of CO_2 in comparison of the mean CO_2 fluxes for all the Argentinean Sea and adjacent open ocean waters of Malvinas Current. This result seems to be counterintuitive having in mind that the main responsible for CO_2 uptake is phytoplankton and it blooms mostly during spring. However, when the dataset is carefully analyzed, it shows up the unbalance in data for each season and region. In particular, the available information for the PSBF in spring is scarce covering a window of time between

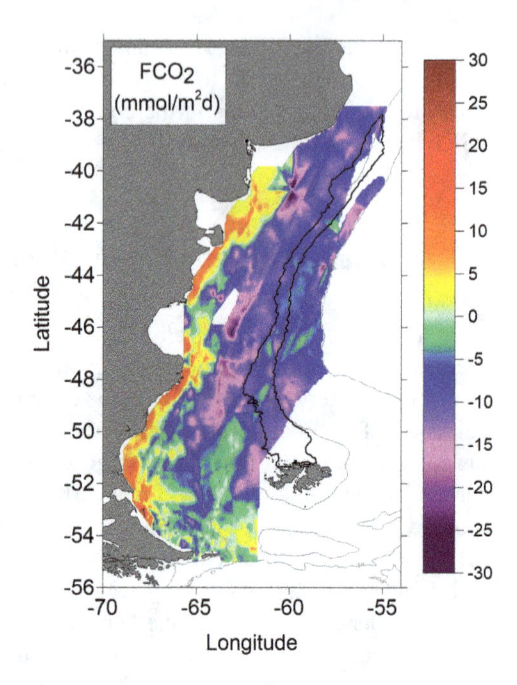

Fig. 9.2 Seasonal climatology of CO_2 fluxes in Argentina Sea based on around 150,000 data of sea water CO_2 fugacity (fCO_2) and the mole fraction of CO_2 in dry air (xCO_2) for the period 2000–2017 from SOCAT2019 database (http://www.socat.info/access.html, Bakker et al. 2016). The black polygon represents the mean position of the Patagonian Shelf-break front estimated from satellite surface Chlorophyll *a* concentration from the period 2002–2017 according to Carranza (2009)

Table 9.1 Seasonal mean values with standard errors of CO_2 fluxes (mmol m^{-2} d^{-1}) estimated for the Argentine Sea, the Patagonian Shelf-break Front region, and the Malvinas Current beyond the Argentine Sea area

	Argentine Sea	Shelf-break front	Malvinas current
Summer	−7.0 (0.3)	−12.6 (0.6)	−5.8 (0.3)
Autumn	−4.6 (0.4)	−13.9 (0.6)	−4.7 (0.3)
Winter	−0.3 (0.3)	−3.0 (0.4)	−1.7 (0.5)
Spring	−10.6 (0.5)	−7.2 (0.7)	−7.0 ± 0.3)
Annual mean annual balance	−5.6 (0.5)	−9.2 (0.7)	−4.8 (0.5)
	−21.3 *TgC year*$^{-1}$	−4.7 *TgC year*$^{-1}$	−9.2 *TgC year*$^{-1}$

Annual mean corresponds to the annual balance in TgC year^{-1}. Calculations were made based on around 150,000 data of sea water CO_2 fugacity (fCO_2) and the mole fraction of CO_2 in dry air (xCO_2) for the period 2000–2017 from SOCAT2019 database (http://www.socat.info/access.html, Bakker et al., 2016). The net CO_2 fluxes were estimated according expressions of Körtzinger et al. (1999), Weiss (1974), Dickson et al. (2007) and Wanninkhof (2014)

November 2 and 8 in 2007. If we look at the satellite product of Chlorophyll *a* for that same time period, it is possible to observe that at those dates the Chlorophyll *a* was also low (<1 mg m^{-3}) in the frontal zone, which could explain the weak CO_2 fluxes observed.

Some areas in the PSBF present high capture fluxes (FCO_2). For example, in the southern sector, FCO_2 is predominantly negative, mainly in summer with FCO_2 of around −20 mmol m^{-2} d^{-1} and even more intense in spring with FCO_2 of around −35 mmol m^{-2} d^{-1}. This intense uptake could be sustained by the high productivity characteristic of the region (Carreto et al., 2016). By separating the thermal and biological effects on pCO_2 in the Patagonia Sea, Kahl et al. (2017) suggest that the biological processes would dominate the CO_2 variability in the shelf region (including PSBF). In winter, both the PSBF and the Argentine Sea present the lowest fluxes. In this season, Chlorophyll *a* concentrations are minimum (<1 mg m^{-3}) in mid and outer Patagonian Sea shelf (Romero et al. 2006). Thus, the decrease in FCO_2 during winter can be understood in terms of the decline in biological activity.

Despite the fact that in the global context, the PSBF only contributes 1.7% of the total marginal seas sink (0.4 $PgCyear^{-1}$, Borges et al. 2005), its relative uptake is high compared to other areas. For instance, the PSBF uptake 2.8 times more than the Benguela upwelling system area. Such upwelling system is bounded in the north and south by warm water current systems and characterized by very high primary production. Notwithstanding its high productivity, the southern Benguela was found to be a very small net CO_2 sink with -1.4 ± 0.6 mol Cm^{-2} per year (1.7 Mt. C $year^{-1}$) (Gregor and Monteiro 2013).

Further examination of the data presented in Table 9.1 reveals that, around the Malvinas Islands, it is observed that from spring to autumn (no data is available in winter) the ocean behaves as an area of intense CO_2 uptake, with very intense flows in spring ($FCO_2 > -40$ mmol m^{-2} d^{-1}). Satellite information shows that in the south of the Malvinas Current, Chlorophyll *a* reaches a maximum in spring and decay to a mean value (~0.5 mg m^{-3}) during the rest of the year (Saraceno et al. 2005). In addition, in situ data show that spring phytoplankton blooms at the PSBF extended in some cases well within the domain of the Malvinas Current (Lutz et al. 2010). This increase in the biological production could explain the intensification in the CO_2 uptake observed during spring in the Malvinas Current region.

Gasses and thermal regulations of the atmosphere by the ocean contribute to global climate regulation. The entire continental shelf and shelf-break with the Malvinas Current area are responsible for capturing approximately 30,500,000 tons of carbon per year, which is equivalent to the total carbon emissions released by residential energy in Argentina, which for 2016 was 34,700,000 tons of carbon (https://www.argentina.gob.ar/sites/default/files/inventario_de_gei_de_2019_de_la_republica_argentina.pdf). Together with few smaller-scale fronts, the PSBF captures more CO_2 per unit area than the mean over the whole shelf region. Even more, the PSBF is a key area for Cant intrusion to the ocean interior in the SW Atlantic Ocean (Orselli et al. 2018).

Spatial and temporal distributions of CO_2 fluxes, as well as their variability, are still poorly understood in this region of the South Atlantic. The works carried out at the moment focus on descriptions of the mean state of the sea. In addition, despite scientific advances in relation to the CO_2 excess problem, there are still areas of oceanographic-fishing importance that have not been explored in relation to this issue yet. In this regard, the PSBF region is still very little explored. It is essential to have time series in different sectors of the Argentine Sea to quantify trends in the absorption of CO_2 from the ocean. Global databases are a useful tool to fill in information gaps; however they are still not enough.

Redistribution of heat by the oceanic currents is also a key for the Earth global climate. In fact, the inclusion of marine fronts in climatic models could improve predictions of its future evolution (Ferrari 2011; Hewitt et al. 2022). The encounter of cold waters from the Malvinas Current with warmer waters from the continental shelf makes possible heat transfer (e.g., Acha et al. 2004. Under the current and future climate change scenario, it has been proposed that this transfer could become relevant and affect the atmosphere-ocean heat exchange (Martinetto et al. 2020). Therefore, and in addition to its capacity to uptake atmospheric CO_2, the role of the PSBF in regional and global climate regulation could be important.

9.4 Seafood from the Patagonian Shelf-Break Front

Several large-scale studies provide evidence of the tight association between fisheries and marine fronts (see Belkin 2021). For instance, some pelagic resources and their fisheries, especially large ones such as the Atlantic bluefin tuna (*Thunnus thynnus*, Druon 2010) and the swordfish (*Xiphias gladius*, Podestá et al. 1993), are coupled with marine fronts. This pattern is found at the PSBF, with several target species showing a strong association with this area (see Chaps. 5 and 7). For instance, the Argentine shortfin squid (*Illex argentinus*) reproduces and feeds during part of its life cycle in the PSBF, and therefore the fishing effort of squid jigging vessels is concentrated there. The Patagonian toothfish (*Dissostichus eleginoides*) is another of the target species caught mostly at the PSBF (Martínez and Wöhler 2016). Likewise, the Argentine hake (*Merluccius hubbsi*) is fished by international fleets in a portion of the PSBF located at international waters (e.g., Spanish fleets, Vilela et al. 2018). In addition, benthic species are also coupled to the PSBF. The most studied example is probably the Patagonian scallop (*Zygochlamys patagonica*) whose larger abundance and fishery coincides with the PSBF area (Bogazzi et al. 2005). Detailed information on the relation between the PSBF and fisheries is examined further in others chapters; here, we focus on fisheries as an ecosystem service providing food and economic income.

In terms of animal protein supply, the global average 2017–2019 contribution of fisheries was <2 g per capita per day (FAO 2022). Argentina is a country with low fish and seafood consumption (average 2010–2019: 6.5 kg per capita per year) in comparison with Latin America and the Caribbean (regional average of 9.9 kg per capita in 2019), and this difference is even bigger in comparison with global values (20.5 kg per capita in 2019; FAO 2022). In fact, according to official Argentinean national reports, most of the products derived from the marine fishing industry are exported (85%, Subsecretaría de Pesca y Acuicultura, Argentina 2021). In terms of economic income, between 2012 and 2021, exports from the fishery sector were ca. 1700 million US dollars per year, which represents about 3% of Argentina total exports (average 2012–2021, INDEC 2024). Beside the large extension of the marine area in Argentina, only 7 species contribute with more than 75% of the marine primary products exported: red shrimp (*Pleoticus muelleri*), Argentine hake,

Argentine shortfin squid, Patagonian scallop, Patagonian hoki (*Macruronus magellanicus*), Patagonian toothfish, and Argentine anchovy (*Engraulis anchoita*) (Martinetto et al. 2020). With exception of the red shrimp, the Argentine anchovy and the Argentine hake, all other species are mainly fished at or near the PSBF (Alemany et al. 2014). The 1700 million US dollars per year mentioned above correspond to an average of 470 thousand tons of seafood per year; almost 40 thousand tons, equivalent to 165 million US dollars, were at least provided by the PSBF. Doing a rough calculation, the PSBF contributes on average with 10% of the export income of seafood production (Fig. 9.3).

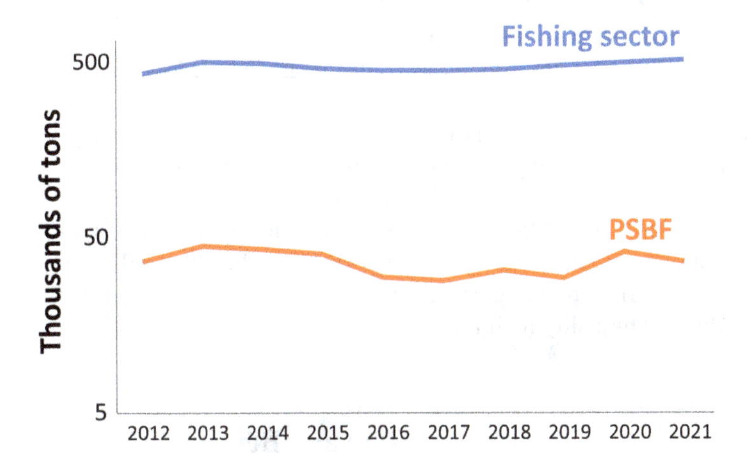

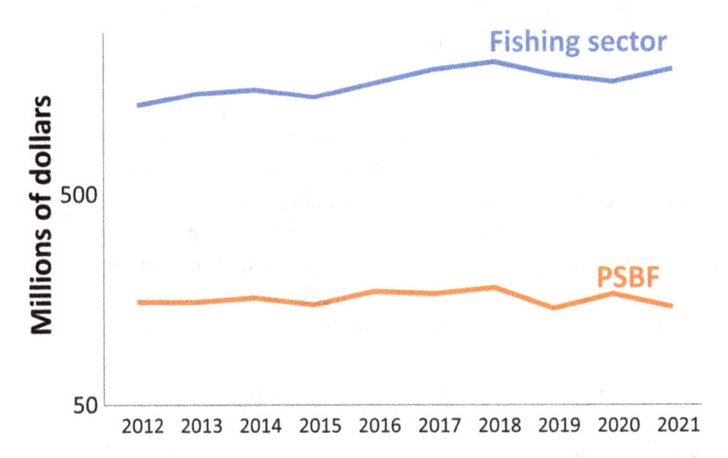

Fig. 9.3 Export of fishery products (in thousands of tons, upper panel) and the economic value (in millions of dollars, lower panel) of the Argentine fishing sector with the particular contribution of the PSBF from 2012 to 2021. *y* axis is in log scale. (Adapted and updated from Martinetto et al. 2020)

There is a relatively small part of the PSBF and the Patagonian Shelf Large Marine Ecosystem located beyond the Argentinean Economic Zone, and some species use these areas during migration. For instance, the Argentine shortfin squid is also being caught around the Malvinas Islands and in the High Seas beyond Argentina's jurisdiction (Vilela et al. 2018). Therefore, when the whole area is considered (including the Argentine Exclusive Economic Zone and the adjacent international waters), about 50% of the Argentine shortfin squid fishery comes from the PSBF, which on average accounts for 213 thousand tons per year equivalent to 286 million US dollars.

Seafood has high nutritional values for human consumption. For instance, ~14% of squids are proteins (Eder and Lewis 2005). Martinetto et al. (2020) calculated that, including both Argentine and international waters, the PSBF provides 30,000 tons of proteins per year from squid only, and the protein exported coming only from the PSBF within the Argentine Exclusive Economic Zone amounts to 3700 tons per year.

It has been hypothesized that the large primary production having place at the PSBF can, in addition, support important fisheries in adjacent areas. It has been be illustrated by the Argentine hake (Martinetto et al. 2020), which is captured in areas close to the PSBF where the trophic web is presumably supported by the primary production in the front (Alemany et al. 2014). Altogether, the PSBF is an important support for fisheries providing benefits not only in economic terms but also in terms of providing high-quality food for humans.

9.5 Supportive Biodiversity Here and Beyond

Under the ES point of view, biodiversity is considered as a service itself but it is also important to face current and future global and climate change scenarios (IPBES 2019; IPCC 2022). The genetic pool, which is composed by all the species in the ecosystem, represents an option value (with future potential) whose ecosystem functions and services may be important under certain environmental scenarios. Thus, the chances to adapt to changes increase with biodiversity.

Besides marine fronts have been highlighted for their high consumer abundances (see Chap. 11), its relationship with biodiversity has been less explored (see Acha et al. 2015). However, there are some cases with such robust evidence that have been even used to design marine protected areas. Such is the case with the UK shelf-sea fronts that were assumed as a surrogate measure for mapping the abundance and diversity of pelagic organisms in the design process for a marine protected area (Miller and Christodoulou 2014). The relationship between biodiversity and the PSBF is still under investigation; however, some studies indicate that biodiversity is high in the PSBF. For example, benthic communities (Mauna et al. 2011) as well as cartilaginous fish (Lucifora et al. 2012) are more diverse in the PSBF than in adjacent areas.

The potential of ocean ecosystems in the discovery of novel chemicals with pharmaceutical and cosmetic uses has also been mentioned (Jaspars et al. 2016). As new marine natural products are being discovered, it allows the development of new drugs (Costa Leal et al. 2012). Marine organisms, from microbes, algae and plants to animals, produce compounds with the potential of being useful for the treatment of human diseases. Future research on these "secondary metabolites" (i.e., chemicals that are not needed by the organism for basic or primary metabolic processes) could eventually help to unlock human medical issues.

Associated to the high primary production in the PSBF, there is an increase in the abundance of zooplankton (e.g., Marrari et al. 2004) with a concomitant increase in the habitat use of consumers such as marine birds and mammals (e.g., Bost et al. 2009). Important commercial species such as the Argentine shortfin squid (e.g., Chen et al. 2007), the Argentine hake (e.g., Ruiz and Fondacaro 1997), the Argentine anchovy (*Engraulis anchoita*, e.g., Sánchez and Ciechomski 1995), and the Patagonian scallop (e.g., Soria et al. 2016) (see Chaps. 7 and 11) are also associated to fronts, with a notorious energy transferred through the food web. After blooms of primary production, zooplankton, mainly composed of small crustaceans and larval stages of other organisms, accumulates along the PSBF. These organisms are the pelagic link between phytoplankton and small fishes like anchovies, myctophids, and juveniles of other fish species or squids, which support some of the top predators (e.g., Ciancio et al. 2008). On the other hand, the benthic pathway is supported by part of the primary production of the PSBF that sinks and is used by scallops and other benthic organisms that in turn are supporting the benthic fish species such as rays and sharks. Therefore, the energy produced by the PSBF is transferred to top consumers through a complex food web. Then this energy is redistributed by large fishes, squids, birds, and marine mammals through migratory movements to other areas of the ocean (van Deurs et al. 2016). In this sense, biodiversity from the PSBF could be also supporting the provision of services, such as seafood, captured in adjacent regions connected via food web interactions.

In addition to supporting target species for fisheries, the PSBF is also a feeding and migratory ground for some of the so called charismatic species (see Chaps. 8 and 10). This is the case of marine birds and mammals, like albatrosses and petrels (Quintana and Dell'Arciprete 2002) and the right whales (Zerbini et al. 2016), which use the PSBF seasonally associated with migratory and reproductive behaviors. Some others, like elephant seals, use the area throughout the year (Falabella et al. 2009). A large part of the elephant seal population breeding at Peninsula Valdés, especially subadult and adult males (Campagna et al. 2020), travels more than 400 km to feed on fish and squids along the shelf-break and the adjacent deep ocean (Aubone et al. 2021). These species constitute the main ecotourism attraction of the Argentine Patagonian coast. Under the ES approach, biodiversity of charismatic species supports cultural services related to ecotourism and recreation, that in somehow are supported by the high productivity of the PSBF.

9.6 Perspectives Facing the Near Future

The hypothesis that the PSBF is the productive "backbone" of the Argentine Sea has been supported so far by the many investigations compiled in this book. However, there are several gaps in key information and a general integrative approach is still pending. We identified key research and tools whose development is paramount to forecasting the provision of ecosystem services under an increasing demand scenario. In particular, the development of a model that allows estimating CO_2 fluxes based on satellite chlorophyll would be extremely useful under the present and future climate change scenarios. Such a model requires increasing the CO_2 fluxes data both in area and time to cover different combinations of environmental variables (e.g., temperature, solar radiation, and nutrient concentrations) that determine phytoplankton blooms. In this sense, it is also important to investigate if there is any relation between the carbon stocks present in the sea floor and the presence of the PSBF. Given that some benthic organisms are coupled to the presence of the PSBF (e.g., scallops are more abundant in the PSBF than in the adjacent continental shelf area), it is possible to hypothesize that C stocks could be also larger in the area influenced by the PSBF.

The development of such models based on satellite chlorophyll fluctuations and other satellite information such as SST, SSH, and SSS (sea surface temperature, height and salinity) could be also useful to estimate or forecast scenarios of provision of other services. For instance, there are several examples in fisheries using satellite information validated with actual fish stocks information (Belkin 2021), particularly for fish species in low steps of the food web. Moreover, it is more often acknowledged that the inclusion of frontal systems in this kind of model increases the precision of the estimations (Belkin 2021; Ferrari 2011; Hewitt et al. 2022). In addition, satellite chlorophyll could be coupled to specific biodiversity studies integrating the information from some taxonomic groups with more general information applying genetic techniques as for example environmental DNA. This kind of research, in addition with trophic web studies, could help to elucidate the role of the PSBF for biodiversity.

Under an accelerated human population growth (doubling over the last 50 years), the world faces two major environmental problems that threaten the provision of ES: species extinction and climate change (Tilman 2022). This situation looks somehow circular: the growing human affluence implies an increasing demand of goods and services which in turn increase pressure on the ecosystems. In fact, the last IPCC report highlights the interconnexion among climate change, nature, and human societies: climate change impacts on human societies and nature, human societies contribute to increase climate change by releasing greenhouse gasses and limit the capacity of nature to adapt to climate change by inducing ecosystem losses and damages; in turns, ecosystems adapt to and mitigates climate change and provide human societies with livelihoods and ES. Human societies also have the capacity to adapt and mitigate climate change, by using specific infrastructure and technology and modifying consumption habits, but also through conservation, restoration, and sustainable management of natural ecosystems. This last is known as nature-based solutions (NbS) and has the advantage that it improves the provision of several ecosystem services besides the ones related with climate change adaptation and mitigation.

Given that the ocean covers 71% of the Earth, marine ecosystems have a large potential for the development of NbS. In fact, marine stakeholders have organized a bottom-up movement, which has ultimately led to the United Nations General Assembly proclaiming a Decade of Ocean Science for Sustainable Development (2021–2030; Visbeck 2018). Under this framework, advances in marine ecological research with a more mechanistic and predictive direction will be essential. The research and tools we mentioned as needed above in this final section are in this line.

Acknowledgements This work was supported by the following grants: IAI-CONICET 3347/14 and CONICET PIP3149. This is INIDEP contribution N° 2370.

References

Acha EM, Mianzan HW, Guerrero RA, Favero M, Bava J (2004) Marine fronts at the continental shelves of austral South America: physical and ecological processes. J Mar Syst 44:83–105

Acha EM, Piola A, IribarneO MH (2015) Ecological processes at marine fronts, oases in the ocean. Springer Brief in Environmental Science

Alemany D, Acha EM, Iribarne O (2014) Marine fronts are important fishing areas for demersal species at the Argentine sea (Southwest Atlantic Ocean). J Sea Res 87:56–67

Armstrong CW, Foley NS, Tinch R, van den Hove S (2012) Services from the deep: steps towards valuation of deep sea goods and services. Ecosyst Serv 2:2–13

Aubone N, Saraceno M, Torres Alberto ML, Campagna J, Le Ster L, Picard B, Hindell M, Campagna C, Guinet CR (2021) Physical changes recorded by a deep diving seal on the Patagonian slope drive large ecological changes. J Mar Syst 223:103612. https://doi.org/10.1016/j.jmarsys.2021.103612

Bakker DCE, Pfeil B, Landa CS et al (2016) A multi-decade record of high-quality fCO2 data in version 3 of the Surface Ocean CO2 atlas (SOCAT). Earth Syst Sci Data 8:383–413. https://doi.org/10.5194/essd-8-383-2016

Bates NR, Astor YM, Church MJ, Currie K, Dore JE, González-Dávila M, Lorenzoni L, Muller-Karger F, Olafsson J, Santana-Casiano JM (2014) A time-series view of changing ocean chemistry due to ocean uptake of anthropogenic CO2 and ocean acidification. Oceanography 27:126–141

Bauer JE, Cai WJ, Raymond PA, Bianchi TS, Hopkinson CS, Regnier PA (2013) The changing carbon cycle of the coastal ocean. Nature 504:61–70. https://doi.org/10.1038/nature12857

Belkin IM (2021) Remote sensing of ocean fronts in marine ecology and fisheries. Remote Sens 13:883. https://doi.org/10.3390/rs13050883

Benoit-Bid KJ, NcManus MA (2012) Bottom-up regulation of a pelagic community through spatial aggregations. Biol Lett 8:813–816

Bianchi AA, Pino DR, Perlender HGI, Osiroff AP, Segura V, Lutz V, Clara ML, Balestrini CF, Piola AR (2009) Annual balance and seasonal variability of sea-air CO2fluxes in the Patagonia Sea: their relationship with fronts and chlorophyll distribution. J Geophys Res: Oceans 114:1–11. https://doi.org/10.1029/2008JC004854

Bogazzi E, Baldoni A, Rivas A, Martos P, Reta R, Orensanz JM, Lasta M, Dell'Arciprete P, Werner F (2005) Spatial correspondence between areas of concentration of Patagonian scallop

(Zygochlamys patagonica) and frontal systems in the southwestern Atlantic. Fish Oceanogr 14:359–376. https://doi.org/10.1111/j.1365-2419.2005.00340.x

Borges AV, Delille B, Frankignoulle M (2005) Budgeting sinks and sources of CO2 in the coastal ocean: diversity of ecosystems counts. Geophys Res Lett 32:L14601. https://doi.org/10.1029/2005GL023053

Bost CA, Cotté C, Bailleul, F, Cherel Y, Charrassin JB, Guinet C, Ainley DG, Weimerskirch H (2009) The importance of oceanographic fronts to marine birds and mammals of the southern oceans. J Mar Syst 78:363–376

Cáceres DM, Tapella E, Quétier F, Díaz S (2015) The social value of biodiversity and ecosystem services from the perspectives of different social actors. Ecol Soc 20:62

Cai WJ (2011) Estuarine and Coastal Ocean carbon paradox: CO2 sinks or sites of terrestrial carbon incineration? Annu Rev Mar Sci 3:123–145. https://doi.org/10.1146/annurev-marine-120709-142723

Cai WJ, Dai M, Wang Y (2006) Air-sea exchange of carbon dioxide in ocean margins: a province-based synthesis. Geophys Res Lett 33:1–4. https://doi.org/10.1029/2006GL026219

Campagna J, Lewis MN, González Carman V, Campagna C, Guinet C, Johnson M, Davis RW, Rodriguez DH, Hindell MA (2020) Ontogenetic niche partitioning in southern elephant seals from Argentine Patagonia. Mar Mam Sci 37:631–651

Carranza M (2009) Indicadores del estado del ambiente marino patagónico en áreas frontales. Tesis de Licenciatura en Oceanografía, Universidad de Buenos Aires, p 115

Carreto JI, Montoya NG, Carignan MO, Akselman R, Acha EM, Derisio C (2016) Environmental and biological factors controlling the spring phytoplankton bloom at the Patagonian shelf-break front – degraded fucoxanthin pigments and the importance of microzooplankton grazing. Progr Oceanogr Elsevier 146:1–21. https://doi.org/10.1016/j.pocean.2016.05.002

Chassot E, Bonhommeau S, Dulvy NK, Mélin F, Watson R, Gascuel D, Le Pape O (2010) Global marine primary production constrains fisheries catches. Ecol Lett 13:495–505

Chen C-TA, Borges AV (2009) Reconciling opposing views on carbon cycling in the coastal ocean: continental shelves as sinks and near-shore ecosystems as sources of atmospheric CO2. Deep-Sea Res II Top Stud Oceanogr. https://doi.org/10.1016/j.dsr2.2009.01.001

Chen CS, Chiu TS, Haung WB (2007) The spatial and temporal distribution patterns of the Argentine short-finned squid, Illex argentinus, abundances in the Southwest Atlantic and the effects of environmental influences. Zool Stud 46:111–122

Chen C-TA, Huang T-H, Chen Y-C, Bai Y, He X, Kang Y (2013) Air-sea exchanges of coin the world's coastal seas. Biogeosciences 10:6509–6544. https://doi.org/10.5194/bg-10-6509-2013

Ciancio JE, Pascual MA, Botto F, Frere E, Iribarne O (2008) Trophic relationships of exotic anadromous salmonids in the southern Patagonian Shelf as inferred from stable isotopes. Limnol Oceanogr 53:788–798

Cognetti G, Maltagliati F (2010) Ecosystem service provision: an operational way for marine biodiversity conservation and management. Mar Pollut Bull 60:1916–1923

Costa Leal M, Puga J, Serôdio J, Gomes NCM, Calado R (2012) Trends in the discovery of new marine natural products from invertebrates over the last two decades – where and what are we bioprospecting? PLoS One 7:e30580

Dai M, Cao Z, Guo X, Zhai W, Liu Z, Yin Z, Xu Y, Gan J, Hu J, Du C (2013) Why are some marginal seas sources of atmospheric CO2? Geophys Res Lett 40:2154–2158. https://doi.org/10.1002/grl.50390

Díaz S, Settele J, Brondízio ES, Ngo HT, Guèze M, Agard J, A. Arneth, Balvanera P, Brauman KA, Butchart SHM, Chan KMA, Garibaldi LA, Ichii K, Liu J, Subramanian SM, Midgley GF, Miloslavich P, Molnár Z, Obura D, Pfaff A, Polasky S, Purvis A, Razzaque J, Reyers B, Roy Chowdhury R, Shin YJ, Visseren-Hamakers IJ, Willis KJ, Zayas CN (eds.). 2019. Summary for policymakers of the global assessment report on biodiversity and ecosystem services of the Intergovernmental Science-Policy Platform on Biodiversity and Ecosystem Services. IPBES secretariat, Bonn, Germany. 56 pages,

Dickson AG, Sabine CL, Christian JR (2007) Guide to best practices for ocean CO2 measurements. PICES Special Publication 3:191. https://doi.org/10.1159/000331784

Druon JN (2010) Habitat zapping of the Atlantic blue fin tuna derived from satellite data: its potential as a tool for the sustainable management of pelagic fisheries. Mar Policy 34:293–297

Eder EB, Lewis MN (2005) Proximate composition and energetic value of demersal and pelagic prey species from the SW Atlantic Ocean. Mar Ecol Prog Ser 291:43–52

Falabella V, Campagna C, Croxall J (2009) Atlas of the Patagonian Sea: species and spaces. Wildlife Conservation Society and BirdLife International, Argentina

Falkowski PG, Barber RT, Smetacek V (1998) Biogeochemical controls and feedbacks on ocean primary production. Science 281:200–206

FAO (2022) The State of World Fisheries and Aquaculture 2022. Towards Blue Transformation. Rome, FAO. https://doi.org/10.4060/cc0461en

Ferrari R (2011) A frontal challenger for climate models. Science 332:316–317

Fisher B, Turner RK, Morling P (2009) Defining and classifying ecosystem services for decision making. Ecol Econ 68:643–653

Friedlingstein P, Jones MW, O'Sullivan M et al (2022) Global carbon budget 2021. Earth Syst Sci Data 14:1917–2005. https://doi.org/10.5194/essd-14-1917-2022

Gattuso JP, Frankignoulle M, Bourge I, Romaine S, Buddemeier RW (1998) Effect of calcium carbonate saturation of seawater on coral calcification. Glob Planet Change 18:37–46. https://doi.org/10.1016/S0921-8181(98)00035-6

Gregor L, Monteiro PMS (2013) Is the southern Benguela a significant regional sink of CO2? S Afr J Sci 109. https://doi.org/10.1590/sajs.2013/20120094

Hartin CA, Bond-Lamberty B, Patel P, Mundra A (2016) Ocean acidification over the next three centuries using a simple global climate carbon-cycle model: projections and sensitivities. Biogeosciences 13:4329–4342. https://doi.org/10.5194/bg-13-4329-2016

Hewitt H, Fox-Kemper B, Pearson B, Roberts M, Klocke D (2022) The small scales of the ocean may hold the key to surprises. Nat Clim Chang 12:496–499. https://doi.org/10.1038/s41558-022-01386-6

INDEC (2024) Instituto Nacional de Estadísticas y Censo, Argentina. https://www.indec.gob.ar/. Accessed 9 Sep 2024

IPBES (2019) Global assessment report on biodiversity and ecosystem services of the Intergovernmental Science-Policy Platform on Biodiversity and Ecosystem Services. In: Brondizio ES, Settele J, Díaz S, Ngo HT (eds) IPBES secretariat, Bonn. https://ipbes.net/global-assessment

IPCC (2022) Climate change 2022: impacts, adaptation, and vulnerability. In: Pörtner H-O, Roberts DC, Tignor M, Poloczanska ES, Mintenbeck K, Alegría A, Craig M, Langsdorf S, Löschke S, Möller V, Okem A, Rama B (eds) Contribution of working group II to the sixth assessment report of the intergovernmental panel on climate change. Cambridge University Press, New York, p 3056. https://doi.org/10.1017/9781009325844

Jaspars M, de Pascalle D, Andersen JH, Reyes F, Crawford AD, Ianora A (2016) The marine biodiscovery pipeline and ocean medicines of tomorrow. J Mar Biol Assoc UK 96:151–158

Kahl LC (2018) Dinámica del CO2 en el Océano Atlántico Sudoccidental. Tesis Doctoral en Meteorología, Universidad de Buenos Aires. Facultad de Ciencias Exactas y Naturales. http://hdl.handle.net/20.500.12110/tesis_n6525_Kahl

Kahl LC, Bianchi AA, Osiroff AP (2017) Distribution of sea-air CO2 fluxes in the Patagonian Sea: seasonal, biological and thermal effects. Continental Shelf Res 143:18–28. https://doi.org/10.1016/j.csr.2017.05.011

Körtzinger A, Rhein M, Mintrop L (1999) Anthropogenic CO2 and CFCs in the North Atlantic Ocean – a comparison of man-made tracers. Geophys Res Lett 26:2065–2068

Kummu M, de MoelH SG, Viviroli D, Ward PJ, Varis O (2016) Over the hills and further away from coast: global geospatial patterns of human and environment over the 20th–21st centuries. Environ Res Lett 1:034010. https://doi.org/10.1088/1748-9326/11/3/034010

Laruelle GG, Lauerwald R, Pfeil B, Regnier P (2014) Regionalized global budget of the CO2 exchange at the air-water interface in continental shelf seas. Global Biogeochem Cycles 28:1199–1214. https://doi.org/10.1002/2014GB004832

Laruelle GG, Cai W-J, Hu X, Gruber N, Mackenzie FT, Regnier P (2018) Continental shelves as a variable but increasing global sink for atmospheric carbon dioxide. Nat Commun 9:454. https://doi.org/10.1038/s41467-017-02738-z

Lucifora L, García V, Menni R, Worm B (2012) Spatial patterns in the diversity of sharks, rays, and chimaeras (Chondrichthyes) in the Southwest Atlantic. Biodivers Conserv 21:407–419

Lutz VA, Segura V, Dogliotti AI, Gagliardini DA, Bianchi AA, Balestrini CF (2010) Primary production in the Argentine Sea during spring estimated by field and satellite models. J Plankton Res 32(2):181–195. https://doi.org/10.1093/plankt/fbp117

Mann KH, Lazier JRN (2006) Dynamics of marine ecosystems. Biological-physical interactions in the oceans. Blackwell Publishing Ltd, Malden, MA

Marrari M, Viñas MD, Martos P, Hernández D (2004) Spatial patterns of mesozooplankton distribution in the Southwestern Atlantic Ocean (34°–41°S) during austral spring: Relationship with the hydrographic conditions. ICES JMar Sci 61:667–679

Martinetto P, Alemany D, Botto F, Mastrángelo M, Falabella V, Acha EM, Antón G, Bianchi A, Campagna C, Cañete G, Filippo P, Iribarne O, Laterra P, Martínez P, Negri R, Piola AR, Romero SI, Santos D, Saraceno M (2020) Linking the scientific knowledge on marine frontal systems with ecosystem services. Ambio 49:541–556. https://doi.org/10.1007/s13280-019-01222-w

Martínez PA, Wöhler OC (2016) Hacia la recuperación de la pesquería de la merluza negra (Dissostichus eleginoides) en el Mar Argentino: un ejemplo de trabajo conjunto entre el sector de la administración, la investigación y la industria. Frente Marítimo 24:115–124

Mauna AC, Acha EM, Lasta ML, Iribarne OO (2011) The influence of a large SW Atlantic shelf-break frontal system on epibenthic community composition, trophic guilds and diversity. J Sea Res 66:39–46

Millennium Ecosystem Assessment (MEA) (2005) Ecosystems and human well-being: synthesis. Island Press, Washington

Miller PI, Christodoulou S (2014) Frequent locations of oceanic fronts as an indicator of pelagic diversity: application to marine protected areas and renewables. Mar Policy 45:318–329

Orselli IBM, Kerr R, Ito RG, Tavano VM, Mendes CRB, Garcia CAE (2018) How fast is the Patagonian shelf-break acidifying? J Mar Syst 178:1–14

Padin XA, Vázquez-Rodríguez M, Castaño M, Velo A, Alonso-Pérez F, Gago J, Gilcoto M, Álvarez M, Pardo PC, de la Paz M, Ríos AF, Pérez FF (2010) Air-Sea CO2 fluxes in the Atlantic as measured during boreal spring and autumn. Biogeosciences 7:1587–1606. https://doi.org/10.5194/bg-7-1587-2010

Pittman SJ, Yates KL, Bouchet PJ, Alvarez-Berastegui D et al (2021) Seascape ecology: identifying research priorities for an emerging ocean sustainability science. Mar Ecol Prog Ser 663:1–29. https://doi.org/10.3354/meps13661

Podestá GP, Browder JA, Hoey JJ (1993) Exploring the relationship between swordfish catch rates and thermal fronts on U.S. longline grounds in the western North Atlantic. Cont Shelf Res 13:253–277

Quintana F, Dell'Arciprete OP (2002) Foraging grounds of southern giant petrels (Macronectes giganteus) on the Patagonian shelf. Polar Biol 25:159–161

Romero SI, Piola AR, Charo M, Garcia CAE (2006) Chlorophyll-a variability off Patagonia based on SeaWiFS data. J Geophys Res Oceans 111: 1–11. doi: https://doi.org/10.1029/2005JC003244

Ruiz AE, Fondacaro RR (1997) Diet of hake (Merluccius hubbsi Marini) in a spawning and nursery area within Patagonian shelf waters. Fish Res 30:157–160

Sagebiel J, Schwartz C, Rhozyel M, Rajmis S, Hirschfeld J (2016) Economic valuation of Baltic marine ecosystem services: blind spot and limited consistency. ICES J Mar Sci 73:991–1003

Sánchez RP, Ciechomski J (1995) Spawning and nursery grounds of pelagic fish species in the sea-shelf off Argentina and adjacent areas. Sci Mar 59:455–478

Saraceno M, Provost C, Piola AR (2005) On the relationship between satellite-retrieved surface temperature fronts and chlorophyll a in the western South Atlantic. J Geophys Res Oceans 110:1–16. https://doi.org/10.1029/2004JC002736
Soria G, Orensanz JM, Morsán EM, Parma AM, Amoroso RO (2016) Chapter 25—Scallops biology, fisheries, and management in Argentina. In Developments in aquaculture and fisheries science, ed. S.E. Shumway and G.J. Parsons, 1019–1046. Amsterdam: Elsevier
Stocker TF (2015) The silent services of the world ocean. Science 350:764–765
Subsecretaría de Pesca y Acuicultura, Informe de Coyuntura (2021). https://www.magyp.gob.ar/sitio/areas/pesca_maritima/informes/coyuntura/_archivos//210000_2021/211201_Informe%20de%20Coyuntura%20-%20Diciembre%202021.pdf
Thurber AR, Sweetman AK, Narayanaswamy BE, Jones DOB, Ingels J, Hansman RL (2014) Ecosystem function and services provided by the deep sea. Biogeosciences 11:3941–3963
Tilman D (2022) Extinction, climate change and the ecology of Homo sapiens. J Ecol 110:744–750. https://doi.org/10.1111/1365-2745.13847
Townsend M, Davies K, Hanley N, Hewitt JE, Lundquist CJ, Lohrer AM (2018) The challenge of implementing the marine ecosystem service concept. Front Mar Sci 5:359. https://doi.org/10.3389/fmars.2018.00359
van Deurs M, Persson A, Lindegren M, Jacobsen C, Neuenfeldt S, Jørgensen C, Nilsson PA (2016) Marine ecosystem connectivity mediated by migrant–resident interactions and the concomitant cross-system flux of lipids. Ecol Evol 6:4076–4087
Vilela R, Conesa D, del Rio JL, López-Quílez A, Portela J, Bellido JM (2018) Integrating fishing spatial patterns and strategies to improve high seas fisheries management. Mar Policy 94:132–142
Visbeck M (2018) Ocean science research is key for a sustainable future. Nat Commun 9:690. https://doi.org/10.1038/s41467-018-03158-3
Wanninkhof R (2014) Relationship between wind speed and gas exchange over the ocean revisited. Limnol Oceanogr Methods 12:351–362. https://doi.org/10.4319/lom.2014.12.351
Weiss RF (1974) Carbon dioxide in water and seawater: the solubility of a non-ideal gas. Mar Chem 2:203–215. https://doi.org/10.1016/0304-4203(74)90015-2
Woodson CB, Litvin SY (2015) Ocean fronts drive marine fishery production and biogeochemical cycling. PNAS 112:1710–1715
Worm B, Sandow M, Oschlies A, Lotze HK, Myers RA (2005) Global patterns of predator diversity in the open oceans. Science 309:1365–1369
Zerbini AN, Rosenbaum H, Mendez M, Sucunza F, Andriolo A, Harris G, Clapham PJ, Sironi M, Uhart M, Ajó AF (2016) Tracking southern right whales through the Southwest Atlantic: an update on movements, migratory routes and feeding destinations. In Paper SC/66b/BRG/26 presented to the International Whale Commission Scientific Committee, June 2016, Bled, Slovenia, p 16

Chapter 10
Species-Dependent Conservation in a SW Atlantic Ecosystem

Claudio Campagna, Valeria Falabella, Pablo Filippo, and Daniela Alemany

Abstract We discuss species conservation in the temperate Extended Shelf System (ESS) of the SW Atlantic, an area the size of the Mediterranean Sea. The ESS is distinctive due to its shallow continental plateau and the predictable Patagonian Shelf-Break Front (PSBF), generated by the interaction of the Brazil and Malvinas currents with the bathymetry of the shelf and slope. The PSBF is a main driving force that sustains species diversity and abundance, including charismatic species of penguins, albatrosses, seals, and whales. From a conservation standpoint, the ESS is vulnerable. An expanding international fishery operates on the shelf and the slope. From 113 regionally red-listed vertebrates, 28% were threatened; sharks and rays being the most affected. Argentina's coastal MPAs protect mainly breeding colonies of birds and mammals. Three oceanic National Parks protect a submerged plateau and its surrounding southern slope (Namuncurá-Burdwood Bank) and some benthic-pelagic habitats in the Drake Passage. Yet, despite protected places and available management tools, threatened species remain exposed to risks of local extinction. The ESS is here envisioned as a conservation unit, where interventions follow

Supplementary Information The online version contains supplementary material available at https://doi.org/10.1007/978-3-031-71190-9_10.

C. Campagna (✉)
Coastal and Marine Conservation, Wildlife Conservation Society, Buenos Aires, Argentina
e-mail: ccdeviaje@gmail.com

V. Falabella
Coastal and Marine Conservation, Wildlife Conservation Society, Buenos Aires, Argentina

P. Filippo
Coastal and Marine Conservation Consultant, Wildlife Conservation Society, Buenos Aires, Argentina

D. Alemany
Instituto de Investigaciones Marinas y Costeras (IIMyC, UNMdP-CONICET), Facultad de Ciencias Exactas y Naturales, Universidad Nacional de Mar del Plata (UNMdP), Consejo Nacional de Investigaciones Científicas y Técnicas (CONICET), Mar del Plata, Argentina

Instituto Nacional de Investigación y Desarrollo Pesquero (INIDEP), Mar del Plata, Argentina

E. M. Acha et al. (eds.), *The Patagonian Shelfbreak Front*, Aquatic Ecology Series 13, https://doi.org/10.1007/978-3-031-71190-9_10

oceanographic function. Conservation effectiveness may be enhanced if focusing on the most endangered groups. Adaptive, dynamic, and ecosystemic management could be combined with seasonal and movable protected sites. Human-forced global warming, but also the sovereignty conflict over the Malvinas Archipelago, pose considerable challenges to this vision.

Acronyms and Abbreviations

AIS	Automatic Identification System
BBNJ	UN Biodiversity Beyond National Jurisdiction Treaty
CDB	Convention on Biological Diversity
COP	Conference of parties
CR	Critically Endangered
DD	Data Deficient
EEZ	Exclusive Economic Zone
EN	Endangered
ESS	Extended Shelf System of the SW Atlantic
FAO	Food and Agriculture Organization (UN)
IBAs	Important Bird Areas
IUCN	International Union for Conservation of Nature
KBAs	Key Biodiversity Areas
LC	Least Concern
LME	Large Marine Ecosystems
MPA	Marine Protected Area
NE	Not Evaluated
NT	Near Threatened
PCA	Permanent Closed Area
PCZMP	Patagonian Coastal Zone Management Plan
PSBF	Patagonian Shelf Break Front
SW	South West
UNCLOS	United Nations Convention on the Law of the Seas
UN	United Nations
UNEP	United Nations Environmental Program
VU	Vulnerable
WCMC	World Conservation Monitoring Centre (UNEP)

10.1 Purpose and Framework

We develop a species-centered conservation vision for a region of the SW Atlantic that we refer to as the Extended Shelf System (ESS; Fig. 10.1a). The ESS encompasses one of the largest, shallowest and most productive of the southern

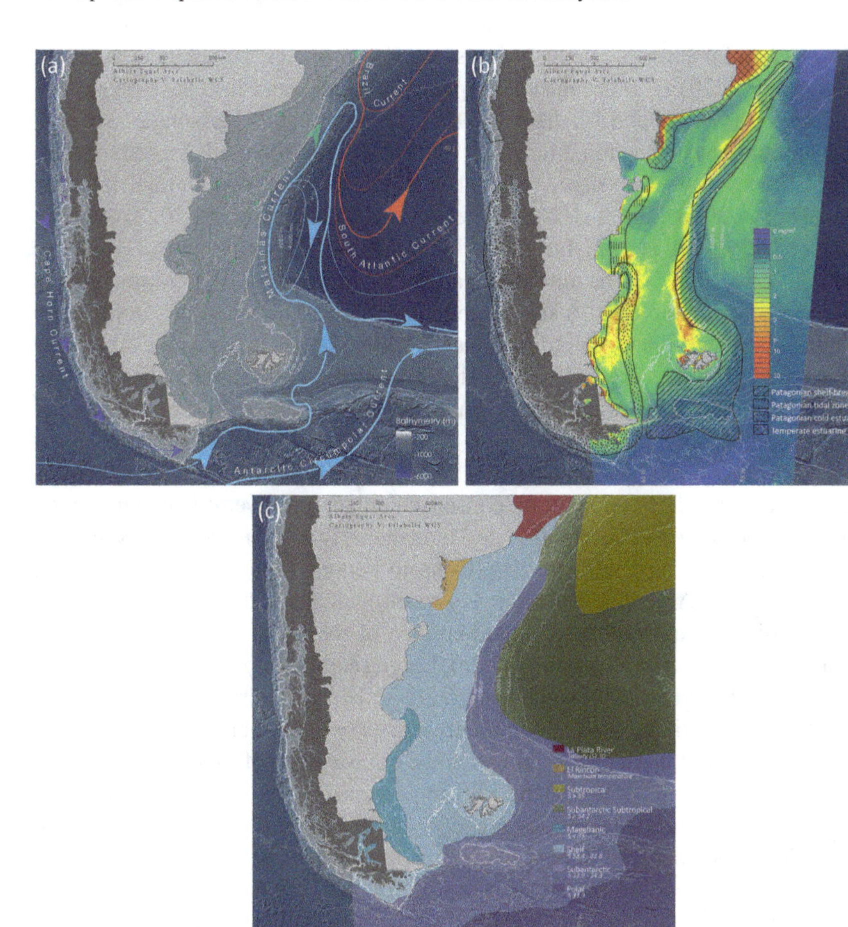

Fig. 10.1 (**a**) Large-scale habitats and currents of the ESS: continental shelf (up to 200 m depth), shelf-break (200–4000 m depth) and a deep basin (> 4000 m depth). The Malvinas (blue) and Brazil (red) currents flow along the shelf-break and part of the shelf. (b) Frontal areas defined by the average satellite chlorophyll *a* concentration for austral summer (January 1998–2009; values in mg/m^3, frontal areas). (Adapted from Acha et al. 2004). (**c**) Oceanographic regimes. (Adapted from Piola 2008)

hemisphere continental shelves. The shelf merges into the continental slope, and the latter into the deep Argentine Basin (Fig. 10.1a). These distinct environments are functionally linked to the circulation of the Malvinas and Brazil Currents (MC, BC; see Chap. 2). The interaction between currents and bathymetry results in predictable and productive frontal areas (Acha et al. 2004). Most important of them is the Patagonian Shelf-break Front (PSBF), that extends for approximately 1500 kilometers along the slope (Fig. 10.1b). This front, that also affects a large part of the shelf, is a main driver of the productivity and species richness of the ESS.

An ensemble of resident and visitor pelagic, top predators forage and migrate through the ESS. This includes several species of marine birds (penguins, petrels, albatrosses) and mammals (elephant seals, sea lions, whales; Falabella et al. 2009; Campagna et al. 2014). More than 100 species of sharks and rays have been cited for the region (Cuevas and Michelson 2023). These are all "charismatic" species, critical for conservation action. Many, as it will be here first reported, are endangered.

The EES also sustains large, national and international, commercial fisheries (Alemany et al. 2016; Sala et al. 2018; Vilela et al. 2018; Giussi et al. 2022). A broad regulatory framework is available for national fisheries operating on the Exclusive Economic Zone (EEZ). The international fleets, however, are operating mostly beyond the limits of the EEZ. A sovereignty dispute over the Malvinas Archipelago precludes integrated management of transboundary species, such as the Argentine shortfin squid, *Illex argentinus* (Chen et al. 2007). Ecosystemic approaches are considered fundamental, yet most management hinges on single-species quotas and spatial-temporal tools.

The ESS lacks a strategic spatial plan (September 2023). No biodiversity conservation scenario has been adopted by consensus either, yet several National Action Plans exist, focusing on conservation and management of chondrichthyans, marine mammals, seabirds, and marine turtles (Consejo Federal Pesquero, https://cfp.gob.ar/publicaciones/). Argentina has relatively recently declared three pelagic national parks, significantly expanding the small surface of the ocean under the numerous coastal-marine protected areas (Fig. 10.2a). Large sectors were also set aside as no-take zones for the operation of some fisheries (Fig. 10.2a). Overall, protected spaces are not functionally integrated, habitat representation is poor, and there are gaps in the representation of biodiversity groups in the conservation efforts.

No doubt, the conservation of the ESS requires an urgent vision for threatened species. A conservation strategy for the ESS could benefit from management and conservation tools that are coupled to the functional ecology of the system. Thus, we will first overview the biophysical aspects of the ESS (other chapters in this book offer more detailed information) and address its conservation status. We will then discuss scenarios integrating oceanographic function, species requirements, use, threats, and the policy context. Our main purpose is to advance discussion on species-driven conservation approaches, particularly for those species found threatened.

10.2 Conceptual Framework

10.2.1 Protecting Spaces May Not Protect Species

Site conservation visions and tools, such as marine protected areas (MPAs), dominate the present ocean conservation agenda, yet MPAs may not necessarily address pressing needs for biodiversity conservation unless they are specifically designed for that purpose. Chondrichthyans, for example, are the most globally threatened of all vertebrates (Klein et al. 2015; Davidson and Dulvy 2017). This generalization

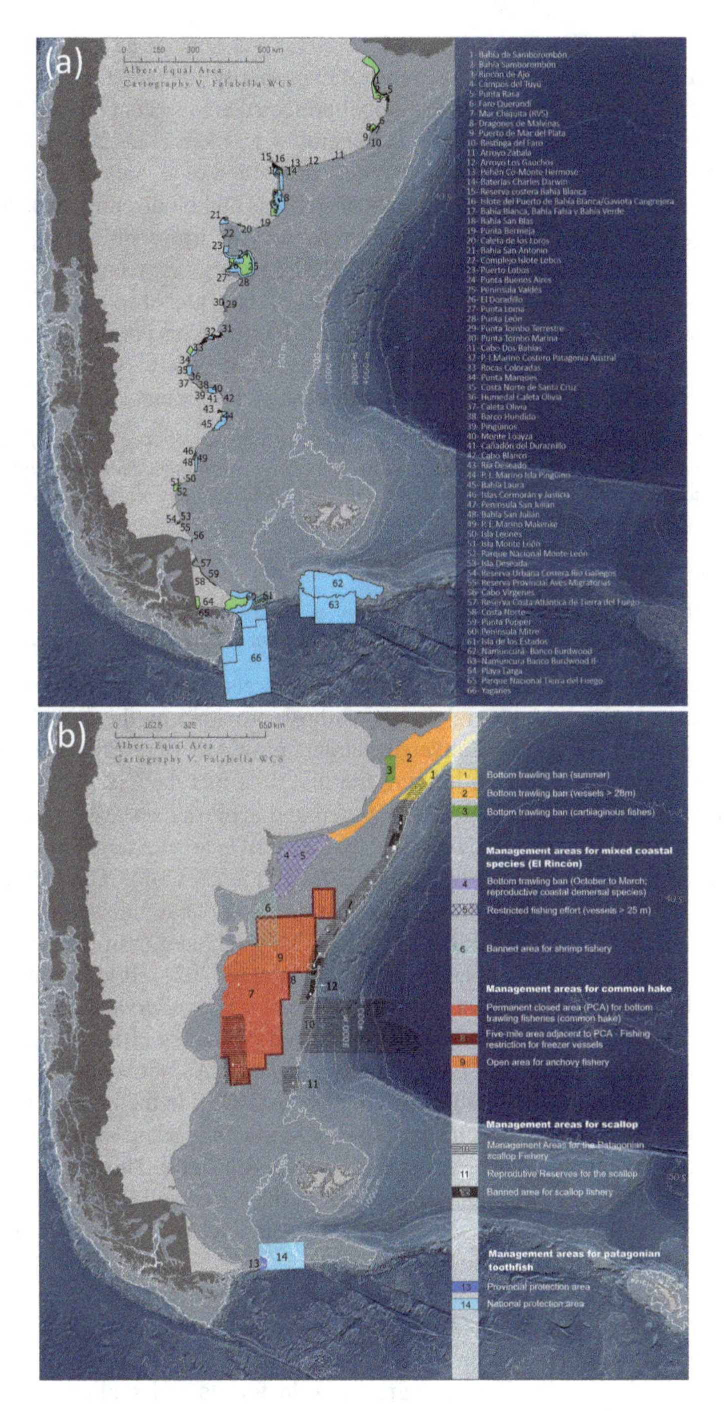

Fig. 10.2 (**a**) Coastal-marine protected areas and pelagic National Parks. (**b**) Important fisheries management areas

applies to the ESS (Cuevas and Michelson 2023), yet none of the existing coastal-marine MPAs, or pelagic parks, were created primarily for safeguarding sharks or rays. Indeed, experts argued that the global protected area system has not been primarily designed to preserve species in general (e.g., Eken et al. 2004). MPAs are justified on many fronts, besides biodiversity protection (e.g., Sala et al. 2021).

As cumulative anthropogenic impacts on habitats and biodiversity expand faster than mitigation efforts, some habitat loss risks may be irreversible (Halpern et al. 2015; Visconti et al. 2019, Jefferson et al. 2021). Facing this scenario, MPAs may be effective to protect both, habitats and species. In fact, global conservation targets are expressed in terms of surface area. In 2010, the Convention on Biological Diversity (CBD) adopted the Strategic Plan for Biodiversity 2011–2020 and the Aichi targets. Two of these targets relate to ocean conservation. Target 11 states:

> *By 2020, at least... 10% of coastal and marine areas, especially areas of particular importance for biodiversity and ecosystem services, are conserved through effectively and equitably managed, ecologically representative and well-connected systems of protected areas and other effective area-based conservation measures...*

Target 12 reads:

> *By 2020 the extinction of known threatened species has been prevented and their conservation status, particularly of those most in decline, has been improved and sustained.*

While Target 11 proactively sets a specific goal, Target 12 requires to be contrasted against conservation status trends, such as those provided by the IUCN Red List indexes. None of the Aichi biodiversity targets proposed were fully met by 2020 (Vaughan 2021), yet, although it is not the first time that the CBD sets unaccomplished objectives, the gap between expected and achieved successes did not discourage experts to recommend, for 2030, that the CBD adopts a target of 30% of the ocean surface under protection (Alberts 2020; CBD 2023; CBD/COP/15/L25 n.d.).

At the present, about 8% of the ocean surface is under some spatial protection category (UNEP-WCMC and IUCN 2023). This includes multiple-use MPAs, where some fishing activity is allowed (Jefferson et al. 2021). It is beyond dispute that no-take MPAs are the most effective of the spatial conservation tools (Lester and Halpern 2008; Lester et al. 2009; Costello and Ballantine 2015; Sciberras et al. 2015; Sala and Giakoumi 2017), yet, if only no-take MPAs were considered, global protection amounts to a negligible 2.7% (MPAtlas 2023). In the attempt to expand these meager results, it has been now proposed to include as protected areas those designated for commercial fisheries management (FAO 2022). This would represent a step away from the goals expressly defined in the Strategic Plan for Biodiversity 2011–2022, and the Kunming-Montreal Global Biodiversity Framework (Laffoley et al. 2017; IUCN WCPA 2019).

Setting a percent surface area as a strategic goal may be arbitrary, yet there is evidence that protecting ca. one third of the marine environment could cover a significant proportion of the marine species diversity (68%), and may also protect fragile biomes such as seagrasses, mangroves, kelp forests and shallow-water coral reefs (Zhao et al. 2020). The challenge then is to set aside areas that are relevant for conserving species and habitats (Jenkins and Van Houtan 2016). This obvious condition may not be fulfilled if protected sites continue to follow a "low-hanging fruit" approach, focusing on sites that are just less relevant to commercial fisheries (Singleton and Roberts 2014; Devillers et al. 2015; Jefferson et al. 2021).

Ocean conservation also requires building stronger bridges between site protection and species needs. An assessment of the overlap between protected areas and priority areas of use for marine species reports that, on average, only 3.6% of the latter's distribution ranges are protected (Jenkins and Van Houtan 2016). The same study shows that "*species of conservation concern (threatened, small-ranged, and data deficient) have less protection than species on average*." Data for the Southern Ocean, for example, suggest a poor overlap between protected spaces and critical foraging areas for top predators (Hindell et al. 2020). Based on a large database of tracked marine birds and mammals foraging in the ocean south of 40 °S, these authors identified Areas of Ecological Significance for top predators and concluded that only one third of them are included in the 7.1% of the ocean under some kind of formal protection. Global MPAs cover less than 10% of the distribution ranges of a large sample of marine species (fishes, mammals, invertebrates; Klein et al. 2015). Only a negligible proportion of priority areas for biodiversity are within marine reserves (Gownaris et al. 2019). Thus, overall, the global protected area system may not reflect species conservation priorities (Eken et al. 2004; Jefferson et al. 2021).

10.2.2 Selecting Spaces Prioritizing Species

Conceptual frameworks and methodologies are in place for the building of stronger bridges between spatial and species-driven conservation. More specifically, sites may be identified that are of differential relevance for strategically targeted species. Important Bird Areas (IBAs), for example, was an early development by the NGO Birdlife International (Osieck et al. 1981) extended to other taxa (Hoyt an di Sciara 2021). It also inspired the notion of Key Biodiversity Areas (KBAs; Eken et al. 2004): "*globally important sites that are large enough, or sufficiently interconnected, to support viable populations of the species for which they are important.*" Marine and terrestrial KBAs are driven by the distribution and abundance of species that require site-level conservation and are selected for their vulnerability and irreplaceability (Eken et al. 2004).

Species-driven, site protection approaches require to evolve towards the designation of important areas specifically for the conservation of endangered species (Coll et al. 2015; Jefferson et al. 2021). When species play a role in MPA design, it is usually richness and endemicity that is primarily taken into account, not necessarily urgency for protection (Jefferson et al. 2021). Thus, priority sites decided based on species richness may overlap poorly with the distribution of threatened biodiversity (e.g., Asaad et al. 2018). In fact, if the most threatened marine species were considered as drivers for area protection, the 30% target of the CBD would be insufficient, and a minimum of 40% of the ocean surface would be required for their effective conservation (Jefferson et al. 2021).

10.2.3 Conservation in the ESS

Mangroves, coral reefs, spawning aggregations, and colonies may be effectively protected by a fixed, spatial tools such as a traditional MPA, but pelagic species, particularly those that make the bulk of the biodiversity of the ESS, have gigantic foraging ranges as are highly migratory (Block et al. 2011; Queiroz et al. 2019; Campagna et al. 2020). For them, fixed, protected sites are partially ineffective or impractical. Considering the natural history of globally threatened species of the ESS, such as some albatrosses, space-driven conservation tools may require specific, science-informed, management considerations. Satellite tracking, for example, may yield information on foraging hot spots distributed over extended ocean surfaces. Site-related tools sensitive to behavioral information may require to move from fixed to dynamic, seasonal, or movable protected areas. Adaptive and dynamic management are not new concepts, except that, in practice, it often responds primarily to the protection of use, not of species.

In sum, a conservation vision for the ESS that prioritizes threatened, pelagic species would hinge on a suite of dynamic and adaptive approaches, reflecting the varying needs and contexts. And need and contexts vary because of the dynamic functioning of frontal areas. The PSBF has a significant effect on the ESS. It is a reliable frontal area, in the sense that is less subject to periodic oscillations than other fronts. The PSBF may then provide a backbone that guides long-term conservation of dependent species. The challenge is to figure conservation approaches built on dynamic tools sensitive to the natural history of the species that require most protection.

10.3 The ESS, an Overview

10.3.1 Oceanographic Regimes as Habitats

Oceanographic regimes may be understood as habitats dependent on the bathymetry and circulation of water masses that create salinity and/or temperature gradients (Figure 10.1c; Guerrero and Piola 1997; Piola et al. 2000). Ocean fronts are part of these regimes (Acha et al. 2004; see also Chap. 2). The continental shelf is the largest of such habitats: an extended, neritic system less than 200 m deep, with areas of differential productivity. The shelf-break works as the outer edge of the plateau, a mesopelagic habitat that coincides roughly with the 200 m isobath. From there, depth increases rapidly to nearly 4000 m. We here refer as "shelf-break" the region where the continental shelf sinks into the basin, or continental margin, differentiating from the PSBF area, that covers the shelf- break but also extends to the continental shelf itself.

10.3.2 *Diversity and Distribution of Megafauna*

Vertebrate diversity of the ESS accounts for at least 700 species, mostly bony fishes (400), cartilaginous fishes (132), marine and coastal birds (83), marine mammals (47) and five of the seven species of marine turtles (e.g., Yorio et al. 2005; Sabadin et al. 2020). Many seabirds, marine mammals, and bony fishes find in the ESS an important proportion of its global population and distribution. As some birds and mammals reproduce in colonies, coastal areas are of highest ecological and conservation value. These occur along the continental shores and the few islands of the ESS, most important: the Malvinas Archipelago. Species abundance and distribution hinge on the physical peculiarities of the ESS, and on the primary and secondary productivity that sustain large fish and invertebrate (squid) populations (Chap. 5), as well as top predators and industrial fisheries (Chaps. 6 and 7). Critical to the ESS food web are the Argentine anchovy (*Engraulis anchoita*) and the Fuegian sprat (*Sprattus fuegensis*). Together with a few other bony fishes and squids, these species are the essential prey for the birds and mammals of the ESS.

10.3.2.1 Marine Birds

At least 17 species nest in the region (Yorio et al. 1998), being the Magellanic penguins, *Spheniscus magellanicus*, the most abundant. This species distributes in some vast colonies in the Southern Cone and beyond. Two colonies, in the shores of continental Argentina, gather about 350,000 breeding pairs for a total population of about 1 million (García-Borboroglu et al. 2022). Southern rockhopper penguins, *Eudyptes chrysocome*, follow in abundance, with about 400,000 pairs in the Malvinas Archipelago and continental shores (Pütz et al. 2013; Gandini et al. 2016). Among the Procellariiformes, the black-browed albatross, *Thalassarche melanophris*, is the most common (Croxall and Gales 1998; Huin and Reid 2007). About 70% of the world population of this species reproduces in one island, in the Malvinas Archipelago.

At least 50 species of seabirds forage in the ESS. It is estimated that about 1 million black-browed albatrosses distribute on the shelf and beyond, as do other eight albatross species (Favero and Silva Rodriguez 2005; Wolfaardt 2012). The shelf and shelf-break are also major seasonal foraging grounds for migratory species coming from South Georgia, the Antarctic Peninsula and even New Zealand (Croxall and Wood 2002). One of them is the wandering albatrosses, *Diomedea exulans* (Nicholls et al. 2002), as well as the albatrosses migrating from Diego Ramirez, Tristan da Cunha, and Gough Islands.

10.3.2.2 Marine Mammals

Three pinniped species are local residents: the South American sea lion, *Otaria flavescens*, with an estimated population of ca. 190,000 animals (Cárdenas-Alayza et al. 2016), the Southern elephant seal, *Mirounga leonina*, with ca. 60,000 animals, and the South American fur seal, *Arctocephalus australis*, also with a population of ca. 60,000 (not including Uruguay), distributed in a few islands off the coast of Argentine Patagonia (Baylis et al. 2019; Vales et al. 2019). A large breeding colony of Southern elephant seals occurs at Península Valdés and nearby stretches of coast, where about 18,000 pups are born per year (data for the 2022 breeding season (Campagna J, personal communication). The population is estimated in ca. 60,000 animals 1 year old or older. Fur seals and sea lion colonies distribute more widely along the coast of Argentina and the Malvinas Archipelago (Fig. 10.3; Campagna and Lewis 1992; Grandi et al. 2008).

From the cetaceans, at least 35 species reside in the ESS, three of which are endemic or have a restricted distribution, and the ESS is an important part of it (e.g., Commerson's dolphins, *Cephalorhynchus commersonii*; La Plata River dolphins, *Pontoporia blainvillei*; Peale's dolphin, *Lagenorhynchus australis*). Southern right whales, *Eubalaena australis*, reproduce in the northern gulfs of Argentine Patagonia, with a population estimated in ca. 6000 animals (https://siguiendoballenas.org/en/whales-in-patagonia/).

10.3.3 Satellite Tracking Studies

Tracking work targeted primarily Magellanic penguins (Pütz et al. 2002; Boersma and Rebstock 2009; Boersma et al. 2009; Barrionuevo et al. 2020), Southern elephant seals (Campagna et al. 1999, 2006; Galimberti and Sanvito 2012; Campagna et al. 2020) and Southern right whales (Zerbini et al. 2016). Despite differences in preys, seals and whales show some interesting similarities in their movements, dispersing widely over the shelf, shelf-break and ocean basin (Campagna et al. 2020; for whales see: https://siguiendoballenas.org/). Elephant seals are unique in their use of the water column, up to a mile deep, as well as the benthic environments. Much less is known about the dispersion at sea of South American sea lions. Early work (e.g., Werner and Campagna 1995) showed that nursing females distribute on the shelf, within short distances from the breeding colonies, while males may reach the shelf-break. Regarding South American fur seals, females also remain close to the colony during lactation but travel to pelagic areas by the end of summer (Thompson et al. 2003). It was estimated that large numbers of Antarctic fur seals, *Arctocephalus gazella*, forage in the ESS waters (Boyd et al. 2002). Southern right whales winter on the shelf and the shelf-break (M. Sironi, personal communication).

An early integration of satellite tracking data for 16 species of marine birds, mammals, and turtles, encompassing about 1300 trips and hundreds of individuals, showed an uneven use of the ESS, with substantial aggregations in frontal areas,

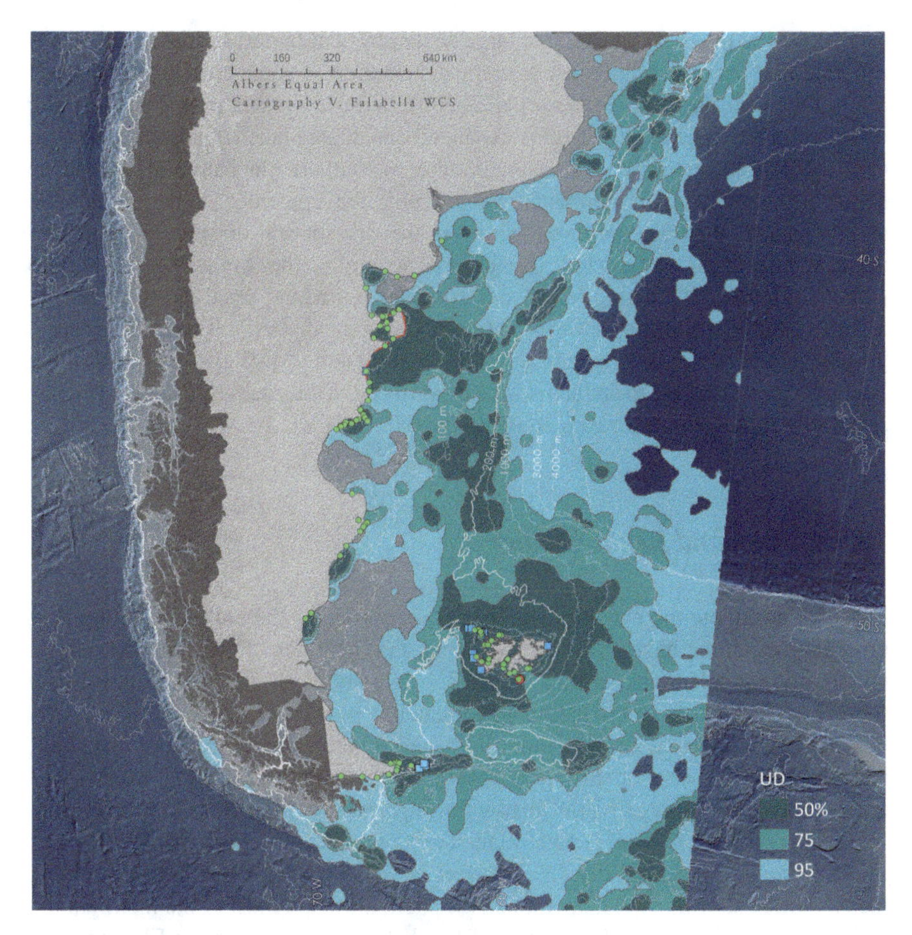

Fig. 10.3 Integrated kernels for 16 species of marine birds and mammals (from Falabella et al. 2009). Most important breeding colonies for South American sea lions (green circles) and fur seals (blue circles). Stretches of coastline where southern elephant seals reproduce (red lines). Breeding area for southern right whales (yellow lines)

particularly in the PSBF (Fig. 10.3; Falabella et al. 2009; see also: Favero and Silva Rodríguez 2005). Tracked individuals from 33 colonies located in costal Patagonia, the Malvinas Archipelago, South Georgia and New Zealand, converged on the ESS, covering all major habitats, from coastal regimes to the high seas. Areas most commonly used by albatrosses, petrels, penguins, sea lions, and elephant seals reveal few seasonal variations. This finding suggests that the same extensive sites, that include the shelf-break, are relevant for these top predators during different periods of the annual and life cycles. The PSBF is important all year round. The sample also represented a variety of life history and foraging strategies, from migratory pelagics (e.g., Northern Royal albatross, *Diomedea sanfordi*), to long-distance swimmer-divers (e.g., Magellanic penguin) and deep divers (e.g., southern elephant seal).

Despite the relative constancy of large-scale physical features for the ESS, it was reported that important shifts occur in the distribution of some critical species such as the Argentine anchovy (Franco et al. 2020b). In the last two decades, the breeding distribution of Magellanic penguins expanded one degree north (García Borboroglu et al. 2022). Conversely, an increasing number of southern elephant seals reproduce about 100 km south of Península Valdés, still the epicenter of the population (Campagna C, unpublished data). Shifts in the distribution of fisheries may also indicate relocation of targeted species. For example, the Argentine red shrimp (*Pleoticus muelleri*) trawling fishery has expanded northward since 2014, suggesting a shift in the distribution of shrimps (de la Garza et al. 2017). From the oceanographic perspective, data suggest an intensification and a shifting southward of the Brazil Current, leading to an intense ocean warming in the south of Brazil, Uruguay, and northern Argentina (Franco et al. 2020a, b).

10.3.4 Use and Legal Framework

Fisheries dominate the extractive industries, in particular the bottom-trawling fleet targeting Argentine hake, *Merluccius hubbsi,* that operates throughout the year on the shelf and shelf-break, within the EEZ (Alemany et al. 2016). Argentine hake represents about 50% of the national fleet catch, yet, it is the Argentine red shrimp that has the highest economic value. Its fishery operates exclusively on the shelf, with outrigger trawlers or shrimpers vessels (de la Garza et al. 2017). Most of the continental shelf is susceptible to trawling operations (Fig. 10.4a), although the spatial distribution of effort is patchy, and the trawling hotspots are small in size and spatially stable inter-annually (Alemany et al. 2016). Argentine jiggers, targeting the Argentine shortfin squid, operate on the shelf, although most of the international fleet of jiggers distributes along the shelf-break, beyond the Argentine EEZ, in association with the PSBF area (Figure 10.4b). The shelf-break, that overlaps both the EEZ and the international waters, is also relevant for the trawler fleet targeting long tail hake, *Macruronus magellanicus*, Argentine hake and Patagonian scallops, *Zygochlamys patagonica*. There is a small anchovy fishery that represents about 2% of the Argentine catch on the EEZ.

The economic relevance of the large commercial fisheries operating in the ESS contrasts with the environmental impact, particularly of bottom trawlers and shrimpers. Trawlers impact benthic environments (Hiddink et al. 2017) and generate large amounts of discards and incidental captures (Zeller et al. 2018). The discard by the Argentine red shrimp fishery includes juveniles of Argentine hake, a species already heavily fished (Bovcon et al. 2013; Irusta et al. 2016). Incidental mortality also affects birds and marine mammals (Marinao and Yorio 2011; Tamini et al. 2015). It was a reason for a drastic reduction in the population of albatrosses (Croxall et al. 2012) although the impact has now been mitigated (Tamini et al. 2015). Fisheries also create large amounts of debris that cause entanglement and mortality of birds, mammals, and turtles (Copello and Quintana 2003; Campagna et al. 2007a; Gregory

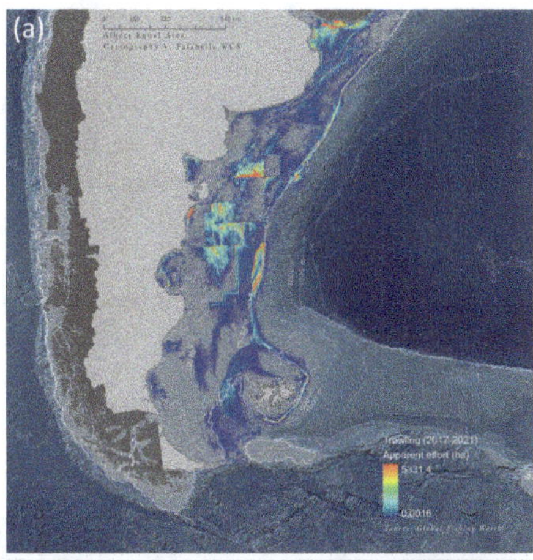

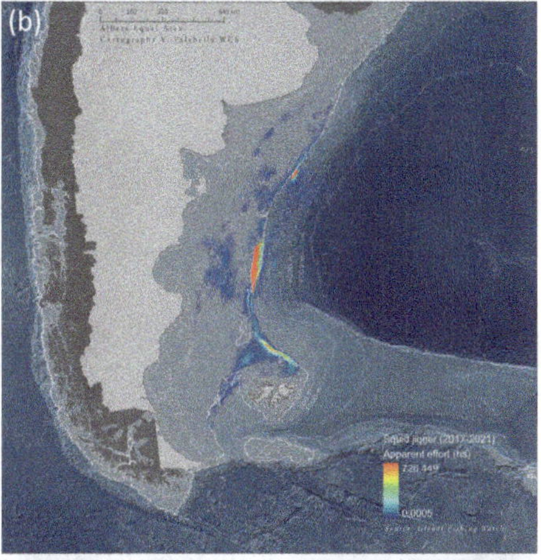

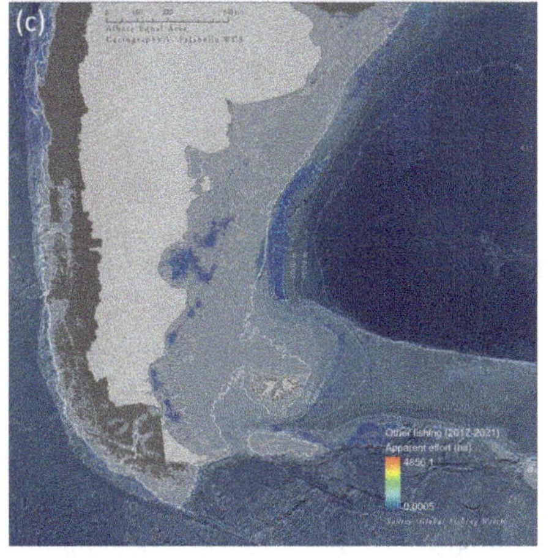

Fig. 10.4 Apparent fishing effort (hours) in the ESS (calculated with the combination of Automatic Identification System AIS data on vessel location and machine learning techniques). Data from 2017 to 2021. Source: Global Fishing Watch. (**a**) Trawlers* (2017–2021). (**b**) Jiggers (2017–2021). *Note that within the "permanent closed area" (PCA, Fig. 10.2b), the fishing effort is explained by the operation of outrigger trawlers or vessels that fish for the Argentine red shrimp

2009; Carroll et al. 2014; Vélez-Rubio et al. 2018). Fishery debris is found in enormous quantities, even in the most protected of coastlines.

Regarding threats beyond fishing operations, plastic pollution from urban sources is a major problem at sea and in coastal areas. Ingestion of plastic debris has been reported in many species (Copello and Quintana 2003; González Carman et al. 2014). Microplastics are recorded in species from coastal environments near large cities (Mandiola et al. 2021), as well as in the water column and in filter-feeding invertebrates from remote and deep oceanic protected environments (e.g., Di Mauro et al. 2022). Oil pollution, originated from ballast waters of oil tankers, killed tens of thousands of penguins per year for decades, until the oil tanker route was moved more than 100 km offshore (Gandini et al. 1994; Garcia-Borboroglu et al. 2006). The number of oiled penguins recorded offshore is now low, but chronic pollution remains a problem (Garcia-Borboroglu et al. 2006; Boersma 2012). There are initiatives to develop hydrocarbon exploration and exploitation activities on the edge of the shelf-break.

Argentine legal regulations include total allowable catches, multi-specific individual transferable quotas, areas under special management considerations for the protection of target species, and National Action Plans to control illegal, unreported and unregulated fishing, manage fisheries targeting chondrichthyans, and reduce the unintended catch of birds, marine mammals and turtles, yet, the institutional, jurisdictional, and legal contexts make the ESS susceptible to overuse due to sovereignty conflicts, unbalanced sectorial interests, and lack of regulations within high seas. An improvement would be to rely less on single species management and advance towards dynamic, ecosystemic decisions, and better extractive and biodiversity policies articulation within national and international jurisdictions.

10.3.5 Species Conservation in the ESS

10.3.5.1 Precedents

Early conservation efforts (1970s and 1980s) hinged around behavioral studies targeting charismatic species, such as southern right whales, Magellanic penguins, southern elephant seals, and South American sea lions. A few coastal protected areas were already in place in the 1970s. Threats were related to oil pollution, illegal whaling, some fisheries practices and disturbance from unregulated tourism. The first project with an ecosystem approach (mid-1990s), the "Patagonian Coastal Zone Management Plan" (PCZMP; Harris 2016) aimed at protecting "wildlife spectacles," particularly colonial species, with a network of MPAs at the regional level. An article titled "*At the edge of a cold, sea river*" (Conway 1992) already referred to the Malvinas Current and the PSBF as critical for coastal conservation. It was grounded on the first satellite tracking studies (e.g., Campagna et al. 1999), as well as by the first satellite images of the squid fishery, along the continental slope

(Waluda et al. 2002), and the first comprehensive overviews of the oceanography of the SW Atlantic (e.g., Guerrero and Piola 1997; Podestá 1997).

The expansion of satellite tracking efforts, in the early 2000s, showed that foraging individuals of many species converged on some areas of the shelf and shelf-break, often overlapping with fishing efforts. In 2004, the "Sea and Sky Project" expanded the PCZMP concept, advancing an ecosystem perspective for the pelagic waters of the ESS. The rationale was that coastal MPAs functioned during a small part of the annual cycle of colonial birds and mammals, as, most of the time, these animals were at sea foraging. Thus, protection of important foraging sites required the format of "Parks of Sea and Sky." Argentina declared three of them since 2010. A fourth, the Blue Hole, on the shelf break, has partial approval in Congress.

A leading concept to inform space conservation during those years was that of Large Marine Ecosystems (LME; Sherman et al. 1990): regional units for the conservation and management of living marine resources in accordance with the legal mandates of UNCLOS (the Law of the Sea). The Patagonian Shelf was among the 64 proposed LMEs (Longhurst 1998). Continental shelves were recognized most relevant for some LMEs, partially because countries have jurisdiction over all or part of them. Frontal areas were also found critical (Belkin 2009), but the main frontal area for the Patagonian LME, the "cold, sea river" that unified the PCZMP and the Sea and Sky Project, was partially beyond national jurisdiction. Thus, function supported the rationale to extend the Patagonian LME into a larger conservation unit: the ESS.

10.3.5.2 Protected Spaces

About 50 years after conservation efforts, a rosary of more than 60 static and permanent areas protects most colonies of seabirds and mammals along the coast of Argentine Patagonia (Fig. 10.2a), yet almost six out of ten coastal MPAs are on land, where colonies are located. Total marine surface under protection is of ca. 25,000 km^2. Most seabird colonies are protected. For example, over 80% of the ca. 67 breeding colonies of Magellanic penguins are included in MPAs. The proportion is important also for some marine mammals: ca. 50% of the 140 breeding and haul-out colonies of the South American sea lion are located in a protected area (https://ampargentina.org/). Most of the coastal areas where elephant seals reproduce are protected. Fur seals reproduce on islands, some of which are under the jurisdiction of national and provincial marine parks. Three "Parks of Sea and Sky" (Figure 10.2a) protect a submerged plateau (Namuncurá-Burdwood Bank I and II, ca. 55,500 km^2) and a large sector of the Drake Passage (Yaganes National Park, 58,200 km^2). The latter includes the water column and benthic environments, with submarine canyons and the only seamount in the Argentine Sea. The pending (October 2022) "Blue Hole" is unique as it would be exclusively a benthic park, legally grounded in the UN extension of the limits of the border of the continental shelf. It would be a park of the PSBF, as a large surface of it is in close relation with the front.

In addition to MPAs, a proportion of the EEZ is under special fisheries management considerations (Fig. 10.2b). These zones have the purpose to protect species under strong fishing pressure, e.g., the Patagonian enclosure to safeguard juveniles of the Argentine hake (Giussi et al. 2022), yet the Argentine red shrimp fishery is allowed to operate within this managed area with some restrictions. Although the CBD recognizes area-based management as effective conservation tools (CBD 2018), these unique spaces of the EZZ should not be considered equivalent to "Parks of Sea and Sky," although they may still contribute to the protection of some biodiversity (Laffoley et al. 2017; IUCN WCPA 2019).

10.3.6 "Seascape" Species and Threatened Marine Vertebrates

Many methodologies are available to advance species-driven conservation. One is grounded in the concept of "landscape species." The approach identifies a suite of species that serves as surrogates for the conservation of the biodiversity of a selected site (Sanderson et al. 2002; Coppolillo et al. 2004). Candidate species selection is based on five criteria: heterogeneity, area distribution, ecological function, vulnerability, and socio-economic importance. This approach was applied to the ESS as a conservation planning tool (Campagna et al. 2007b). At the time, the conservation status of the candidate species was unknown at the regional level. A red listing of vertebrate species was conducted for the region during the period 2016–2019. We will summarize results and integrate them with results from the seascape species approach.

10.3.7 Overview of a Regional Red-Listing Exercise

A total of 113 vertebrate species that distribute in the EES were analyzed following the IUCN Red List methodology (Tables 10.1 and 10.2). Overall, 28% (32 species) were threatened; 11% (12) were Critically Endangered. Sharks and rays were the most threatened: 57% (20) of 35 species; two were Critically Endangered. The list of threatened species for the ESS is completed with one of the three assessed sea turtles, 17% (4) of the bony fishes, 25% (5) of the marine mammals, and 13% (5) of the birds.

Among the 113 sampled species, 46 species were endemic, or had restricted distribution and the ESS was critical ground. The Patagonian Sea horse, *Hippocampus patagonicus*, for example, is a CR species of restricted distribution in most urgent need for conservation action. Most threatened species were shelf users, and the largest number of them concentrated in the northern shelf and coastal areas, where water masses are heterogeneous in salinity, temperature, and productivity (Fig. 10.5).

Table 10.1 Number of species assessed and threatened by taxon

	Assessed	Threatened
Bony fishes	23	4
Sharks and rays	35	20
Sea turtles	3	1
Seabirds	35	5
Marine mammals	17	2
Total	113	32

Table 10.2 Conservation status by taxon. Threatened species integrate the categories of Critically Endangered (CR), Vulnerable (VU) and Endangered (EN)

	CR	VU	EN	NT	DD	LC	Total	% Threatened
Bony fishes	**1**	**3**			7	12	23	17
Sharks and rays	**8**	**6**	**6**	2	7	6	35	57
Sea turtles		**1**		1		1	3	33
Seabirds	**3**	**2**		2		28	35	14
Marine mammals		**1**	**1**		2	13	17	12
Total	**12**	**13**	**7**	5	16	60	113	28

The shelf-break followed in relevance, with 65% (74) of the species using it. The PSBF was priority habitat for six species: hairy conger eel (*Bassanago albescens*), grenadier (*Coelorinchus fasciatus*), longtail southern cod (*Patagonotothen ramsayi*), Cousseau's skate (*Bathyraja cousseauae*), Graytail skate (*Bathyraja griseocauda*), and Cuphead skate (*Bathyraja scaphiops*) (Table 10.3). One in three threatened species distributed on the PSBF; a few species were strictly shelf-slope dependent (Table 10.3).

10.3.8 Threats

Sharks and rays were highly impacted by overfishing and mismanagement of recreational, artisanal, and industrial fisheries. At least five species of sharks are caught by commercial fisheries: Argentine angel shark (*Squatina argentina*), Tope shark (*Galeorhinus galeus*), Striped smooth-hound (*Mustelus fasciatus*), Copper shark (*Carcharhinus brachyurus*), Porbeagle shark (*Lamna nasus*). Caught rays and skates are landed as part of a multi-species bulk, being Argentina a top exporting country (Niedermüller et al. 2021).

The 23 species of bony fishes assessed were not targets of the most important fisheries. Thus, the assessment of this group is incomplete. Two ecologically critical species, Argentine anchovy and Fuegian sprat, were assessed and found Least Concern, but fisheries also affect many species of marine mammals and birds,

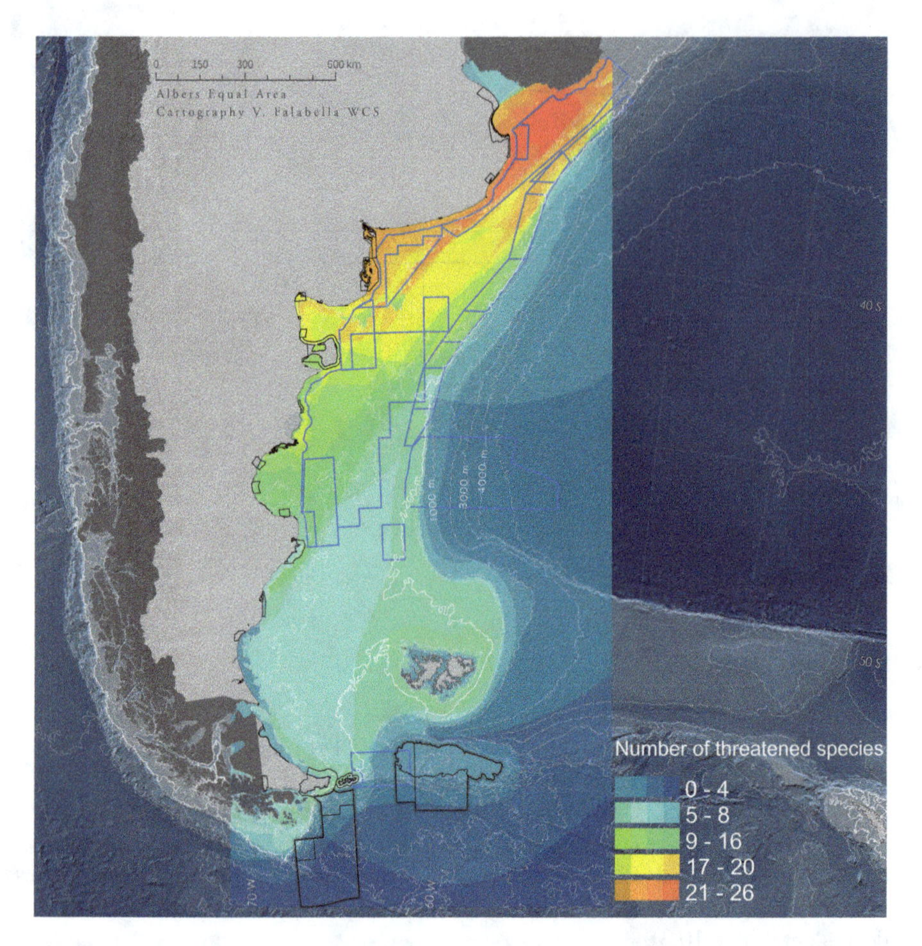

Fig. 10.5 Overlap of coastal-marine protected areas (black polygons) and the main management fisheries zones (blue polygons) with the distribution of threatened vertebrates for the ESS

caught as unintended catch or entanglement (Dans et al. 2003; Favero et al. 2003; Gandini and Frere 2006; Campagna et al. 2007a; Mendez et al. 2008; Pütz et al. 2011; Negri et al. 2012). The Vulnerable La Plata River dolphin, *Pontoporia blainvillei,* has a small population impacted by artisanal fisheries throughout its distribution range (Mendez et al. 2008; Negri et al. 2012).

10.3.8.1 Integrating Red Listed and Seascape Species

As surrogates, seascape species should serve to protect threatened species. Seven species were selected as surrogates from the 33 candidate species for the region (Table 10.4; see also Tables 10.3 and 10.4 in Campagna et al. 2007b). Among the 33, 21 species were subsequently red listed, and 24% (5) were categorized as

Table 10.3 Threatened species of the ESS organized according to their conservation status (following the Red List categories)

Common name	Species	ESS	Global 2020	Slope user
Tristan albatross	*Diomedea dabbenena*	CR	CR	Yes
Patagonian Sea-horse	*Hippocampus patagonicus*	CR	VU	No
Great shearwater	*Ardenna gravis*	CR	LC	Yes
Guanay cormorant	*Leucocarbo bougainvilliorum*	CR	NT	No
Tope shark	*Galeorhinus galeus*	CR	VU	No
Striped smooth-hound	*Mustelus fasciatus (*)*	CR	CR	No
Patagonian smooth-hound	*Mustelus schmitti (*)*	CR	EN	No
Sand tiger shark	*Carcharias taurus*	CR	VU	Little
Spiny butterfly ray	*Gymnura altavela*	CR	VU	No
Southern eagle ray	*Myliobatis goodei*	CR	DD	No
Brazilian guitarfish	*Pseudobatos horkelii (*)*	CR	CR	No
Shortnose guitarfish	*Zapteryx brevirostris (*)*	CR	VU	No
La Plata River dolphin	*Pontoporia blainvillei (*)*	EN	VU	No
Roughskin skate	*Dipturus trachyderma (*)*	EN	VU	Yes
Spotback skate	*Atlantoraja castelnaui (*)*	EN	EN	No
Broadnose skate	*Bathyraja brachyurops (*)*	EN	LC	Yes
Graytale skate	*Bathyraja griseocauda (**)*	EN	EN	Yes, almost exclusively
Argentine angel shark	*Squatina argentina (*)*	EN	CR	No
Spiny angel shark	*Squatina guggenheim (*)*	EN	EN	Little
Copper/bronze shark	*Carcharhinus brachyurus*	VU	NT	Little
Narrowmouth catshark	*Schroederichthys bivius (*)*	VU	DD	Yes
Yellownose skate	*Zearaja chilensis (*)*	VU	VU	Yes
Pike icefish	*Champsocephalus esox*	VU	NE	No
Black drum	*Pogonias courbina (*)*	VU	NE	No
Patagonian flounder	*Paralichthys patagonicus (**)*	VU	NE	Little
Steamerduck	*Tachyeres leucocephalus (**)*	VU	VU	No
White-chinned petrel	*Procellaria aequinoctialis*	VU	VU	Yes
Bottlenose dolphin	*Tursiops truncatus*	VU	LC	Yes
Green turtle	*Chelonia mydas*	VU	EN	No or little
Eyespot skate	*Atlantoraja cyclophora*	VU	VU	No
Rio skate	*Rioraja agassizi (*)*	VU	VU	No
Bignose fanskate	*Sympterygia acuta (*)*	VU	VU	No

Marine mammals, seabirds, and sea turtles were assessed in 2016, sharks and rays in 2017 and bony fishes in 2019. **: Endemic. *: The ESS covers most of the distribution. *NE* not evaluated

threatened. The Tope shark was a high ranked, seascape candidate categorized as Critically Endangered (Table 10.4). At the time of the seascape species analysis, lack of regional data precluded focalizing on threatened candidates, and figuring the best list of seascape species to protect them. Such an exercise may be helpful in the future.

Table 10.4 Conservation status of the seven seascape species (italics) chosen from 33 candidates (Table 10.3 in Campagna et al. 2007a, b)

Species	Rank as seascape species	Rank as seascape candidate	Regional Red List	Shelf user	Shelf slope user
Magellanic penguin	1	1	LC	Yes	Yes
Black-browed albatross	2	2	LC	Yes	Yes
Argentine shortfin squid	3	7	NE	Yes	Yes
Yellownose skate	4	5	VU	Yes	Yes
Rockhopper penguin	5	3	NT	Yes	Yes
Southern right whale	6	10	LC	Yes	Yes
Southern blue whiting	7	25	NE	Yes	Yes
Tope shark	–	4	CE	Yes	No
Patagonian or narrownose Smoothhound	–	12	EN	Yes	No
La Plata river dolphin	–	24	EN	Yes	No
White-chinned petrel	–	28	VU	Yes	Yes

In bold: seascape and non-chosen candidate species categorized as threatened (see legend of Table 10.3 for abbreviations). *LC* least concern, *NE* not evaluated. The rank of a species as candidate may be high and yet it may not make the short list of seascape species. This is because a previously ranked seascape species is already functioning as surrogates for the same features (see methodology in Campagna et al. 2007a, b)

10.3.8.2 Integrating Red Listed Species and Protected Spaces

The ESSs follow the generalities regarding the disjunction between protected spaces and endangered species. For example, no MPA was ever declared exclusively to revert the status of a Critically Endangered species. Present MPAs offer virtually no protection for threatened vertebrates. Threats operate mainly far from the coast and at a scale that coastal MPAs do not match. Also, most endangered species distribute beyond the limits of the oceanic parks, which were not designed to suit their needs. Important foraging hot spots for pelagic birds and mammals, as indicated by satellite tracking studies, overlap partially with the oceanic protected areas (Fig. 10.5; see also Campagna et al. 2020). The fisheries sectors for special management cover a proportion of core habitat for some endangered species (Fig. 10.5).

Gaps also affect habitats. A recent analysis identified 11 bioregions for the ESS and regulated intensive fishing occurs in six of them (Falabella V. y Alemany D, unpublished data), yet only three of the 11 have more than 10% of their surface area protected, and six have less than 3%.

10.3.9 A Conservation Plan for the ESS

A conservation plan for the ESS could revert the endangered status of some of its species and achieve better representation of habitats-oceanographic regimes and bioregions. Such a plan will require a variety of interventions. We illustrate the point

Table 10.5 Set of species selected for their relevance as surrogates or their highly threatened status

Criteria	Species	Red list	Description	Interventions
Seascape or candidate seascape species	Yellownose skate	NE	Seascape species (ranked 5th). It is threatened by the demersal trawl fishery.	Develop species-specific conservation programs for immediate results, facing threats. MPAs could be useful. Requires significant improvements of fishing practices.
	Tope shark	CR	Ranked fourth as seascape species candidate. Transboundary species. Overfished. Targeted also by sport fishing.	Moratorium of the fishery. Enforce the law. Otherwise, same as Yellownose skate.
	Magellanic penguin	LC	Ranked first as seascape species. Extended home ranges over all oceanographic regimes. High exposure to threats susceptible to changes in distribution of prey.	Targeted satellite tracking program to spot changes in preferred foraging places, which, in turn, would indicate changes in distribution of prey. Important as platforms to collect oceanographic data.
	Argentine shortfin squid	NE	Seascape species. Transboundary. Most important for the shelf-break habitat. Targeted by an international, high-profile fishery. Integrated management affected by sovereignty dispute over the Malvinas archipelago.	Insight could be gained on the functioning of the SB-F by monitoring the distribution pattern of the fishery dependent on this species.
	Argentine anchovy	LC	Candidate seascape species. Critical for the food web of the ESS. Impacting its abundance or distribution may cause cascading effects reaching many dependent natural predators.	Priority monitoring programs
	Fuegian sprat	LC	Same as argentine anchovy, but for the southern shelf.	Priority monitoring programs
	Southern elephant seal	LC	Seascape species candidate: Ranked 9 of 33. Deep diver, makes a differential use of the water column. Broad distribution in the ESS and beyond. Southward shift in the breeding distribution.	Similar to Magellanic penguin, but mainly for the SB-F and the deep waters of the argentine basin.
Threatened species not ranked as seascape candidates	Graytail skate	EN	A distributions that overlaps with the shelf break front. Targeted by the demersal trawlers.	Same as Yellownose skate
	Sand Tiger shark	CR	Very low reproductive output. Slow maturation. Caught accidentally in industrial fisheries. Target of sport fishing.	Enforce the law.
	Patagonian smooth-hound shark	CR	Overfished in the core area of distribution until the population crashed.	Develop conservation plan while gathering more data.

NE not evaluated

by selecting a set of ten species included on both, the red listing and the seascape species studies (Table 10.5). Conservation interventions were derived based on their natural history and conservation profile (Table 10.5). These are some broad conclusions based on this set:

1. Endangered species with restricted distribution would benefit from site-conservation when designed specifically for their needs.
2. Species of differential ecological relevance may be closely monitored to spot changes in abundance and distribution, no matter their conservation status. Indeed, some least concern species are the best indicators for a variety of ocean regimes and should be the target of scientific research and monitoring programs.
3. Monitoring of highly specialized fisheries (e.g., squid jiggers) may serve to infer abundance and distribution of their targets, and may be more easily monitored.

These conclusions are by no means exhaustive, but suggest that spatial conservation tools are insufficient for an integrated plan. Fixed protected areas would be useful for few of the threatened species. Best candidates for fixed MPAs could be the Patagonian Sea horse, a set of sharks and rays that concentrate in a relatively small coastal areas, including the Sand tiger shark, *Carcharias taurus* (see Cuevas and Michelson 2023) and, perhaps, the La Plata River dolphin. Multi-species, seasonal MPAs are most required and would be most effective in the northern shelf, and in the area of the Río de la Plata salinity front, where many threatened species converge. Coastal MPAs may serve for monitoring population data regarding colonial species. Behavioral research projects should be focused on diving species with wide distributions that could also serve as platforms for gathering oceanographic data. A set of satellite tracked species combined with remote sensing data may contribute real time information to guide fishing effort and avoid unnecessary costs impinged on biodiversity. An integrated satellite tracking program of indicator species would also help identify shifts in foraging hot spots that may be related to forced changes in the climate. Results from these studies would be of interest to managers, as it is in these hot spots where negative interactions with extractive industries may take place. Areas of high foraging relevance could then be turned into dynamic, no-take, short-term protected sites for periods of days to an entire season. Dynamic space management will be required, as these priority areas may change from year to year.

10.4 General Discussion

Space-driven conservation guided by environmental diversity is receiving considerable attention as a global strategy. It may be argued that the approach is linked to biodiversity conservation, but habitat diversity and functional-ecological data may not be surrogates for species conservation (Brooks et al. 2004), and it does not work in the case of threatened species. MPAs may serve many purposes besides species conservation, such as the protection of habitats and human activities (Lascelles et al. 2012). Thus, space conservation strategies require better synergy with efforts to prevent local species extinctions. In fact, focusing too much on protected areas might delay progress on other fronts, management included. Despite being also their beneficiaries, the extractive industries often resist MPAs declarations. As a consequence, space protection often targets areas of secondary relevance. Thus, the establishment of protected spaces may risk creating the illusion of sufficiency in conservation interventions.

The ESS requires a conservation plan that prioritizes the rescuing of most endangered species from local extinction. Today, the region is better conceived as a valuable fishing zone, or an area to be more developed for oil extraction, than as an ocean in need to preserve its biodiversity, yet about one third of its vertebrates are threatened, a similar proportion to global data (McCauley et al. 2015). Despite legislation, sharks and rays are in urgent need of protection from threats that deepen rather than disappearing. From the 100 species of chondrichthyans cited for the ESS, 62% are globally threatened (Cuevas and Michelson 2023), a proportion reflected for those species that were red listed regionally.

It can be predicted that site-related solutions will continue to dominate as priority conservation interventions worldwide. Thus, future site-driven conservation interventions in the ESS should focus on threatened biodiversity.

10.4.1 The Relevance of Frontal Areas in Conservation

Any conservation strategy for the ESS should depart from considering its oceanographic profile. This translates as the need to couple function and management. The ESS is differential due to the shallow continental shelf and the PSBF. The PSBF is critical for function, the shelf for management. The PSBF is intrinsically valuable as it integrates the dynamic physical structure of the water masses with spatial predictability and annual constancy. Also, the best evidence, so far, of climate-related change in the ESS is pertinent to this system (Franco et al. 2020a, b). The PSBF is the functional driver of the ecological profile of every habitat of the ESS, being the shelf the most relevant. The northern portion of the shelf is priority habitat for most regional biodiversity, including endangered species. A few species use many habitats, and here the Magellanic penguin plays an important indicator role, yet, if the deep ocean basin and the deep benthic habitats are to be included, then the southern elephant seals are unique in their role as indicator species.

In sum, conservation actions focused on threatened species should consider the shelf as a priority conservation unit. The PSBF is to be monitored in terms of its physical variables as it seems to be most sensitive to climate change and exerts an influence on all other regimes. Thus, the continental slope plays two roles, a functional one related to the PSBF, and as a unique conservation site related to the deep-water canyons that connect the shelf with the deep basin. Oceanic protected areas in this region may safeguard unique environments and species.

10.4.2 Tools for Dynamic Conservation and Management Strategies

The size of the ESS, the scale of industrial fisheries, the plans to expand offshore oil exploitation, the sovereignty conflict over the Malvinas archipelago, the lack of consensus over a conservation vision, and the unpredictability of climate-change scenarios are but some examples of the complex scenario that faces any conservation vision, yet the most urgent need is to improve the conservation status of endangered species. Considering the functional signature of the ESS, the fundamental quality of any conservation tool would be its capacity to adapt dynamically to ecological conditions. Thus, fixed site protection approaches would require to be replaced with dynamic habitat protection. Adaptive interventions should also combine tools, as protecting spaces are often impractical for pelagic species.

Marine spatial planning may help optimize the allocation of conservation efforts, yet the value of zoning is relativized by unpredictable scenarios. Two sources of unpredictability prevail: the global economy, dependent on political stability, and the forced climate change, that operates on the core function of the system. Fisheries are challenged by both, yet, it is our view that management should not be sensitive to global economics at the expense of ecological costs. Variations in productivity impact the entire system, and use should be guided by ecosystemic principles. Dynamic fisheries management is already in place in some places of the world (Hobday and Hartmann 2006; Hobday et al. 2014) such as in some Argentinian fisheries, although this does not guarantee an ecosystem-based management framework.

Monitoring becomes of paramount relevance in unpredictable scenarios, and management plans should incorporate more insights gained by studying the behavior of species that play a key ecological role in the systems. Combining satellite tracking of indicator species with remote sensing will yield real-time descriptions of the location of frontal areas (Scales et al. 2014). Spatial conservation tools would also be more effective if conceived dynamically: it is necessary to identify mechanisms for dynamic and real-time management of MPAs, including zoning that can vary seasonally. Most useful MPAs would target benthic habitats, the water column, the deep sea or particular features such as seamounts, canyons, or submerged plateaus.

Although Argentina has an extensive array of fisheries regulations and environmental related fisheries national plans in place, these have been developed from an extractive perspective and mainly for the fisheries industry impact mitigation. A relevant challenge, still in 2023, is how an effective ecosystem approach to fisheries will be implemented under the mandates of its National General Environmental Law within its ZEE.

Following the approval, in 2017, of the extension of the continental shelf limits of Argentina, by the United Nations Continental Shelf Limits Commission, the country incorporated, in 2020, approximately 1,782,500 km^2 under its sovereignty control. A bill to create a Marine Benthic Protected area (the "Blue Hole"), partially included within the extended shelf limits, is under consideration by the National Congress (June 2023). This bill, when implemented, will contribute to fisheries benthic environmental impact mitigation. Furthermore, in 2023, the UN member states agreed to a text on the first international treaty to protect the high seas, known as "Biodiversity Beyond National Jurisdiction" (BBNJ); this treaty aims to enhance the international legal regime concerning the conservation and sustainable use of biological diversity in the oceans beyond the EEZs and continental shelves of states. It is expected that when this treaty enters into force, it will complement the prospective benthic conservation area in the High Seas, beyond the Argentine extended shelf.

10.4.3 Forced Climate Change

No conservation or management approach will be completed if the forced climate change factor is not included. For example, large-scale, climate driven effects (e.g., Franco et al. 2020a, b) may relativize one of the most relevant features of the ESS: the constancy and predictability of some of its largest frontal areas. The single, most important physical variable under the effect of forced climate change in the system seems to be the surface wind patterns (see Piola et al. 2018 and references therein). Alterations in the intensity of the winds will impact on the circulation of water masses and the vertical stratification of coastal waters, generating changes in the availability of nutrients (Carranza et al. 2017). The cascading effects may reach the distribution of food for the higher predators, impacting on the energetics of foraging trips, the migratory path and the location of colonies along the coast.

State of the art, oceanographic modeling would require effective translation into management recommendations (Cavanagh et al. 2021), but the challenges of climate change transcend the science. Built for predictability, present bureaucratic institutional frameworks are poorly adapted to respond to global warming. Jurisdictional overlapping, for example, occurs when the authority on a particular issue is divided by concurrent agencies of the government. The administration of the ocean worldwide, but also at the national level, suffers from overlapping bureaucracies.

As a general reflection, the future of marine conservation requires a paradigmatic shift. Space and species-driven conservation are reactive to threats, rather than proactive. What is obviously required is a move towards precaution, that we call "species-dependent conservation." Species-dependent conservation considers the natural history of living things. It understands fisheries as targeting wildlife, altering their behavior and interfering with their biology. Only when traditional approaches to management and conservation are overcome, species-dependent conservation may have a chance. A really ambitious vision would be to propose the ESS as a candidate to advance the approach.

References

Acha EM, Mianzan HW, Guerrero RA, Favero M, Bava J (2004) Marine fronts at the continental shelves of austral South America: physical and ecological processes. J Mar Syst 44:83–105

Alberts EC (2020) Scientists agree on the need to protect 30% of the seas. But which 30%? https://news.mongabay.com/2020/06/scientists-agree-on-the-need-to-protect-of-30-of-the-sea-but-which-30/. Accessed on 25 Jan 2023

Alemany D, Acha EM, Iribarne OO (2016) Distribution and intensity of bottom trawl fisheries in the Patagonian shelf large marine ecosystem and its relationship with marine fronts. Fish Oceanogr 25(2):183–192

Asaad I, Lundquist CJ, Erdmann MV, Costello MJ (2018) Delineating priority areas for marine biodiversity conservation in the coral triangle. Biol Conserv 222:198–211

Barrionuevo M, Ciancio J, Steinfurth A, Frere E (2020) Geolocation and stable isotopes indicate habitat segregation between sexes in Magellanic penguins during the winter dispersion. J Avian Biol 51(2):e02325

Baylis AMM, Orben RA, Arkhipkin AA, Barton J, Brownell RL, Staniland IJ, Brickle P (2019) Re-evaluating the population size of South American fur seals and conservation implications. Aquat Conserv: Mar Freshw Ecosyst 29:1988–1995

Belkin IM (2009) Observational studies of oceanic fronts. J Mar Syst 78(3):317–318

Block BA, Jonsen ID, Jorgensen SJ, Winship AJ, Shaffer SA, Bograd SJ, Hazen EL, Foley DG, Breed GA, Harrison AL et al (2011) Tracking apex marine predator movements in a dynamic ocean. Nature 475(7354):86–90

Boersma PD (2012) Penguins and petroleum: lessons in conservation ecology. Front Ecol Evol 10:218–219

Boersma PD, Rebstock GA (2009) Foraging distance affects reproductive success in Magellanic penguins. Mar Ecol Progr Ser 375:263–275

Boersma PD, Rebstock GA, Frere E, Moore SE (2009) Following the fish: penguins and productivity in the South Atlantic. Ecol Monogr 79(1):59–76

Bovcon ND, Góngora ME, Marinao C, Gonzalez-Zevallos D (2013) Catches composition and discards generated by hake *Merluccius hubbsi* and shrimp *Pleoticus muelleri* fisheries: a case of study in the high-sea ice trawlers of San Jorge Gulf, Chubut, Argentina. Rev Biol Mar Oceanogr 48(2):303–319

Boyd IL, Staniland IJ, Martin AR (2002) Distribution of foraging by female Antarctic fur seals. Mar Ecol Progr Ser 242:285–294

Brooks TM, da Fonseca GAB, Rodrigues ASL (2004) Protected areas and species. Conserv Biol 18(3):616–618

Campagna C, Lewis M (1992) Growth and distribution of a southern elephant seal colony. Mar Mamm Sci 8(4):387–396

Campagna C, Fedak MA, McConnell BJ (1999) Post-breeding distribution and diving behavior of adult male southern elephant seals from Patagonia. J Mammal 80(4):1341–1352

Campagna C, Piola AR, Marin MR, Lewis M, Fernandez T (2006) Southern elephant seal trajectories, fronts and eddies in the Brazil/Malvinas confluence. Deep-Sea Res I Oceanogr Res Pap 53(12):1907–1924

Campagna C, Falabella V, Lewis M (2007a) Entanglement of southern elephant seals in squid fishing gear. Mar Mamm Sci 23(2):414–418

Campagna C, Sanderson EW, Coppolillo PB, Falabella V, Piola AR, Strindberg S, Croxall JP (2007b) A species approach to marine ecosystem conservation. Aquat Conserv: Mar Freshw Ecosyst 17:S122–S147

Campagna C, Falabella V, Zavattieri MV (2014) Species-driven conservation of Patagonian seascapes. In: Carleton R, McCormick-Ray J (eds) Marine conservation. Science, policy, and management. Wiley-Blackwell, p 368

Campagna J, Lewis MN, González Carman V, Campagna C, Guinet C, Johnson M, Davis RW, Rodríguez DH, Hindell MA (2020) Ontogenetic niche partitioning in southern elephant seals from Argentine Patagonia. Mar Mamm Sci 37(2):631–651

Cárdenas-Alayza S, Crespo E, Oliveira L (2016) *Otaria byronia*. The IUCN Red List of Threatened Species 2016: e.T41665A61948292 https://doi.org/10.2305/IUCN.UK.2016-1.RLTS.T41665A61948292.en

Carranza MM, Gille ST, Piola AR, Charo M, Romero SI (2017) Wind modulation of upwelling at the shelf-break front off Patagonia: observational evidence. J Geophys Res: Oceans 122(3):2401–2421

Carroll CR, Sousa J, Thevenon F (2014) Plastic debris in the ocean: the characterization of marine plastics and their environmental impacts. Situation Analysis Report, IUCN, p 52. https://doi.org/10.2305/IUCN.CH.2014.03.en

Cavanagh RD, Trathan PN, Hill SL, Melbourne-Thomas J, Meredith MP, Hollyman P, Krafft BA, Mc Muelbert M, Murphy EJ, Sommerkorn M, Turner J, Grant SM (2021) Utilising IPCC assessments to support the ecosystem approach to fisheries management within a warming Southern Ocean. Mar Pol 131:104589

CBD (2018) Convention on biological diversity. Decision adopted by the conference of the parties to the convention on biological diversity: 14/8 – Protected areas and other effective area- based conservation measures. CBD/COP/DEC/14/8 https://www.cbd.int/doc/decisions/cop-14/cop-14-dec-08-en.pdf

CBD (2023) Convention on biological diversity. Aichi Target Pages (11 and 12). https://www.cbd.int/aichi-targets/. Accessed on 25 Jan 2023

CBD/COP/15/L25 (n.d.). https://www.cbd.int/conferences/2021-2022/cop-15/documents

Chen CS, Chiu TS, Haung WB (2007) The spatial and temporal distribution patterns of the Argentine short-finned squid, *Illex argentinus*, abundances in the Southwest Atlantic and the effects of environmental influences. Zool Stud 46(1):111–122

Coll M, Steenbeek J, Ben Rais Lasram F, Mouillot D, Cury P (2015) 'Low-hanging fruit' for conservation of marine vertebrate species at risk in the Mediterranean Sea. Glob Ecol Biogeogr 24(2):226–239

Conway W (1992) A orillas de un helado río de mar. In: Lichter A (ed) Huellas en la arena, sombras en el mar: los mamíferos marinos de la Argentina y la Antártida. Ediciones Terra Nova, Buenos Aires

Copello S, Quintana F (2003) Marine debris ingestion by Southern Giant Petrels and its potential relationships with fisheries in the Southern Atlantic Ocean. Mar Poll Bull 46(11):1513–1515

Coppolillo P, Gomez H, Maisels F, Wallace R (2004) Selection criteria for suites of landscape species as a basis for site-based conservation. Biol Conserv 115(3):419–430

Costello MJ, Ballantine B (2015) Biodiversity conservation should focus on no-take marine reserves: 94% of marine protected areas allow fishing. Trends Ecol Evol 30(9):507–509

Croxall J, Gales R (1998) An assessment of the conservation status of albatrosses. In: Robertson G, Gales R (eds) Albatross biology and conservation. Surrey Beatty, pp 46–65

Croxall JP, Wood AG (2002) The importance of the Patagonian Shelf for top predator species breeding at South Georgia. Aquat Conserv: Mar Freshw Ecosyst 12:101–118

Croxall JP, Butchart SHM, Lascelles BEN, Stattersfield AJ, Sullivan BEN, Symes A, Taylor P (2012) Seabird conservation status, threats and priority actions: a global assessment. Bird Conserv Int 22(1):1–34

Cuevas JM, Michelson AM (2023) Estado actual del conocimiento sobre condrictios en la Reserva Natural de Usos Múltiples Bahía San Blas, provincia de Buenos Aires, p 120

Dans SL, Alonso MK, Pedraza SN, Crespo EA (2003) Incidental catch of dolphins in trawling fisheries off Patagonia, Argentina: can populations persist? Ecol Appl 13(3):754–762

Davidson LNK, Dulvy NK (2017) Global marine protected areas to prevent extinctions. Nat Ecol Evol 1(2):0040. https://doi.org/10.1038/s41559-016-0040

de la Garza J, Moriondo Danovaro P, Fernández M, Ravalli C, Souto V, Waessle J (2017) An overview of the Argentine red shrimp (*Pleoticus muelleri*, Decapoda, Solenoceridae) fishery in Argentina: biology, fishing, management and ecological interactions. Instituto Nacional de Investigación y Desarrollo Pesquero (INIDEP), p 42

Devillers R, Pressey RL, Grech A, Kittinger JN, Edgar GJ, Ward T, Watson R (2015) Reinventing residual reserves in the sea: are we favoring ease of establishment over need for protection? Aquat Conserv: Mar Freshw Ecosyst 25:480–504

Di Mauro R, Castillo S, Perez A, Iachetti CM, Silva L, Tomba JP, Chiesa IL (2022) Anthropogenic microfibers are highly abundant at the Burdwood Bank seamount, a protected sub-Antarctic environment in the Southwestern Atlantic Ocean. Environ Pollut 306:119364

Eken G, Bennun L, Brooks TM et al (2004) Key biodiversity areas as site conservation targets. Bioscience 54(12):1110–1118

Falabella V, Campagna C, Croxall JP (2009) Atlas del Mar Patagónico. Especies y espacios. Wildlife Conservation Society and BirdLife International, Buenos Aires, p 304

FAO (2022) A handbook for identifying, evaluating and reporting other effective area-based conservation measures in marine fisheries. Rome, p 100

Favero M, Silva Rodríguez MP (2005) Estado actual y conservación de aves pelágicas que utilizan la plataforma continental Argentina como área de alimentación. Hornero 20(1):95–110

Favero M, Khatchikian CE, Arias A, Silva R, Cañete G, Mariano-Jelicich R (2003) Estimates of seabird by-catch along the Patagonian shelf by Argentine longline fishing vessels, 1999–2001. Bird Conserv Int 13(4):273–281

Franco BC, Combes V, González Carman V (2020a) Subsurface Ocean warming hotspots and potential impacts on marine species: the southwest South Atlantic Ocean case study. Front Mar Sci 7(824). https://doi.org/10.3389/fmars.2020.563394

Franco BC, Defeo O, Piola AR, Barreiro M, Yang H, Ortega L, Gianelli I, Castello JP, Vera C, Buratti C, Pájaro M, Pezzi LP, Möller OO (2020b) Climate change impacts on the atmospheric circulation, ocean, and fisheries in the southwest South Atlantic Ocean: a review. Clim Chang 162:2359–2377

Galimberti F, Sanvito S (2012) Tracking at sea the elephant seals of sea lion island. ESRG Technical reports, Elephant Seal Research Group. Available from www.eleseal.org, p 44

Gandini P, Frere E (2006) Spatial and temporal patterns in the bycatch of seabirds in the Argentinian longline fishery. Fish Bull 104(3):482–485

Gandini P, Boersma D, Frere E, Gandini M, Holik T, Lichtschein V (1994) Magellanic penguins (*Spheniscus magellanicus*) affected by chronic petroleum pollution along coast of Chubut, Argentina. Auk 111(1):20–27

Gandini P, Millones A, Morgenthaler A, Frere E (2016) Population trends of the southern rockhopper penguin (*Eudyptes chrysocome chrysocome*) at the northern limit of its breeding range: Isla Pingüino, Santa Cruz. Argentina Polar Biol 40(5):1023–1028

Garcia-Borboroglu P, Boersma PD, Ruoppolo V, Reyes L, Rebstock GA, Griot K, Heredia SR, Adornes AC, da Silva RP (2006) Chronic oil pollution harms Magellanic penguins in the Southwest Atlantic. Mar Poll Bull 52(2):193–198

García-Borboroglu P, Pozzi LM, Parma AM, Dell'Arciprete P, Yorio P (2022) Population distribution shifts of Magellanic penguins in northern Patagonia, Argentina: implications for conservation and management strategies. Ocean Coastal Manag 226:106259

Giussi AR, Prosdocimi L, Carozza CR, Navarro G (2022) Estado de los recursos pesqueros bajo administración exclusiva de la República Argentina. Aportes para el informe SOFIA 2022. Informe de Asesoramiento y Transferencia INIDEP 012–22

Gonzalez Carman V, Acha EM, Maxwell SM, Albareda D, Campagna C, Mianzan H (2014) Young green turtles, *Chelonia mydas*, exposed to plastic in a frontal area of the SW Atlantic. Mar Poll Bull 78(1–2):56–62

Gownaris NJ, Santora CM, Davis JB, Pikitch EK (2019) Gaps in protection of important ocean areas: a spatial meta-analysis of ten global mapping initiatives. Front Mar Sci 6. https://doi.org/10.3389/fmars.2019.00650

Grandi MF, Dans SL, Crespo EA (2008) Social composition and spatial distribution of colonies in an expanding population of south American sea lions. J Mammal 89(5):1218–1228

Gregory MR (2009) Environmental implications of plastic debris in marine settings-entanglement, ingestion, smothering, hangers-on, hitch-hiking and alien invasions. Philos Trans R Soc Lond Ser B Biol Sci 364(1526):2013–2025
Guerrero RA, Piola AR (1997) Masas de agua en la plataforma continental. In: Boschi EE (ed) INIDEP El Mar Argentino y sus Recursos Pesqueros 1:107–118
Halpern BS, Frazier M, Potapenko J, Casey KS, Koenig K, Longo C, Lowndes JS, Rockwood RC, Selig ER, Selkoe KA, Walbridge S (2015) Spatial and temporal changes in cumulative human impacts on the world's ocean. Nat Commun 6:7615
Harris G (2016) Conserving the Cold Sea River. Civil Society and government working together in Patagonia. In Voices of Impact: Speaking for the global commons. Stories from 25 years of environmental innovation for sustainable development. Mandy Cadman, editor. UNDP- Global Environmental Finance, p 55–57
Hiddink JG, Jennings S, Sciberras M, Szostek CL, Hughes KM, Ellis N, Rijnsdorp AD, McConnaughey RA, Mazor T, Hilborn R, Collie JS, Pitcher CR, Amoroso RO, Parma AM, Suuronen P, Kaiser MJ (2017) Global analysis of depletion and recovery of seabed biota after bottom trawling disturbance. Proc Natl Acad Sci USA 114(31):8301–8306
Hindell MA, Reisinger RR, Ropert-Coudert Y et al (2020) Tracking of marine predators to protect Southern Ocean ecosystems. Nature 580(7801):87–92
Hobday AJ, Hartmann K (2006) Near real-time spatial management based on habitat predictions for a longline bycatch species. Fish Manag Ecol 13:365–380
Hobday AJ, Maxwell SM, Forgie J et al (2014) Dynamic Ocean management: integrating scientific and technological capacity with law, policy and management. Stanf Environ Law J 33:125–165
Hoyt E, di Sciara GN (2021) Important marine mammal areas: a spatial tool for marine mammal conservation. Oryx 55:330–330
Huin N, Reid T (2007) Census of the black-browed albatross population of the Falkland Islands: 2000 and 2005. Falklands Conservation, Stanley/Puerto Argentino, Falkland Islands/Malvinas Islands, p 41
Irusta G, Macchi GJ, Louge E, Rodrigues KA, D'Atri L et al (2016) Biology and fishery of the Argentine hake (*Merluccius hubbsi*). Rev Invest Des Pesq INIDEP 28:9–36
IUCN WCPA (2019) Guidelines for recognizing and reporting other effective area-based conservation measures. IUCN, Switzerland, p 46
Jefferson T, Costello MJ, Zhao Q, Lundquist CJ (2021) Conserving threatened marine species and biodiversity requires 40% ocean protection. Biol Conserv 264:109368
Jenkins CN, Van Houtan KS (2016) Global and regional priorities for marine biodiversity protection. Biol Conserv 204:333–339
Klein C, Brown C, Halpern B et al (2015) Shortfalls in the global protected area network at representing marine biodiversity. Sci Rep 5:17539
Laffoley D, Dudley N, Jonas H, MacKinnon D, MacKinnon K, Hockings M, Woodley S (2017) An introduction to 'other effective area-based conservation measures' under Aichi target 11 of the convention on biological diversity: origin, interpretation and emerging ocean issues. Aquat Conserv: Mar Freshw Ecosys 27:130–137
Lascelles BG, Langham GM, Ronconi RA, Reid JB (2012) From hotspots to site protection: identifying marine protected areas for seabirds around the globe. Biol Conserv 156:5–14
Lester SE, Halpern BS (2008) Biological responses in marine no-take reserves versus partially protected areas. Mar Ecol Progr Ser 367:49–56
Lester SE, Halpern BS, Grorud-Colvert K, Lubchenco J, Ruttenberg BI, Gaines SD, Airame S, Warner RR (2009) Biological effects within no-take marine reserves: a global synthesis. Mar Ecol Progr Ser 384:33–46
Longhurst A (1998) Ecological geography of the sea. Academic Press, San Diego, p 560
Mandiola MA, Bagnato R, Gana JCM, De León MC, Dassis M, Albareda D, Denuncio P (2021) Baseline data of the presence of meso and microplastics in digestive tract of a commercially important teleost fish from the Rio de la Plata estuary system (Southwest Atlantic Ocean). Mar Fish Sci (MAFIS) 35(1). https://doi.org/10.47193/mafis.3512022010101
Marinao CJ, Yorio P (2011) Fishery discards and incidental mortality of seabirds attending coastal shrimp trawlers at Isla Escondida, Patagonia, Argentina. Wilson J Ornith 123(4):709–719

McCauley DJ, Pinsky ML, Palumbi SR, Estes JA, Joyce FH, Warner RR (2015) Marine defaunation: animal loss in the global ocean. Science 347(6219):1255641

Mendez M, Rosenbaum HC, Bordino P (2008) Conservation genetics of the Franciscana dolphin in northern Argentina: population structure, by-catch impacts, and management implications. Conserv Genet 9(2):419–435

MPAtlas (2023). https://mpatlas.org/. Accessed on 25 Jan 2023

Negri MF, Denuncio P, Panebianco MV, Cappozzo HL (2012) Bycatch of Franciscana dolphins *Pontoporia blainvillei* and the dynamic of artisanal fisheries in the species' southernmost area of distribution. Braz J Oceanogr 60(2):149–158

Nicholls DG, Robertson CJR, Prince PA, Murray MD, Walker KJ, Elliot GP (2002) Foraging niches of three *Diomedea albatrosses*. Mar Ecol Progr Ser 231:260–277

Niedermüller S, Ainsworth G, de Juan S, García R, Ospina-Alvarez A, Pita P, Villasante S (2021) The shark and ray meat network. A deep dive into a global affair. WWF, p 34

Osieck ER, Bruyns MFM, Hallmann B (1981) Important bird areas in the European Community. International Council for Bird Preservation, Cambridge

Piola AR (2008) Oceanografía Física. In: Estado de conservación del Mar Patagónico y áreas de influencia, vol versión electrónica - en línea. Foro para la Conservación del Mar Patagónico y Áreas de Influencia, Puerto Madryn, pp 1–21. http://www.marpatagonico.org

Piola AR, Campos EJD, Moller OO, Charo M, Martinez C (2000) Subtropical shelf front off eastern South America. J Geophys Res 105(C3):6565–6578

Piola AR, Palma ED, Bianchi AA, Castro BM, Dottori M, Guerrero RA, Marrari M, Matano RP, Möller Jr OO, Saraceno M (2018) Physical oceanography of the SW Atlantic shelf: a review. In: Plankton ecology of the Southwestern Atlantic. https://doi.org/10.1007/978-3-319-77869-3_2

Podestá G (1997) Utilización de datos satelitarios en investigaciones oceanográficas y pesqueras en el océano Atlántico sudoccidental. In: Boschi EE; ed. Antecedentes históricos de las exploraciones en el mar y las características ambientales. INIDEP, El Mar Argentino y sus Recursos Pesqueros 1:195–222

Pütz K, Ingham RJ, Smith JG (2002) Foraging movements of Magellanic penguins *Spheniscus magellanicus* during the breeding season in The Falkland Islands. Aquat Conserv Mar Freshw Ecosyst 12(1):75–87

Pütz K, Hiriart-Bertrand L, Simeone A, Riquelme V, Reyes-Arriagada R, Luthi B (2011) Entanglement and drowning of a Magellanic penguin (*Spheniscus magellanicus*) in a gill net recorded by a time-depth recorder in south-central Chile. Waterbirds 34(1):121–125

Pütz K, Raya Rey A, Otley H (2013) Southern Rockhopper penguin. In: García Borboroglu P, Boersma PD (eds) Penguins: natural history and conservation. University of Washington Press, Seattle

Queiroz N, Humphries NE, Couto A et al (2019) Global spatial risk assessment of sharks under the footprint of fisheries. Nature 572(7770):461–466

Sabadin DE, Lucifora L, Barbini SA, Figueroa DE, Kittlein MJ (2020) Towards regionalization of the chondrichthyan fauna of the Southwest Atlantic: a spatial framework for conservation planning. ICES J Mar Sci 77(5):1893–1905

Sala E, Giakoumi S (2017) No-take marine reserves are the most effective protected areas in the ocean. ICES J Mar Sci 75(3):1166–1168

Sala E, Mayorga J, Costello C, Kroodsma D, Palomares MLD, Pauly D, Sumaila UR, Zeller D (2018) The economics of fishing the high seas. Sci Adv 4(6):eaat2504

Sala E, Mayorga J, Bradley D et al (2021) Protecting the global ocean for biodiversity, food and climate. Nature 592(7854):397–402

Sanderson EW, Redford KH, Vedder A, Coppolillo PB, Ward SE (2002) A conceptual model for conservation planning based on landscape species requirements. Landsc Urban Plan 58:41–56

Scales KL, Miller PI, Hawkes LA, Ingram SN, Sims DW, Votier SC (2014) Review: on the front line: frontal zones as priority at-sea conservation areas for mobile marine vertebrates. J Appl Ecol 51:1575–1583

Sciberras M, Jenkins SR, Mant R, Kaiser MJ, Hawkins SJ, Pullin AS (2015) Evaluating the relative conservation value of fully and partially protected marine areas. Fish Fish 16(1):58–77

Sherman K, Alexander LM, Gold BD (1990) Large marine ecosystems: Patterns, processes and yields. American Association for the Advancement of Science: Washington D.C, XIII, p 242

Singleton RL, Roberts CM (2014) The contribution of very large marine protected areas to marine conservation: giant leaps or smoke and mirrors? Mar Pollut Bull 87:7–10

Tamini LL, Chavez LN, Góngora ME, Yates O, Rabuffetti FL, Sullivan B (2015) Estimating mortality of black-browed albatross (*Thalassarche melanophris*, Temminck, 1828) and other seabirds in the Argentinean factory trawl fleet and the use of bird-scaring lines as a mitigation measure. Pol Biol 38:1867–1879

Thompson D, Moss SEW, Lovell P (2003) Foraging behaviour of south american fur seals *Arctocephalus australis*: extracting fine scale foraging behaviour from satellite tracks. Mar Ecol Progr Ser 260:285–296

UNEP-WCMC and IUCN (2023) Protected planet: the world database on protected areas (WDPA) and world database on other effective area-based conservation measures (WD-OECM) Available at: www.protectedplanet.net [July 2023]

Vales DG, Mandiola A, Romero MA, Svendsen G, Túnez JI, Negrete J, Grandi MF (2019) *Arctocephalus australis*. In: SAyDS–SAREM (eds) Categorización 2019 de los mamíferos de Argentina según su riesgo de extinción. Lista Roja de los mamíferos de Argentina

Vaughan A (2021) The world has missed its target for protecting oceans to save species. https://www.newscientist.com/article/2277886-the-world-has-missed-its-target-for-protecting-oceans-to-save-species/

Vélez-Rubio GM, Teryda N, Asaroff PE, Estrades A, Rodriguez D, Tomás J (2018) Differential impact of marine debris ingestion during ontogenetic dietary shift of green turtles in Uruguayan waters. Mar Pollut Bull 127:603–611

Vilela R, Conesa D, del Rio JL, López-Quflez A, Portela J, Bellido JM (2018) Integrating fishing spatial patterns and strategies to improve high seas fisheries management. Mar Pol 94:132–142

Visconti P, Butchart SHM, Brooks TM, Langhammer PF, Marnewick D, Vergara S, Yanosky A, Watson JEM (2019) Protected area targets post-2020. Science 364(6437):239–241

Waluda CM, Trathan PN, Elvidge CD, Hobson VR, Rodhouse PG (2002) Throwing light on straddling stocks of *Illex argentinus*: assessing fishing intensity with satellite imagery. Can J Fish Aquat Sci 59(4):592–596

Werner R, Campagna C (1995) Diving behaviour of lactating southern sea lions (*Otaria flavescens*) in Patagonia. Can J Zool Rev Can Zool 73(11):1975–1982

Wolfaardt A (2012) An assessment of the population trends and conservation status of Black-browed Albatrosses in the Falkland Islands. Joint Nature Conservation Committee (JNCC), July 2012

Yorio P, Frere E, Gandini P, Harris G (1998) Atlas de la distribución de aves marinas en el litoral patagónico argentino. Instituto Salesiano de Artes Gráficas, Buenos Aires, Buenos Aires, p 221

Yorio P, Quintana F, Lopez de Casenave J (2005) Ecología y conservación de las aves marinas del litoral marítimo argentino. Hornero 20(1):1–3

Zeller D, Cashion T, Palomares M, Pauly D (2018) Global marine fisheries discards: a synthesis of reconstructed data. Fish Fish 19:30–39

Zerbini AN, Rosenbaum H, Mendez M, Sucunza F, Andriolo A, Harris G, Clapham P, Sironi M, Uhart M, Fernández Ajó AA (2016) Tracking southern right whales through the southwest Atlantic: An update on movements, migratory routes and feeding grounds. International Whaling Commission SC/66b/BRG/26

Zhao Q, Stephenson F, Lundquist C, Kaschner K, Jayathilake D, Costello MJ (2020) Where marine protected areas would best represent 30% of ocean biodiversity. Biol Conserv 244:108536

Chapter 11
Food Web Topology Associated with the Patagonian Shelf-Break Front

Florencia Botto, Paulina Martinetto, Daniela Alemany, and Clara Díaz de Astarloa

Abstract This chapter provides a summary of trophic interactions within the Patagonian Shelf-break Front (PSBF) ecosystem under the food web conceptual framework. We collected the published diet information of the organisms inhabiting the PSBF and constructed the trophic interaction network. We considered 55 trophic species (i.e., phytoplankton, POM and debris as basal species, 19 species/groups of invertebrates, 23 species of fish, 7 sea birds, and 2 marine mammals) with 342 links among them. The connectance of the PSBF food web was 0.11 and the maximum trophic levels was 4.7 for the elephant seal. A pelagic and a benthic pathway were detected in the PSBF, with large demersal generalist fish species (e.g., Argentine hake, long tail hake), and top predators (e.g., elephant seals) playing an important role linking both pathways. The analysis reveals gaps in knowledge, particularly regarding interactions involving basal species and microorganisms with early life stages of fishes and invertebrates. The PSBF food web is dynamic, influenced by spatial and temporal variability in the encounters between species, given their movements and diet shifts during their life cycles. Understanding the food web dynamic and functioning is crucial for predicting the impact of human activities, and the topological description is a first step toward that goal.

Supplementary Information The online version contains supplementary material available at https://doi.org/10.1007/978-3-031-71190-9_11.

F. Botto (✉) · P. Martinetto · C. Díaz de Astarloa
Instituto de Investigaciones Marinas y Costeras (IIMyC, UNMdP-CONICET), Facultad de Ciencias Exactas y Naturales, Universidad Nacional de Mar del Plata (UNMdP), Consejo Nacional de Investigaciones Científicas y Técnicas (CONICET), Mar del Plata, Argentina

D. Alemany
Instituto de Investigaciones Marinas y Costeras (IIMyC, UNMdP-CONICET), Facultad de Ciencias Exactas y Naturales, Universidad Nacional de Mar del Plata (UNMdP), Consejo Nacional de Investigaciones Científicas y Técnicas (CONICET), Mar del Plata, Argentina

Instituto Nacional de Investigación y Desarrollo Pesquero (INIDEP), Mar del Plata, Argentina

E. M. Acha et al. (eds.), *The Patagonian Shelfbreak Front*, Aquatic Ecology Series 13, https://doi.org/10.1007/978-3-031-71190-9_11

Acronyms and Abbreviations

POM Particulate organic matter
PSB Patagonian Shelf-break front

11.1 Introduction

Food webs describe trophic interactions among species in a community to comprehensively understand the ecological roles of species, the structure of the communities, and ecosystem functions (Layman et al. 2015 and references therein). Food web theory provides valuable tools to evaluate the stability and persistence of communities (Dunne et al. 2005), to predict the development of populations (Morin and Lawler 1995), and to understand ecosystem services (Hines et al. 2015). All these tools are useful for the management and conservation of natural systems (McDonald-Madden et al. 2016). Gaining a better understanding of the structure and functioning of food webs in marine systems is of particular importance given the increasing interest of implementing an ecosystem approach to manage fisheries (Link et al. 2010; Tam et al. 2017). Specifically, a management approach that considers not only the single target species, but also their multiple ecological interactions, with the objective to maintain the natural structure, balance and functioning of marine ecosystems (e.g., Tam et al. 2017).

Most of the food web theory was first developed in terrestrial or non-marine aquatic systems where spatial boundaries are usually well-defined (Layman et al. 2015). In marine communities, the difficulty of studying food webs is due to the spatio-temporal scale in which interactions occur (Link 2002). Some populations of marine species disperse over large areas and some can even perform large distance movements daily. That is why the first studies describing marine food webs were performed in defined areas such as bays (e.g., Baird and Ulanowicz 1989), or estuaries (e.g., Menge and Sutherland 1976), or associated with some bottom structures (e.g., rocky shores: Paine 1966, reefs: Opitz 1993), which spatially determines the community. However, since then, the description of the structure of marine food webs increased (e.g., Link 2002) and accelerated with the use of stable isotopes analysis (SIA) as an important tool to infer trophic relationships (Layman et al. 2012). Since the 2000s, network theory developed as a new framework for food web theory and was used to describe and analyze properties of food webs (e.g., Dunne et al. 2002; Dunne 2006) including robustness (e.g., Dunne et al. 2004) and spatial and temporal dynamics of marine food webs (Kortsch et al. 2019, 2021).

Particularly, in the SW Atlantic, the marine food web of the Argentine Sea was described in the pioneer study of Angelescu and Prenski (1987). They focused on the role of the main fishery target species, the Argentine hake *Merluccius hubbsi*, proposing it as the main link between demersal-pelagic and demersal-benthic food web compartments. Other food web studies in the region focused on estuarine coastal environments (e.g., Mar Chiquita: Botto et al. 2005; Río de la Plata: Botto et al. 2011); the Southern Patagonian shelf (Ciancio et al. 2008; Gaitán 2012; Riccialdelli et al. 2017), and the Namuncurá/Burdwood Bank (Riccialdelli et al. 2017). In addition, a network analysis was performed in San Jorge Gulf showing the effect of fishery (Funes et al. 2022). However, food web studies in the offshore zone are less common and are usually focused on a single compartment of the food web (i.e., benthos: Botto et al. 2006; fishes: Saporiti et al. 2015; Zhu et al. 2017). In oceanic systems, frontal areas are of particular interest, given that they are hotspots of primary production and therefore considered as areas that support food webs. This is the case of the Patagonian Shelf-break Front (herein PSBF), which is well described in previous chapters. From an ecological point of view, this area can be considered as a boundary system with influence from the deep ocean and from the continental shelf. Some species distributions are highly associated with the frontal zone, which through trophic pathways can transfer material between these systems. The diet of many of the components of the food web that inhabit the PSBF has been well studied and described through stomach content or stable isotope analyses (see Table 11.1 for references) but until now has not been analyzed under a topological food web framework. Although most novel food web theory attempts to understand fluxes of energy and biomass (e.g., Kortsch et al. 2021), the first step is to have a basic topological description of the trophic interactions in the food web. Therefore, the goal of this chapter is to describe the general characteristics of the food web associated with the PSBF.

11.2 General Description of the Patagonian Shelf Break Front Food Web

In order to describe the food web of the PSBF, we constructed a network of all known trophic interactions among species that inhabit the area at least during some stage of their life cycle. After selecting species or groups, we reviewed the published diet information, which is summarized in Table 11.1. Then, we identified trophic links considering all prey items found in each species diet. We prioritized studies with sampling points in the area of interest and detailed stomach content analysis. For those species or groups without stomach content information, we used studies based on SIA when available, considering trophic links inferred by mixing models (Botto et al. 2006). For some invertebrates (i.e., ctenophores. Polychaetes, Chaetognata) diet information was only found for other areas. When two or more studies were found with comparable detailed diet information, some data dating back 20 years or more were discarded. To describe the PSBF food web, we calculated commonly used metrics (e.g., Pimm et al. 1991; Williams and Martinez 2000; Kortsch et al. 2019), using packages igraph for the number of species or trophic "species" (S = groups of organisms at taxonomically feasible and functionally related levels), the interaction among those groups (L = links), the realized number

Table 11.1 Food web parameters calculated (TL = trophic level, V = vulnerability, G = generality and main prey items (ordered according to importance) of each species or group. References indicate the studies from which trophic links were taken to construct the food web. The diet here is described in a very broad way according the references; s and l indicate small and large organisms respectively

	TL	V	G	Main prey items	References
Pelagic invertebrates:					
Munida gregaria	2	13	2	Phytoplankton and POM	Romero et al. (2004), Pérez-Barros et al. (2010)
Copepods	2.17	8	3	Phytoplankton and dinoflagelletes	Carreto et al. (2016)
Euphausids	2.17	22	3	Phytoplankton and dinoflagelletes	Carreto et al. (2016), Viñas et al. (2016)
Ctenophora	2.33	10	2	Copepods and microzooplankton	Purcell (1991), Briz et al. (2017)
Chaetognatha	2.44	4	3	Copepods and microzooplankton	Sabatini et al. (2012)
Themisto gaudichaudii	2.57	19	4	Copepods, euphausids chaetognatha	Viñas et al. (2016)
Doryteuthis gahi	3.62	22	5	s: euphausiids and other crustaceans, l: Fishes (*Salilota australis*) and squids (cannibalism)	Rosas-Luis et al. (2016), Büring et al. (2021)
Martialia hyadesi	3.64	3	4	Euphausids, myctophids l: cannibalism	Brunetti et al. (1998)
Illex argentinus	3.79	16	8	Euphausids, *T. gaudichaudii*; l: fishes (Myctophids) and squids (*D. gahi)*	Ivanovic and Brunetti (1994), Ivanovic (2010), Rosas-Luis et al. (2016)
Onykia ingens	3.85	5	8	Squids (*D. gahi, I. argentinus*), fishes (Myctophids) and *Munida gregaria*	Rosas-Luis et al. (2014, 2016)
Benthic invertebrates:					
Zygochlamys patagónica	2.16	3	3	Diatoms dinoflagelletes and POM	Schejter (2000)
Crabs	2.25	10	2	Benthic small organisms and debris	
Gastropods	3.05	11	3	Bivalves, infauna	
Sea stars	3.24	0	3	Benthic invertebrates, gastropods	Botto et al. (2006)
Pelagic fishes					
Sprattus fueguensis	2.58	7	3	Zooplankton	Ciancio et al. (2008), Montecinos Garrido (2015)
Engraulis anchoita	2.6	3	4	Zooplankton,	Pájaro (2002)
Myctophids	3.16	12	2	Zooplankton	
Demersal-pelagic fishes					
Macruronus magellanicus	3.43	11	12	s: zooplankton (Euphausids, *Munida gregaria T. gaudichaudii* l: zooplankton and fishes (*Sprattus fueguensis*, Myctophids)	Mari and Sanchez (2002), Brickle et al. (2009), Alvarez et al. (2022)

(continued)

Table 11.1 (continued)

	TL	V	G	Main prey items	References
Micromesistius australis	3.51	11	9	Euphausids, *T. gaudichaudi*, Myctophids	Brickle et al. (2009)
Merluccius hubbsi	3.7	13	12	s: Euphausids, *T. gaudichaudi, M. gregaria*; l: squids and fishes (*P. ramsayi, E. anchoita*, cannibalism)	Belleggia et al. (2014)
Dissostichus eleginoides	4	7	14	Fishes (*P. ramsayi, Macrourus spp*) and squids (*D. gahi, O. ingens*). S: Euphausids	Troccoli et al. (2020) de la Rosa et al. (1997)
Merluccius australis	4.57	0	5	Fishes (*M. magellanicus*), squids (*I. argentinus, D. gahi, O. ingens*)	Giussi et al. (2016)
Demersal-benthic fishes:					
M. holotrachys	3	1	4	Benthic organisms: Amphipods gammarids, polychaetes, isopods	Romanelli (2017)
Macrourus carinatus,	3.19	0	8	Benthic organisms: Amphipods gammarids, polychaetes, isopods	Romanelli (2017)
Coelorinchus fasciatus	3.19	2	8	Benthic invertebrates: Amphipods gammarids, polychaetes, euphausids	Romanelli (2017)
Patagonotothen ramsayi	3.35	21	12	Benthic invertebrates, jellyfishes; l: polychaetes s: isopods	Laptikhovsky and Arkhipkin (2003), Laptikhovsky (2004), Fischer et al. (2022)
Bassanago albescens	3.42	2	7	s: amphipods, l: isopods and squids	Izzo (2010)
Schroederichthys bivius	3.54	1	14	s: crustaceans and polychaetes, l: squids and fishes	Laptikhovsky et al. (2001)
Salilota australis	3.63	5	16	s: gammarid amphipods, isopods, squids; l: *P. ramsayi*	Arkhipkin et al. (2001)
Squalus acanthias	3.64	0	12	Squids (*I. argentinus, D. gahi*) and fishes (*M. hubbsi, S. fueguensis*), ctenophors	Laptikhovsky et al. (2001), Koen-Alonso et al. (2002)
Genypterus blacodes	3.78	5	17	Fishes (*P. ramsayi, M. magellanicus, Mi. australis*), s: isopods	Nyegaard et al. (2004)
Benthic fishes:					
Bathyraja macloviana	3.08	0	4	Polychaetes, crustaceans	Mabragaña et al. (2005)
Bathyraja albomaculata	3.42	1	10	Polychaetes, crustaceans	Brickle et al. (2003)
Cottoperca gobio	3.64	6	7	Squids *Doryteuthis gahi*, fishes *Patagonotothen ramsayi*	Laptikhovsky and Arkhipkin (2003)
Bathyraja griseocauda	3.67	1	15	Benthic crustaceans: Amphipods and isopods	Brickle et al. (2003)
Bathyraja brachiurops	3.68	2	14	S: isopods, l: crabs and benthic fishes	Belleggia et al. (2008)

(continued)

Table 11.1 (continued)

	TL	V	G	Main prey items	References
Zearaja chilensis	3.69	0	19	Fishes (*M. hubbsi, P. ramsayi*); s: isopods	Belleggia et al. (2016)
Birds:					
Pygoscelis papua	3.55	0	8	Crustaceans (*M. gregaria*) Fishes (*S. fueguensis*), squids (*O. ingens*)	Clausen et al. (2005)
Spheniscus magellanicus	3.77	1	5	Fishes (*P. ramsayi, S. fueguensis, Mi. australis*), crustaceans (*M. gregaria*)	Thompson (1993), Pütz et al. (2000)
Eudyptes chrysocome	3.85	0	8	Crustaceans (*T. gaudichaudii*, Euphausids, *M. gregaria*), squids (*I. argentines, O. ingens*), fishes (*S. fueguensis*, Myctophids)	Rey and Schiavini (2005), Dehnhard et al. (2016)
Aptenodytes patagonicus	4.6	0	6	Fishes (Myctophids) and squids (*O. ingens, M. hyadesi*)	Piatkowski et al. (2001), Cherel et al. (2002), Pütz (2002)
Diomedea exulans	4.6	0	5	Squids, fishes, fishery discards	Xavier et al. (2004), Bugoni et al. (2010)
Thalassarche melanophris	4	0	9	Squids, fishes, fishery discards	Thompson (1992), Mariano-Jelicich et al. (2014)
Macronectes giganteus	4.67	0	6	Squids, fishes, fishery discards	Copello et al. (2008)
Mammals:					
Eubalaena australis	3.11	0	3	Zoooplankton	Best and Schell (1996), Valenzuela et al. (2018)
Mirounga leonina	4.7	0	4	Fishes	Campagna et al. (2021)

of links in relation to maximum possible links (Connectance; $C = L*S^{-2}$, trophic level, % of cannibalism, omnivory index (the standard deviation of the trophic levels of a species' prey), generality (number of prey per predator), and vulnerability (number of predators per prey).

The PSBF network contained 55 trophic species (S): three highly aggregated basal groups (i.e., phytoplankton, POM and debris), 19 species/groups of invertebrates, 23 species of fish, 7 seabirds, and 2 marine mammals (Table 11.1). All components were classified as pelagic, benthic, demersal-benthic or demersal-pelagic (for fishes and cephalopods, we followed Angelescu and Prenski 1987). Although most abundant components of the food web are represented, like all food webs, this would be by default incomplete. A total of 342 links were identified considering all prey consumed by each predator (gray lines in Fig. 11.1). For fish species, main links (black lines Fig. 11.1) were considered when frequency of occurrence of a prey item reached values greater than 10%. Trophic level was calculated for each species (Table 11.1), being mean TL = 3, with top predators reaching values of TL = 4.7. Mean generality (number of prey items per predator) was 6, and omnivory was 0.6. Connectance in the PSBF food web was C = 0.113, which is similar to average values of compilations of food webs (e.g., Martinez 1991; Dunne et al.

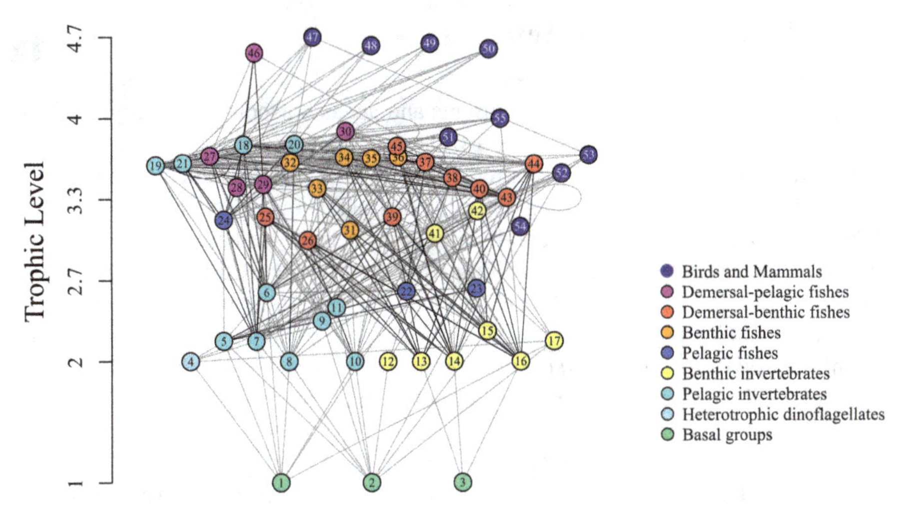

Fig. 11.1 Trophic interactions network of the Patagonian Shelf-break front, considering all trophic links reported in literature. Each circle represents a node and each line a trophic link. 1. Phytoplankton, 2. POM, 3. Debris, 4. Heterotrophic dinoflagellates, 5. Copepods, 6. *Themisto gaudichaudii,* 7. Euphausiids, 8. *Munida gregaria,* 9. Ctenophores, 10. Salps, 11. Chaetognaths, 12. Holothuroids, 13. Gammarid Amphipods, 14. Isopods, 15. Crabs, 16. Polychaetes, 17. *Zygochlamys patagonica,* 18. *Illex argentinus,* 19. *Doryteuthis gahi,* 20. *Onykia ingens,* 21. *Martialia hyadesi,* 22. *Sprattus fueguensis,* 23. *Engraulis anchoita,* 24. Myctophids, 25. *Macrourus carinatus,* 26. *Macrourus holotrachys,* 27. *Merluccius hubbsi,* 28. *Macruronus magellanicus,* 29. *Micromesistius australis,* 30. *Dissostichus eleginoides,* 31. *Bathyraja macloviana,* 32. *Cottoperca gobio,* 33. *Bathyraja albomaculata,* 34. *Bathyraja brachyurops,* 35. *Bathyraja griseocauda,* 36. *Zearaja chilensis,* 37. *Squalus acanthias,* 38. *Schroederichthys bivius,* 39. *Coelorinchus fasciatus,* 40. *Bassanago albescens,* 41. Gastropods, 42. Sea stars, 43. *Patagonotothen ramsayi,* 44. *Salilota australis,* 45. *Genypterus blacodes,* 46. *Merluccius australis,* 47. *Macronectes giganteus,* 48. *Diomedea exulans,* 49. *Mirounga leonina,* 50. *Aptenodytes patagonicus,* 51. *Eudyptes chrysocome,* 52. *Pygoscelis papua,* 53. *Spheniscus magellanicus,* 54. *Eubalaena australis,* 55. *Thalassarche melanophris*

2004). However, C was found to be higher in marine food webs than other environments (between 0.3 and 0.4 considering S > 40, Link 2002). Nevertheless, some food webs have shown very low connectance (e.g., C = 0.01 in Antarctic food webs; Marina et al. 2018). Connectance value of the PSBF food web could indicate intermediate sensitivity to removal of species (threshold of 20–30% species removal to display extreme sensitivity according to Dunne et al. (2002), although specific studies in the area are necessary to evaluate this. Also, to better predict the stability of food webs, it is important to weight metrics considering the interaction strengths among species such as it was performed elsewhere (e.g., van Altena et al. 2016; Kortsch et al. 2021; Marina et al. 2023). Given the large extension of the studied area, some spatial limitation of species interactions may occur and therefore compartmentation is probable, that is the presence of subsystems with minimal linkage to other subsystems (Mougi 2018). Indeed, we are aware that the PSBF food web is not isolated and together with the temporal and spatial variability, the interconnections with the continental shelf and deep ocean deserve further study.

11.3 Base of the Food Web

Marine fronts usually concentrate nutrients and act as hotspots of primary production (Acha et al. 2015). In fact, the PSBF is characterized by a band of consistent high satellite-chlorophyll concentration (Romero et al. 2006; Bianchi et al. 2009). These blooms are observed during austral spring and summer and along more than 1000 km (Carreto et al. 2016). Fieldwork supports the hypothesis that these algal blooms are caused by nutrient upwelling from the Malvinas Current waters reaching the surface (Carreto et al. 2007; Garcia et al. 2008). The main components (90% of biomass of Chl a) of phytoplankton in this area are small diatoms (8.5–12.0 μm) mainly *Thalassiosira bioculata* (Ferrario et al. 2013). The diatom bloom is initiated with minimal stratification (Carreto et al. 2016) and progresses from north to south along the front (Rivas et al. 2006).

Phytoplankton is not only the primary source for the pelagic components of the food web, but also supports benthic species by downwelling as suggested by theoretical models (e.g., Franco et al. 2017) and empirical data (Mauna et al. 2008). During the late phase of phytoplankton blooms, a strong stratification limits the availability of light, inducing diatom sinking. The sinking of these cells acts as seed for future upwelling blooms; while some cells accumulate as source of nutrients and carbon sequestration, others act as the main source of food for benthic filter feeders. This material known as Particulate Organic Matter (herein POM) that reaches the bottom is probably composed of dead phytoplankton but also of dead cells of microzooplankton, and other components of organisms living in the water column such as eggs and feces among others. These particles link the planktonic and benthic pathways in the food web. Particularly, organic matter of phytoplanktonic origin supports the Patagonian scallop, *Zygochlamys patagonica* (Schejter 2000), an important component of the benthic compartment associated with the PSBF even supporting a well-established fishery (Bogazzi et al. 2005). Benthic debris composed by organic matter in sediment is the other component considered as the base of the food web. Models tracking passive particles in the shelf-break suggest the importance of the physical processes in the suspension, transport, and support of epibenthic organisms (Franco et al. 2017) that could affect the dynamics of components supporting food webs.

11.4 A Big World of Small Organisms

The main grazers of diatoms are microplankton, mostly heterotrophic or phagotrophic dinoflagellates, which are abundant in the western band of the spring diatom bloom of the PSBF (Carreto et al. 2016). Other potential grazers are copepods, which are an important component of the zooplankton community along the PSBF in summer with dominance of species typical of shelf waters, but also with some oceanic species typical of Malvinas Current waters (Ramírez 2007). Some species of copepods can feed on microzooplankton and therefore form a trophic chain of diatoms-microzooplankton-copepods along the PSBF (Carreto et al. 2016).

In addition, euphausiids are found in high abundance associated with diatom blooms (Carreto et al. 2016) feeding on phytoplankton (Viñas et al. 2016). In a next trophic level, the most abundant zooplanktonic components are the amphipods, of which *Themisto gaudichaudii* is the most abundant species. This species was suggested to play a key role as a "wasp waist" species supporting fishes in the Argentine continental shelf (Padovani et al. 2012). Likewise, the squat lobster *Munida gregaria* is expanding its geographical distribution (Diez et al. 2016) and gaining importance as food source for many fishes (Belleggia et al. 2017; Alvarez et al. 2022). In conclusion, the evidence suggests that, although with spatial and temporal variability in their relative importance, copepods, euphausids, amphipods, and squat lobsters are all important groups supporting the PSBF food web (Figs. 11.1 and 11.2).

Other components of the base of the food web are early life stages of fishes, such as larvae and small juveniles. During these early stages, fish have a high-energy demand for growing and are also susceptible to predation, even by their conspecifics (cannibalism). In our system, the trophic role of most species in their larval or early juvenile stages is unknown with the exception of studies in few commercial species (e.g., anchovy: Brown and Sánchez 2010; Argentine hake: Temperoni et al. 2013, 2018a, b; Botto et al. 2019). The role of fish early life stages in oceanic food webs is usually ignored given the complexity and dynamics of this system and the difficulty of their study in the open ocean. Moreover, the temporal and spatial variability of these components surely add complexity to this part of the food web, which deserves further investigation.

11.5 Pelagic and Benthic Pathways

The high abundance of zooplankton that concentrates in the PSBF is likely the main source of energy for small pelagic and large demersal fishes (Fig. 11.2). The three small (15–18 cm maximum length) pelagic fishes, the Argentine anchovy *Engraulis anchoita*, the Patagonian sprat *Sprattus fueguensis,* and myctophids, seem to have similar roles in the food web, but with a clear spatial segregation in their distributions. The Argentine anchovy is the most abundant pelagic species that spreads along the continental shelf, with the Patagonian stock reaching a distribution up to 48 °S (Hansen et al. 2001; Pájaro 2002). The Patagonian sprat is characteristic of the southern Patagonian shelf (Sánchez et al. 1995) and only reaches part of the PSBF around the Malvinas Islands (Agnew 2002) and the Namuncurá/Burdwood Bank (García Alonso et al. 2018). Myctophids, on the other hand, are characteristic of the deep ocean along the SW Atlantic (Fig. 11.3a). There is little information regarding the ecology of myctophids in the zone close to the PSBF but more than 20 species were reported, being the most abundant *Gymnoscopelus* spp. and *Protomyctophum* spp. (Figueroa et al. 1998). All these pelagic fishes move in dense schools, and although small in body size, their high abundance makes them suitable food for

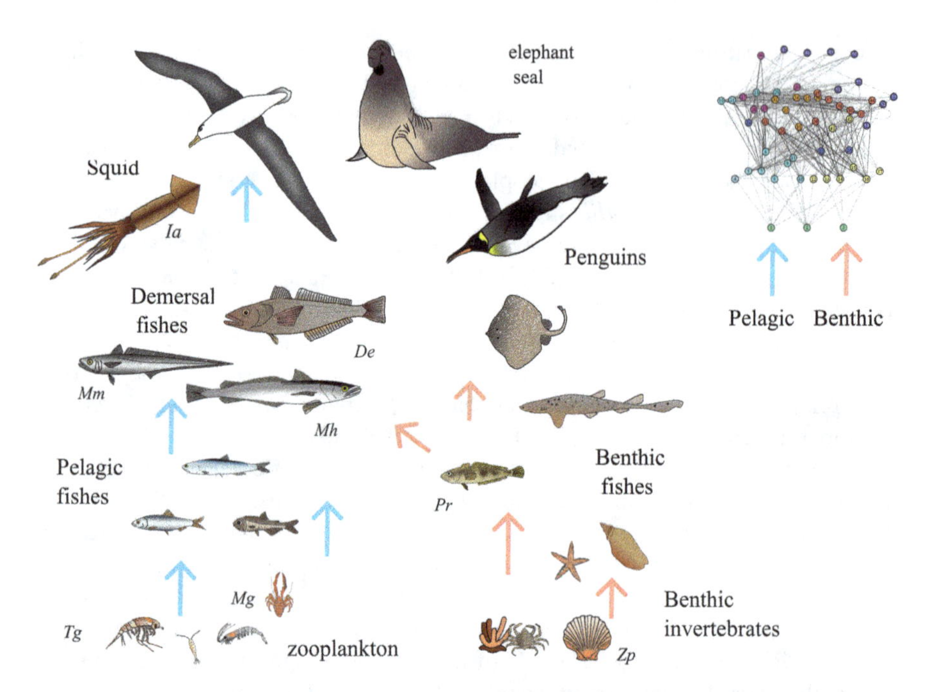

Fig. 11.2 Schematic representation of the trophic pathways of the Patagonian Shelf-break front, considering the main groups of the food web. The figure highlights the role of the following species: the amphipod *Themisto gaudichaudii* (*Tg*), the squat lobster *Munida gregaria* (*Mg*), the scallop, *Zygochlamys patagonica* (*Zp*), the large demersal fishes, *Merluccius hubbsi* (*Mh*), *Macruronus magellanicus* (*Mm*) and *Dissostichlus eleginoides* (*De*), the shortfin squid *Illex argentinus* (*Ia*), the notothenioid *Patagonotethen ramsayi* (*Pr*), and the elephant seal *Mirounga leonina* (*Ml*). In the upper right figure, arrows indicate the benthic and pelagic pathways in the complete network (see Fig. 11.1 for details)

some ichthyophagous organisms. Therefore, these species act as the link between the zooplankton and top predators in the food web.

The PSBF is also characterized by high abundance of the Argentine shortfin squid *Illex argentinus* (Brunetti et al. 1998; Acha et al. 2004). This is reflected by the effort of squid jigging vessels, which is concentrated in these areas (e.g., Acha et al. 2015; Martinetto et al. 2020). Globally, squids occupy a high range of trophic levels and exploit a high diversity of trophic resources with clear differences among oceans and ecosystem types (Coll et al. 2013). In the PSBF, the Argentine shortfin squid shows a high trophic level feeding on a variety of organisms (Table 11.1, Fig. 11.1). It is a semelparous species, with a life span of 1 year. Four stocks were identified in the PSBF, the most important being distributed between 45° and 47 °S. Larvae and juveniles concentrate over the outer shelf, near the Brazil-Malvinas Confluence and move to the intermediate shelf. Then, juveniles move southward along the continental shelf, preadults and adults reach the reproductive zones in the PSBF and finally move to the area where spawning and death occurs, which location remains controversial (Torres-Alberto et al. 2021). It was proposed

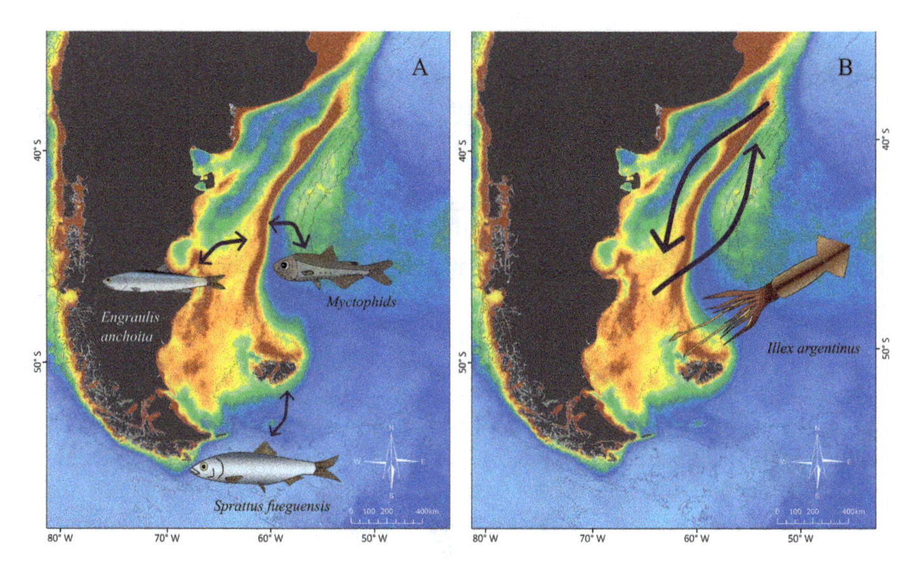

Fig. 11.3 Localization of the PSBF according satellite Chlorophyll *a* distribution during austral summer. (Taken from Martinetto et al. 2020). (**a**) Differential spatial distribution of the three small pelagic fishes with similar roles in the PSBF food web: *Engraulis anchoita* in the shelf area, *Sprattus fueguensis* in the southern area and Myctophids in the offshore area. (**b**) The possible energy distribution of nutrients and energy along the PSBF by migration movements of the shortfin squid *Illex argentinus*

a north spawning ground (i.e., southern Brazil; Arkhipkin 2013), which implies back northward migrations, or a Patagonian spawning ground over the shelf-break, which implies egg transport by oceanic currents to the hatch ground (Brunetti et al. 1998; Torres-Alberto et al. 2021). Therefore, shortfin squids exert an important role yearly re-distributing a high biomass of organic matter and nutrients along the PSBF (Fig. 11.3b), being a crucial pathway in the exchange of resources among marine habitats (Arkhipkin 2013). Because this species reaches a high biomass and has a high protein content, it is considered a food item of good nutrient quality (Eder and Lewis 2005), which is consumed by demersal-pelagic fishes, seabirds, and marine mammals (Figs. 11.1 and 11.2).

Another species highly associated with the PSFB, but inhabiting the sea bottom is the Patagonian scallop *Zygochlamys patagonica* (e.g., Bogazzi et al. 2005). As mentioned before, the Patagonian scallop is a primary consumer, filtering phytoplankton, microzooplankton, and other small organic particles that reach the bottom. This species is a prey item mainly for other benthic organisms such as sea stars and gastropods (Botto et al. 2006). The pink cusk eel *Genypterus blacodes* is the only fish species that has been reported to be consuming scallops, although in low abundance (Nyegaard et al. 2004). This suggests that scallops accumulate energy from the pelagic realm that remains within the benthic compartment for at least one or two trophic levels. Although scallops are not eaten by many predators, they can have ecosystem engineering effects, indirectly increasing the abundance and biomass of other benthic species (Schejter and Bremec 2007).

Despite that more than 250 species encompass the benthic assemblage (Chap. 6), only highly aggregated groups were considered in this study. This is partly because benthic fauna is rarely determined to the species or genera category in stomach content analyses of their predators, yet some important predators higher up in the food web such as the notothenioid longtail southern cod *Patagonotethen ramsayi*, some sharks (e.g., the narrow-mouthed sharkcat *Schroederichthys bivius*) and skates (e.g., *Bathyraja* species) feed on the benthos that inhabit the scallop beds along the PSBF. Some of these predators are specialist feeders of particular groups of benthic fauna such as *Bathyraja macloviana* (polychaete feeder, Mabragaña et al. 2005) or *B. albimaculata* (polychaetes and amphipods Brickle et al. 2003). The most abundant species is the longtail southern cod, which is also the species with the most diverse predators (Table 11.1, Figs. 11.1 and 11.2). Together with grenadiers (*Coelorhincus fasciatus* and *Macrourus spp*) and other notothenioid *Cottoperca gobio*, longtail southern cod are the fishes at intermediate trophic levels linking the benthic production to higher trophic level species, in particular skates (e.g., *Zearaja brevicaudata*) and demersal fishes (e.g., pink cusk-eel *Genypterus blacodes* and Patagonian toothfish *Dissostichus eleginoides*).

11.6 Size Matters for Generalist Fishes

Except for a few skates, most fishes in the PSBF were classified as generalists in the literature, and they consumed on average 11 prey items (Table 11.1).). However, it must be taken into account that this value integrates the complete life cycle of organisms. Some fishes can change their trophic niche while growing, with a substantial shift in diet, or an increase in their trophic niche by incorporating new items to their diet (Hammerschlag-Peyer et al. 2011).

In the PSBF system, many species showed differences in the preference for prey items among individuals of different size ranges (Table 11.1), in general, increasing their trophic spectra with size. For example, the most abundant demersal-pelagic fishes (i.e., Argentine hake *Merluccius hubbsi* and long tail hake *Macruronus magellanicus*) feed mainly on planktonic crustaceans during their life cycle, but large individuals incorporate fishes in their diet. Anatomic limitation of the mouth size to the incorporation of some prey items is one of the main explanations of this change; in particular considering that these fishes can enhance their body size up to 20 times. Large teleost species in boreal and temperate regions are typically slower growing, demersal and generalist feeders, coupling both pelagic and benthic pathways (van Denderen et al. 2018).

Trophic level of most fishes of the PSBF ranged between 3 and 4, and some were found feeding in more than one trophic level (mean omnivory index is 0.6). The fish with the highest trophic level (TL = 4.7) in the food web was the southern hake *Merluccius australis,* which has been recorded to feed almost exclusively on fishes (Giussi et al. 2016). Other species at high trophic level are the pink cusk-eel (TL = 4), the Patagonian toothfish (TL = 3.8), and the Argentine hake (TL = 3.7,

Table 1.11). Given their ontogenetic changes in diet, probably large individuals can reach even higher trophic levels. Geographic range and movements of large consumers also increase their ability to exploit different prey spots and therefore affect trophic niche breadth (Hayden et al. 2019). Spatial variability in stomach contents of some species were found among different areas of the ocean (e.g., Brickle et al. 2009; Belleggia et al. 2014, 2016) and in some cases associated to frontal systems (e.g., Alemany et al. 2018). For example, the long tail hake can move among distant feeding grounds and therefore can take advantage of spatially variable food items (Alvarez et al. 2022). In general, the final outcome of the diet is probably determined by both the quality and quantity of the different prey available, which often exhibit trade-offs (e.g., Zheng et al. 2021). Therefore, there is spatial variability in the relative importance of the different trophic links, which may determine different compartments in the PSBF food web, which may contribute to its stability.

11.7 Birds and Mammals: Are They Top Predators?

Top predators are particularly abundant in ocean fronts or water mixing zones such as the continental slope (Campagna et al. 2006; Zerbini et al. 2006; Bost et al. 2009). Particularly the elephant seal *Mirounga leonina* takes advantage of marine fronts to feed (Siegelman et al. 2019). In the Southern Atlantic, they breed and molt in oceanic islands and in Península Valdes, from where they make long trips to feed. Most satellite tracking shows that they reach the PSBF, where some individuals stay there to eat while others make longer trips to deeper waters (Campagna et al. 2021). Individual specialization is suggested for this species with high fidelity to foraging areas. They are top predators with high energetic demand, and few organisms may have local top-down impact on the food web.

Also, the large amount of food concentrated in the PSBF is exploited by a large number of several pelagic seabirds such as the black-browed albatross *Thalassarche melanophris* (Huin 2002), the wandering albatross *Diomedea exulans* (Nicholls et al. 2002; Xavier et al. 2004), and the Southern giant petrel *Macronectes giganteus* (Quintana and Dell'Arciprete 2002; Copello et al. 2011). These species often travel very long distances from their breeding sites to the PSBF (Xavier et al. 2004). An important proportion of the diet of these pelagic seabirds is byproduct of fishery activity, which makes available a high amount of food (e.g., Xavier et al. 2004; Copello et al. 2008; Mariano-Jelicich et al. 2014). This association of fisheries with bird foraging has been a matter of concern given the negative effect of trampling mortality. However, fisheries make a high amount of food available with a low searching effort, which could compensate for the negative effects. Regardless, there is no doubt that humans through fishery activity facilitate the trophic link between demersal fishes and pelagic seabirds.

Another bird species commonly associated with fronts is the King penguin *Aptenodytes patagonicus.* This species is particularly associated with the Antarctic Polar Front (Pütz 2002; Scheffer et al. 2010) but some individuals from the Malvinas

Islands colonies make use of the PSBF as far north as 38 °S in winter (Pütz 2002; Baylis et al. 2015). The Magellanic penguin, *Spheniscus magellanicus* inhabiting continental colonies in Argentina feed mostly on the ample Patagonian shelf (Ciancio et al. 2015; Blanco et al. 2022) and only some males reach the PSBF during winter (Barrionuevo et al. 2020; Dodino et al. 2021). However, the PSBF is included as part of the foraging areas of Magellanic penguins of Malvinas Islands colonies (Pütz et al. 2000), Gentoo penguins, *Pygoscelis papua* (Baylis et al. 2021), and Rockhoppers penguins *Eudyptes chrysocome* (Pütz et al. 2006), with similar diet but variable in space and time (Pütz 2002). All these penguin species showed trophic level (between 3.5 and 3.8, Table 11.1) similar to most demersal and benthic fishes, as found in the Patagonian shelf (Ciancio et al. 2008).

We cannot ignore, however, that humans are also major predators in marine systems. There is evidence of coupling between marine fronts and fishing (Alemany et al. 2014, 2016). In the PSBF in particular, there is an aggregation of fishing vessels in the frontal area. Species targeted by fisheries in the continental shelf and close to the shelf break, and that are part of this food web, are the Argentine shortfin squid, the Patagonian scallop, the Argentine anchovy and the demersal fishes, such as the Argentine hake and the long tail hake, all with different topological roles in the food web. Therefore, the fishery, together with other human activities that affect marine ecosystems, such as climate change, may affect the entire food web.

11.8 Conclusion

This chapter summarizes the information of trophic interactions studied in the PSBF analyzed under the food web conceptual framework. As pointed out in the different sections, some gaps of information appear. For example, trophic interactions of basal species are poorly studied, including those between microorganisms and early life stages of fishes and invertebrates. In addition, as in most marine food web studies, information of trophic links is biased towards important commercial species, whereas the role of some groups, such as myctophids or jellyfishes, may be underestimated. The present food web represents a metaweb given that it considers potential trophic interactions, yet the realized interactions depend on the actual encounter in space and time of the different species, which is dynamic. For example, the migration patterns of squids along the food web imply that links involving them change spatially and seasonally. Moreover, the different distribution of pelagic fish species along the PSBF implies local changes in components of the food web. The same occurs with some penguins that nest in Malvinas Islands and only interact in the southern portion of the PSBF food web. Understanding the food web dynamic is crucial to predict the outcomes of human impact. It has been suggested to understand food webs as open and flexible Jenga-like systems that can reach different stable conditions depending on changes in species attributes, composition, and dynamics (de Ruiter et al. 2005). Maintenance of such stable states in marine food webs is a goal for ecosystem management of marine systems. This chapter aims to be a starting point to better understand the structure and dynamics of the PSBF food web towards achieving this goal.

Acknowledgements We thank the editorial team and two anonymous reviewers for their valuable comments on an early version of this chapter. This is INIDEP contribution no. 2368.

References

Acha EM, Mianzan HW, Guerrero RA et al (2004) Marine fronts at the continental shelves of austral South America: physical and ecological processes. J Mar Syst 44:83–105. https://doi.org/10.1016/j.jmarsys.2003.09.005

Acha EM, Piola A, Iribarne O, Mianzan H (2015) Ecological processes at marine fronts: oases in the ocean. Springer

Agnew D (2002) Critical aspects of The Falkland Islands pelagic ecosystem: distribution, spawning and migration of pelagic animals in relation to oil exploration. Aquat Conserv Mar Freshw Ecosyst 12:39–50. https://doi.org/10.1002/aqc.474

Alemany D, Acha EM, Iribarne OO (2014) Marine fronts are important fishing areas for demersal species at the Argentine sea (Southwest Atlantic Ocean). J Sea Res 87:56–67. https://doi.org/10.1016/j.seares.2013.12.006

Alemany D, Acha EM, Iribarne OO (2016) Distribution and intensity of bottom trawl fisheries in the Patagonian shelf large marine ecosystem and its relationship with marine fronts. Fish Oceanogr 25:183–192. https://doi.org/10.1111/fog.12144

Alemany D, Iribarne OO, Acha EM (2018) Marine fronts as preferred habitats for young Patagonian hoki *Macruronus magellanicus* on the southern Patagonian shelf. Mar Ecol Prog Ser 588:191–200

Alvarez CD, Giussi AR, Botto F (2022) Trophic variability of long tail hake *Macruronus magellanicus* in the Southwestern Atlantic: movements evidenced by stomach content and stable isotope analysis. Polar Biol 45:1131–1143. https://doi.org/10.1007/s00300-022-03063-y

Angelescu VA, Prenski LB (1987) Ecología trófica de la merluza común (*Merluccius hubbsi*) del Mar Argentino: Parte 2. Dinámica de la alimentación analizada sobre la base de las condiciones ambientales, la estructura y las evaluaciones de los efectivos en su área de distribución. Contribución INIDEP 561:1–205

Arkhipkin AI (2013) Squid as nutrient vectors linking Southwest Atlantic marine ecosystems. Deep Sea Res Part II Top Stud Oceanogr 95:7–20. https://doi.org/10.1016/j.dsr2.2012.07.003

Arkhipkin A, Brickle P, Laptikhovsky V et al (2001) Variation in the diet of the red cod, *Salilota australis* Pisces: Moridae, with size and season at The Falkland Islands south-west Atlantic. J Mar Biol UK 81:6

Baird D, Ulanowicz RE (1989) The seasonal dynamics of the Chesapeake Bay ecosystem. Ecol Monogr 59:329–364. https://doi.org/10.2307/1943071

Barrionuevo M, Ciancio J, Steinfurth A, Frere E (2020) Geolocation and stable isotopes indicate habitat segregation between sexes in Magellanic penguins during the winter dispersion. J Avian Biol 51:eo2325. https://doi.org/10.1111/jav.02325

Baylis AMM, Orben RA, Pistorius P et al (2015) Winter foraging site fidelity of king penguins breeding at The Falkland Islands. Mar Biol 162:99–110. https://doi.org/10.1007/s00227-014-2561-0

Baylis AMM, Tierney M, Orben RA et al (2021) Non-breeding movements of Gentoo penguins at The Falkland Islands. Ibis 163:507–518. https://doi.org/10.1111/ibi.12882

Belleggia M, Mabragaña E, Figueroa DE et al (2008) Food habits of the broad nose skate, *Bathyraja brachyurops* (Chondrichthyes, Rajidae), in the south-west Atlantic. Sci Mar 72:701–710

Belleggia M, Figueroa DE, Irusta G, Bremec C (2014) Spatio-temporal and ontogenetic changes in the diet of the Argentine hake *Merluccius hubbsi*. J Mar Biol Assoc U K 94:1701–1710. https://doi.org/10.1017/S0025315414000629

Belleggia M, Andrada N, Paglieri S et al (2016) Trophic ecology of yellownose skate *Zearaja chilensis*, a top predator in the south-western Atlantic Ocean. J Fish Biol 88:1070–1087. https://doi.org/10.1111/jfb.12878

Belleggia M, Giberto D, Bremec C (2017) Adaptation of diet in a changed environment: increased consumption of lobster krill *Munida gregaria* (Fabricius, 1793) by Argentine hake. Mar Ecol 38:e 12445. https://doi.org/10.1111/maec.12445

Best PB, Schell DM (1996) Stable isotopes in southern right whale (*Eubalaena australis*) baleen as indicators of seasonal movements, feeding and growth. Mar Biol 124:483–494. https://doi.org/10.1007/BF00351030

Bianchi AA, Pino DR, Perlender HGI et al (2009) Annual balance and seasonal variability of sea-air CO2 fluxes in the Patagonia Sea: their relationship with fronts and chlorophyll distribution. J Geophys Res Oceans 114:1–11. https://doi.org/10.1029/2008JC004854

Blanco GS, Gallo L, Pisoni JP et al (2022) At-sea distribution, movements and diving behavior of Magellanic penguins reflect small-scale changes in oceanographic conditions around the colony. Mar Biol 169:29. https://doi.org/10.1007/s00227-021-04016-5

Bogazzi E, Baldoni A, Rivas A et al (2005) Spatial correspondence between areas of concentration of Patagonian scallop (*Zygochlamys patagonica*) and frontal systems in the southwestern Atlantic. Fish Oceanogr 14:359–376. https://doi.org/10.1111/j.1365-2419.2005.00340.x

Bost CA, Cotté C, Bailleul F, Cherel Y, Charrassin JB, Guinet C, Ainley DG, Weimerskirch H (2009) The importance of oceanographic fronts to marine birds and mammals of the southern oceans. J Mar Syst 78:363–376. https://doi.org/10.1016/j.jmarsys.2008.11.022

Botto F, Valiela I, Iribarne O et al (2005) Impact of burrowing crabs on C and N sources, control, and transformations in sediments and food webs of SW Atlantic estuaries. Mar Ecol Prog Ser 293:155–164

Botto F, Bremec C, Marecos A et al (2006) Identifying predators of the SW Atlantic Patagonian scallop *Zygochlamys patagonica* using stable isotopes. Fish Res 81:45–50

Botto F, Gaitán E, Mianzan H et al (2011) Origin of resources and trophic pathways in a large SW Atlantic estuary: an evaluation using stable isotopes. Estuar Coast Shelf Sci 92:70–77

Botto F, Gaitán E, Iribarne OO, Acha EM (2019) Trophic niche changes during settlement in the Argentine hake *Merluccius hubbsi* reveal the importance of pelagic food post metamorphosis. Mar Ecol Prog Ser 619:125–136. https://doi.org/10.3354/meps12947

Brickle P, Laptikhovsky V, Pompert J, Bishop A (2003) Ontogenetic changes in the feeding habits and dietary overlap between three abundant rajid species on The Falkland Islands' shelf. J Mar Biol Assoc U K 83:1119–1125. https://doi.org/10.1017/S0025315403008373h

Brickle P, Arkhipkin AI, Laptikhovsky V et al (2009) Resource partitioning by two large planktivorous fishes *Micromesistius australis* and *Macruronus magellanicus* in the Southwest Atlantic. Estuar Coast Shelf Sci 84:91–98. https://doi.org/10.1016/j.ecss.2009.06.007

Briz LD, Sánchez F, Marí N et al (2017) Gelatinous zooplankton (ctenophores, salps and medusae): an important food resource of fishes in the temperate SW Atlantic Ocean. Mar Biol Res 13:630–644. https://doi.org/10.1080/17451000.2016.1274403

Brown DR, Sánchez RP (2010) Larval and juvenile growth of two Patagonian small pelagic fishes: *Engraulis anchoita* and *Sprattus fuegensis*. Rev Investig Desarro Pesq 20:35–50

Brunetti NE, Ivanovic ML, Elena B (1998) Calamares omastréfidos (Cephalopoda, Ommastrephidae). In: El Mar Argentino y sus recursos pesqueros. Tomo 2. Los moluscos de interés pesquero. Cultivos y estrategias reproductivas de bivalvos y equinoideos. In: Boschi EE (ed) Instituto Nacional de Investigación y Desarrollo Pesquero, Mar del Plata, Argentina, p 37–68

Bugoni L, McGill RAR, Furness RW (2010) The importance of pelagic longline fishery discards for a seabird community determined through stable isotope analysis. J Exp Mar Biol Ecol 391:190–200. https://doi.org/10.1016/j.jembe.2010.06.027

Büring T, Schroeder P, Jones JB, Pierce G, Rocha F, Arkhipkin AI (2021) Size-related, seasonal and interdecadal changes in the diet of the Patagonian longfin squid *Doryteuthis gahi* in the South-Western Atlantic. J Mar Biol Assoc UK 101:11–28. https://doi.org/10.1017/S0025315422000194

Campagna C, Piola AR, Marin MR, Lewis M, Fernández T (2006) Southern elephant seal trajectories, fronts and eddies in the Brazil/Malvinas confluence. Deep-Sea Res I Oceanogr Res Pap 53:1907–1924. https://doi.org/10.1016/j.dsr.2006.08.015

Campagna J, Lewis MN, González Carman V et al (2021) Ontogenetic niche partitioning in southern elephant seals from Argentine Patagonia. Mar Mamm Sci 37:631–651. https://doi.org/10.1111/mms.12770

Carreto JI, Carignan MO, Montoya NG, Cucchi Colleoni AD (2007) Ecología del fitoplancton en los sistemas frontales del Mar Argentino. In: El Mar Argentino y sus recursos pesqueros, Tomo 5. El ecosistema marino, Carreto J. I., Bremec C, p 11–31

Carreto JI, Montoya NG, Carignan MO et al (2016) Environmental and biological factors controlling the spring phytoplankton bloom at the Patagonian shelf-break front – degraded fucoxanthin pigments and the importance of microzooplankton grazing. Prog Oceanogr 146:1–21. https://doi.org/10.1016/j.pocean.2016.05.002

Cherel Y, Pütz K, Hobson KA (2002) Summer diet of king penguins (*Aptenodytes patagonicus*) at The Falkland Islands, southern Atlantic Ocean. Polar Biol 25:898–906. https://doi.org/10.1007/s00300-002-0419-2

Ciancio JE, Pascual MA, Botto F et al (2008) Trophic relationships of exotic anadromous salmonids in the southern Patagonian Shelf as inferred from stable isotopes. Limnol Oceanogr 53:788–798

Ciancio J, Botto F, Frere E (2015) Combining a geographic information system, known dietary, foraging and habitat preferences, and stable isotope analysis to infer the diet of Magellanic penguins in their austral distribution. Emu Austral Ornithol 115:237–246. https://doi.org/10.1071/MU14032

Clausen AP, Arkhipkin AI, Laptikhovsky VV, Huin N (2005) What is out there: diversity in feeding of gentoo penguins (*Pygoscelis papua*) around The Falkland Islands (Southwest Atlantic). Polar Biol 28:653–662. https://doi.org/10.1007/s00300-005-0738-1

Coll M, Navarro J, Olson RJ, Christensen V (2013) Assessing the trophic position and ecological role of squids in marine ecosystems by means of food-web models. Deep Sea Res Part II Top Stud Oceanogr 95:21–36. https://doi.org/10.1016/j.dsr2.2012.08.020

Copello S, Quintana F, Pérez F (2008) Diet of the southern giant petrel in Patagonia: fishery-related items and natural prey. Endanger Species Res 6:15–23. https://doi.org/10.3354/esr00118

Copello S, Dogliotti AI, Gagliardini DA, Quintana F (2011) Oceanographic and biological landscapes used by the Southern Giant Petrel during the breeding season at the Patagonian Shelf. Mar Biol 158:1247–1257. https://doi.org/10.1007/s00227-011-1645-3

de la Rosa SG, Sanchez F, Figueroa D (1997) Comparative feeding ecology of Patagonian toothfish (*Dissostichus eleginoides*) in the southwestern Atlantic. CCAMLR Sci 4:105–124

de Ruiter PC, Wolters V, Moore JC, Winemiller KO (2005) Food web ecology: playing Jenga and beyond. Science 309:68–71. https://doi.org/10.1126/science.1096112

Dehnhard N, Ludynia K, Masello JF et al (2016) Plasticity in foraging behaviour and diet buffers effects of inter-annual environmental differences on chick growth and survival in southern rockhopper penguins *Eudyptes chrysocome chrysocome*. Polar Biol 39:1627–1641. https://doi.org/10.1007/s00300-015-1887-5

Diez MJ, Cabreira AG, Madirolas A, Lovrich GA (2016) Hydroacoustical evidence of the expansion of pelagic swarms of *Munida gregaria* (Decapoda, Munididae) in the Beagle Channel and the Argentine Patagonian Shelf, and its relationship with habitat features. J Sea Res 114:1–12. https://doi.org/10.1016/j.seares.2016.04.004

Dodino S, Lois NA, Riccialdelli L et al (2021) Sex-specific spatial use of the winter foraging areas by Magellanic penguins and assessment of potential conflicts with fisheries during winter dispersal. PLoS One 16:e0256339. https://doi.org/10.1371/journal.pone.0256339

Dunne J (2006) The network structure of food webs. In: Ecological networks: linking structure to dynamics in food webs, pp 27–86

Dunne JA, Williams RJ, Martinez ND (2002) Network structure and biodiversity loss in food webs: robustness increases with connectance. Ecol Lett 5:558–567. https://doi.org/10.1046/j.1461-0248.2002.00354.x

Dunne JA, Williams RJ, Martinez ND (2004) Network structure and robustness of marine food webs. Mar Ecol Prog Ser 273:291–302. https://doi.org/10.3354/meps273291

Dunne JA, Brose U, Williams RJ, Martinez ND (2005) Modeling food-web dynamics: complexity-stability implications. Aquat Food Webs Ecosyst Approach:117–129

Eder EB, Lewis MN (2005) Proximate composition and energetic value of demersal and pelagic prey species from the SW Atlantic Ocean. Mar Ecol Prog Ser 291:43–52

Ferrario ME, Almandoz GO, Cefarelli AO, Sastre V, Santinelli N (2013) Ultrastructure *Thalassiosira bioculata* var. rariprora in the Argentine sea slope. First record for the southern hemisphere. In: X international Phycological congress, Orlando, USA

Figueroa DE, de Astarloa JMD, Martos P (1998) Mesopelagic fish distribution in the Southwest Atlantic in relation to water masses. Deep Sea Res Part Oceanogr Res Pap 45:317–332. https://doi.org/10.1016/S0967-0637(97)00076-9

Fischer L, Covatti Ale M, Deli Antoni M et al (2022) Feeding ecology of the longtail southern cod, *Patagonotothen ramsayi* (Regan, 1913) (Notothenioidei) in the marine protected area Namuncurá-Burdwood Bank, Argentina. Polar Biol 45:1483–1494. https://doi.org/10.1007/s00300-022-03082-9

Franco BC, Palma ED, Combes V, Lasta ML (2017) Physical processes controlling passive larval transport at the Patagonian Shelf Break Front. J Sea Res 124:17–25. https://doi.org/10.1016/j.seares.2017.04.012

Funes M, Saravia LA, Cordone G et al (2022) Network analysis suggests changes in food web stability produced by bottom trawl fishery in Patagonia. Sci Rep 12:10876. https://doi.org/10.1038/s41598-022-14363-y

Gaitán E (2012) Tramas tróficas en sistemas frontales del Mar Argentino: estructura, dinámica y complejidad analizada mediante isótopos estables. Tesis de Doctorado, Universidad Nacional de Mar del Plata

García Alonso VA, Brown D, Martín J et al (2018) Seasonal patterns of Patagonian sprat *Sprattus fuegensis* early life stages in an open sea sub-Antarctic marine protected area. Polar Biol 41:2167–2179. https://doi.org/10.1007/s00300-018-2352-z

Garcia VMT, Garcia CAE, Mata MM et al (2008) Environmental factors controlling the phytoplankton blooms at the Patagonia shelf-break in spring. Deep Sea Res Part Oceanogr Res Pap 55:1150–1166. https://doi.org/10.1016/j.dsr.2008.04.011

Giussi AR, Gorini F, Marco ED et al (2016) Biology and fishery of the Southern Hake (*Merluccius australis*) in the Southwest Atlantic Ocean. Mar Fish Sci MAFIS 28:37–53

Hammerschlag-Peyer CM, Yeager LA, Araújo MS, Layman CA (2011) A hypothesis-testing framework for studies investigating ontogenetic niche shifts using stable isotope ratios. PLoS One 6:e27104. https://doi.org/10.1371/journal.pone.0027104

Hansen JE, Martos P, Madirolas A (2001) Relationship between spatial distribution of the Patagonian stock of Argentine anchovy, *Engraulis anchoita*, and sea temperatures during late spring to early summer. Fish Oceanogr 10:193–206. https://doi.org/10.1046/j.1365-2419.2001.00166.x

Hayden B, Palomares MLD, Smith BE, Poelen JH (2019) Biological and environmental drivers of trophic ecology in marine fishes – a global perspective. Sci Rep 9:11415. https://doi.org/10.1038/s41598-019-47618-2

Hines J, van der Putten WH, De Deyn GB et al (2015) Chapter four-towards an integration of biodiversity–ecosystem functioning and food web theory to evaluate relationships between multiple ecosystem services. Adv Ecol Res 53:161–199

Huin N (2002) Foraging distribution of the black-browed albatross, *Thalassarche melanophris*, breeding in The Falkland Islands. Aquat Conserv Mar Freshw Ecosyst 12:89–99. https://doi.org/10.1002/aqc.479

Ivanovic ML (2010) Alimentación del calamar *Illex argentinus* en la región patagónica durante el verano de los años 2006, 2007 y 2008. Rev Investig Desarro Pesq 20:51–63

Ivanovic ML, Brunetti NE (1994) Food and feeding of *Illex argentinus*. Antarct Sci 6:185–193. https://doi.org/10.1017/S0954102094000295

Izzo LP (2010) Ecología trófica del congrio de profundidad *Bassanago albescens* (Barnard, 1923) en el Atlántico Sudoccidental (35°'S-45°'S). Tesis de Licenciatura, Universidad Nacional de Mar del Plata

Koen-Alonso M, Crespo EA, Garcıa NA, Pedraza SN, Mariotti PA, Mora NJ (2002) Fishery and ontogenetic driven changes in the diet of the spiny dogfish, *Squalus acanthias*, in Patagonian waters, Argentina. Environ Biol Fish 63:193–202

Kortsch S, Primicerio R, Aschan M et al (2019) Food-web structure varies along environmental gradients in a high-latitude marine ecosystem. Ecography 42:295–308. https://doi.org/10.1111/ecog.03443

Kortsch S, Frelat R, Pecuchet L et al (2021) Disentangling temporal food web dynamics facilitates understanding of ecosystem functioning. J Anim Ecol 90:1205–1216. https://doi.org/10.1111/1365-2656.13447

Laptikhovsky VV (2004) A comparative study of diet in three sympatric populations of *Patagonotothen* species (Pisces: Nototheniidae). Polar Biol 27:202–205. https://doi.org/10.1007/s00300-003-0573-1

Laptikhovsky VV, Arkhipkin AI (2003) An impact of seasonal squid migrations and fishing on the feeding spectra of subantarctic notothenioids *Patagonotothen ramsayi* and *Cottoperca gobio* around The Falkland Islands. J Appl Ichthyol 19:35–39. https://doi.org/10.1046/j.1439-0426.2003.00340.x

Laptikhovsky VV, Arkhipkin AI, Henderson AC (2001) Feeding habits and dietary overlap in spiny dogfish *Squalus acanthias* (Squalidae) and narrowmouth catshark *Schroederichthys bivius* (Scyliorhinidae). J Mar Bioll Assoc UK 81:1015–1018

Layman CA, Araujo MS, Boucek R et al (2012) Applying stable isotopes to examine food-web structure: an overview of analytical tools. Biol Rev 87:545–562. https://doi.org/10.1111/j.1469-185X.2011.00208.x

Layman CA, Giery ST, Buhler S et al (2015) A primer on the history of food web ecology: fundamental contributions of fourteen researchers. Food Webs 4:14–24. https://doi.org/10.1016/j.fooweb.2015.07.001

Link J (2002) Does food web theory work for marine ecosystems? Mar Ecol Prog Ser 230:1–9

Link JS, Yemane D, Shannon LJ et al (2010) Relating marine ecosystem indicators to fishing and environmental drivers: an elucidation of contrasting responses. ICES J Mar Sci 67:787–795. https://doi.org/10.1093/icesjms/fsp258

Mabragaña E, Giberto DA, Bremec CS (2005) Feeding ecology of *Bathyraja macloviana* (Rajiformes, Arhynchobatidae): a polychaete-feeding skate from the south-west Atlantic. Sci Mar 69:405–413. https://doi.org/10.3989/scimar.2005.69n3405

Mari NR, Sanchez F (2002) Espectros troficos especificos de varias especies de peces demersales de la region austral y sus variaciones anuales entre 1994 y 2000. Documento Cientifico No. 88. Instituto Nacional de Investigacion y Desarrollo Pesquero, Mar del Plata

Mariano-Jelicich R, Copello S, Pon JPS, Favero M (2014) Contribution of fishery discards to the diet of the black-browed albatross (*Thalassarche melanophris*) during the non-breeding season: an assessment through stable isotope analysis. Mar Biol 161:119–129. https://doi.org/10.1007/s00227-013-2320-7

Marina TI, Salinas V, Cordone G et al (2018) The food web of potter cove (Antarctica): complexity, structure and function. Estuar Coast Shelf Sci 200:141–151. https://doi.org/10.1016/j.ecss.2017.10.015

Marina TI, Saravia LA, Kortsch S (2023) New insights into the Weddell Sea ecosystem applying a quantitative network approach. EGUsphere 1–20. https://doi.org/10.5194/egusphere-2022-1518

Martinetto P, Alemany D, Botto F et al (2020) Linking the scientific knowledge on marine frontal systems with ecosystem services. Ambio 49:541–556. https://doi.org/10.1007/s13280-019-01222-w

Martinez ND (1991) Artifacts or attributes? Effects of Resolution on the Little Rock Lake Food Web Ecol Monogr 61:367–392. https://doi.org/10.2307/2937047

Mauna AC, Franco BC, Baldoni A et al (2008) Cross-front variations in adult abundance and recruitment of Patagonian scallop (*Zygochlamys patagonica*) at the SW Atlantic Shelf Break Front. ICES J Mar Sci 65:1184–1190. https://doi.org/10.1093/icesjms/fsn098

McDonald-Madden E, Game ET, Possingham HP et al (2016) Using food-web theory to conserve ecosystems. Nat Commun 7:10245. https://doi.org/10.1038/ncomms10245

Menge BA, Sutherland JP (1976) Species diversity gradients: synthesis of the roles of predation, competition, and temporal heterogeneity. Am Nat 110:351–369. https://doi.org/10.1086/283073

Montecinos Garrido SO (2015) Composición dietaria de *Sprattus fuegensis* y determinación del nivel trófico mediante isótopos estables de $\delta^{13}C$ y $\delta^{15}N$ en la zona sur austral

Morin PJ, Lawler SP (1995) Food web architecture and population dynamics: theory and empirical evidence. Annu Rev Ecol Syst 26:505–529

Mougi A (2018) Spatial compartmentation and food web stability. Sci Rep 8. https://doi.org/10.1038/s41598-018-34716-w

Nicholls DG, Robertson CJR, Prince PA et al (2002) Foraging niches of three *Diomedea* albatrosses. Mar Ecol Prog Ser 231:269–277. https://doi.org/10.3354/meps231269

Nyegaard M, Arkhipkin A, Brickle P (2004) Variation in the diet of *Genypterus blacodes* (Ophidiidae) around The Falkland Islands. J Fish Biol 65:666–682. https://doi.org/10.1111/j.0022-1112.2004.00476.x

Opitz S (1993) A quantitative model of the trophic interactions in a Caribbean coral reef ecosystem. In: Trophic models of aquatic ecosystems. ICLARM Conf. Proc. 26, p 259–267

Padovani LN, Viñas MD, Sánchez F, Mianzan H (2012) Amphipod-supported food web: *Themisto gaudichaudii*, a key food resource for fishes in the southern Patagonian shelf. J Sea Res 67:85–90. https://doi.org/10.1016/j.seares.2011.10.007

Paine RT (1966) Food web complexity and species diversity. Am Nat 100:65–75

Pájaro M (2002) Alimentación de la anchoíta Argentina (*Engraulis anchoita* Hubbs y Marini, 1935) (Pisces: Clupeiformes) durante la época reproductiva. Rev Investig Desarro Pesq 15:111–125

Pérez-Barros P, Romero MC, Calcagno JA, Lovrich GA (2010) Similar feeding habits of two morphs of *Munida gregaria* (Decapoda) evidence the lack of trophic polymorphism. Rev Biol Mar Oceanogr 45:461–470

Piatkowski U, Pütz K, Heinemann H (2001) Cephalopod prey of king penguins (*Aptenodytes patagonicus*) breeding at Volunteer Beach, Falkland Islands, during austral winter 1996. Fish Res 52:79–90. https://doi.org/10.1016/S0165-7836(01)00232-6

Pimm SL, Lawton JH, Cohen JE (1991) Food web patterns and their consequences. Nature 350:669–674. https://doi.org/10.1038/350669a0

Purcell JE (1991) A review of cnidarians and ctenophores feeding on competitors in the plankton. Hydrobiologia 216:335–342. https://doi.org/10.1007/BF00026483

Pütz K (2002) Spatial and temporal variability in the foraging areas of breeding king penguins. Condor 104:528–538. https://doi.org/10.1093/condor/104.3.528

Pütz K, Ingham RJ, Smith JG (2000) Satellite tracking of the winter migration of Magellanic penguins *Spheniscus magellanicus* breeding in The Falkland Islands. Ibis 142:614–622. https://doi.org/10.1111/j.1474-919X.2000.tb04461.x

Pütz K, Rey AR, Huin N et al (2006) Diving characteristics of southern rockhopper penguins (*Eudyptes c. chrysocome*) in the Southwest Atlantic. Mar Biol 149:125–137. https://doi.org/10.1007/s00227-005-0179-y

Quintana F, Dell'Arciprete PO (2002) Foraging grounds of southern giant petrels (*Macronectes giganteus*) on the Patagonian shelf. Polar Biol 25:159–161. https://doi.org/10.1007/s003000100313

Ramírez F (2007) Distribución y alimentación del zooplancton. In: El mar argentino y sus recursos pesqueros. Tomo 5. El ecosistema marino, Carret J. I, Bremec C. Instituto Nacional de Investigación y Desarrollo Pesquero, Mar del Plata, Argentina, p 45–69

Rey AR, Schiavini A (2005) Inter-annual variation in the diet of female southern rockhopper penguin (*Eudyptes chrysocome chrysocome*) at Tierra del Fuego. Polar Biol 28:132–141. https://doi.org/10.1007/s00300-004-0668-3

Riccialdelli L, Newsome SD, Fogel ML, Fernández DA (2017) Trophic interactions and food web structure of a subantarctic marine food web in the Beagle Channel: Bahía Lapataia, Argentina. Polar Biol 40:807–821. https://doi.org/10.1007/s00300-016-2007-x

Rivas AL, Dogliotti AI, Gagliardini DA (2006) Seasonal variability in satellite-measured surface chlorophyll in the Patagonian Shelf. Cont Shelf Res 26:703–720. https://doi.org/10.1016/j.csr.2006.01.013

Romanelli J (2017) Ecología trófica de tres especies de granaderos, *Coelorhinchus fasciatus*, *Macrourus carinatus* y *Macrourus holotrachys*, presentes en el Océano Atlántico Sudoccidental. Tesis de Licenciatura, Universidad Nacional de Mar del Plata

Romero MC, Lovrich GA, Tapella F, Thatje S (2004) Feeding ecology of the crab *Munida subrugosa* (Decapoda: Anomura: Galatheidae) in the Beagle Channel, Argentina. J Mar Biol Assoc U K 84:359–365

Romero SI, Piola AR, Charo M, Garcia CAE (2006) Chlorophyll-a variability off Patagonia based on SeaWiFS data. J Geophys Res Oceans 1-11. https://doi.org/10.1029/2005JC003244

Rosas-Luis R, Sánchez P, Portela JM, del Rio JL (2014) Feeding habits and trophic interactions of *Doryteuthis gahi*, *Illex argentinus* and *Onykia ingens* in the marine ecosystem off the Patagonian Shelf. Fish Res 152:37–44. https://doi.org/10.1016/j.fishres.2013.11.004

Rosas-Luis R, Navarro J, Sánchez P, Río JLD (2016) Assessing the trophic ecology of three sympatric squid in the marine ecosystem off the Patagonian Shelf by combining stomach content and stable isotopic analyses. Mar Biol Res 12:402–411. https://doi.org/10.1080/17451000.2016.1142094

Sabatini ME, Akselman R, Reta R et al (2012) Spring plankton communities in the southern Patagonian shelf: hydrography, mesozooplankton patterns and trophic relationships. J Mar Syst 94:33–51. https://doi.org/10.1016/j.jmarsys.2011.10.007

Sánchez RP, Remeslo A, Madirolas A, de Ciechomski JD (1995) Distribution and abundance of post-larvae and juveniles of the Patagonian sprat, *Sprattus fuegensis*, and related hydrographic conditions. Fish Res 23:47–81. https://doi.org/10.1016/0165-7836(94)00339-X

Saporiti F, Bearhop S, Vales DG et al (2015) Latitudinal changes in the structure of marine food webs in the Southwestern Atlantic Ocean. Mar Ecol Prog Ser 538:23–34. https://doi.org/10.3354/meps11464

Scheffer A, Trathan PN, Collins M (2010) Foraging behaviour of King Penguins (*Aptenodytes patagonicus*) in relation to predictable mesoscale oceanographic features in the polar front zone to the north of South Georgia. Prog Oceanogr 86:232–245. https://doi.org/10.1016/j.pocean.2010.04.008

Schejter L (2000) Alimentación de la vieira patagónica *Zygochlamys patagonica* (King y Broderip, 1832) en el banco Reclutas (39ºS-55ºW) durante un período anual. Tesis de Licenciatura, Universidad Nacional de Mar del Plata

Schejter L, Bremec C (2007) Benthic richness in the Argentine continental shelf: the role of *Zygochlamys patagonica* (Mollusca: Bivalvia: Pectinidae) as settlement substrate. J Mar Biol Assoc U K 87:917–925. https://doi.org/10.1017/S0025315407055853

Siegelman L, O'Toole M, Flexas M et al (2019) Submesoscale Ocean fronts act as biological hotspot for southern elephant seal. Sci Rep 9:5588. https://doi.org/10.1038/s41598-019-42117-w

Tam JC, Link JS, Rossberg AG et al (2017) Towards ecosystem-based management: identifying operational food-web indicators for marine ecosystems. ICES J Mar Sci 74:2040–2052. https://doi.org/10.1093/icesjms/fsw230

Temperoni B, Viñas MD, Buratti CC (2013) Feeding strategy of juvenile (age-0+ year) Argentine hake *Merluccius hubbsi* in the Patagonian nursery ground. J Fish Biol 83:1354–1370. https://doi.org/10.1111/jfb.12238

Temperoni B, Massa A, Viñas MD (2018a) Fatty acids composition as an indicator of food intake in *Merluccius hubbsi* larvae. J Mar Biol Assoc U K 1–8. https://doi.org/10.1017/S002531541800070X

Temperoni B, Massa AE, Derisio C et al (2018b) Effect of nursery ground variability on condition of age 0+ year *Merluccius hubbsi*. J Fish Biol 93:1090–1101. https://doi.org/10.1111/jfb.13816

Thompson KR (1992) Quantitative analysis of the use of discards from squid trawlers by black-browed albatrosses *Diomedea melanophris* in the vicinity of The Falkland Islands. Ibis 134:11–21. https://doi.org/10.1111/j.1474-919X.1992.tb07223.x

Thompson KR (1993) Variation in Magellanic penguin *Spheniscus magellanicus* diet in The Falkland Islands. Mar Ornithol 21:57–67

Torres Alberto ML, Bodnariuk N, Ivanovic M et al (2021) Dynamics of the confluence of Malvinas and Brazil currents, and a southern Patagonian spawning ground, explain recruitment fluctuations of the main stock of *Illex argentinus*. Fish Oceanogr 30:127–141. https://doi.org/10.1111/fog.12507

Troccoli GH, Aguilar E, Martínez PA, Belleggia M (2020) The diet of the Patagonian toothfish *Dissostichus eleginoides*, a deep-sea top predator off Southwest Atlantic Ocean. Polar Biol 43:1595–1604. https://doi.org/10.1007/s00300-020-02730-2

Valenzuela LO, Rowntree VJ, Sironi M, Seger J (2018) Stable isotopes ($\delta^{15}N$, $\delta^{13}C$, $\delta^{34}S$) in skin reveal diverse food sources used by southern right whales *Eubalaena australis*. Mar Ecol Prog Ser 603:243–255. https://doi.org/10.3354/meps12722

van Altena C, Hemerik L, de Ruiter PC (2016) Food web stability and weighted connectance: the complexity-stability debate revisited. Theor Ecol 9:49–58. https://doi.org/10.1007/s12080-015-0291-7

Valenzuela LO, Rowntree VJ, Sironi M, Seger J (2018) Stable isotopes ($\delta^{15}N$, $\delta^{13}C$, $\delta^{34}S$) in skin reveal diverse food sources used by southern right whales Eubalaena australis. Mar Ecol Prog Ser 603:243–255. https://doi.org/10.3354/meps12722

Viñas MD, Álvarez Colombo GL, Padovani LN (2016) Anfípodos hiperideos. In: El Mar Argentino y sus recursos pesqueros TOMO 6 Los crustáceos de interés pesquero y otras especies relevantes en los ecosistemas marinos, Boschi, E. E. Instituto Nacional de Investigación y Desarrollo Pesquero (INIDEP), Mar del Plata, Argentina

Williams RJ, Martinez ND (2000) Simple rules yield complex food webs. Nature 404:180–183. https://doi.org/10.1038/35004572

Xavier JC, Trathan PN, Croxall JP et al (2004) Foraging ecology and interactions with fisheries of wandering albatrosses (*Diomedea exulans*) breeding at South Georgia. Fish Oceanogr 13:324–344. https://doi.org/10.1111/j.1365-2419.2004.00298.x

Zerbini A, Andriolo A, Heide-Jørgensen MP, Pizzorno J, Geyer Y, Van Blaricom G, De Master D, Simões-Lopes P, Moreira S, Bethlem C (2006) Satellite-monitored movements of humpback whales *Megaptera novaeangliae* in the Southwest Atlantic Ocean. Mar Ecol Prog Ser 313:295–304. https://doi.org/10.3354/meps313295

Zheng J, Brose U, Gravel D et al (2021) Asymmetric foraging lowers the trophic level and omnivory in natural food webs. J Anim Ecol 90:1444–1454. https://doi.org/10.1111/1365-2656.13464

Zhu G, Zhang H, Yang Y et al (2017) Upper trophic structure in the Atlantic Patagonian shelf break as inferred from stable isotope analysis. Chin J Oceanol Limnol 1–9. https://doi.org/10.1007/s00343-018-6340-5

Printed in the USA
CPSIA information can be obtained
at www.ICGtesting.com
CBHW071915181124
17601CB00002B/8

9 783031 711893